Urban Ecology and Global Climate Change

Urban Ecology and Global Climate Change

Edited by

Rahul Bhadouria
University of Delhi

Shweta Upadhyay
Banaras Hindu University

Sachchidanand Tripathi
University of Delhi

Pardeep Singh
PGDAV College, University of Delhi

WILEY Blackwell

This edition first published 2022

Registered Offices
John Wiley & Sons, Inc., 111 River Street, Hoboken, NJ 07030, USA
John Wiley & Sons Ltd, The Atrium, Southern Gate, Chichester, West Sussex, PO19 8SQ, UK

Editorial Office
9600 Garsington Road, Oxford, OX4 2DQ, UK

For details of our global editorial offices, customer services, and more information about Wiley products visit us at www.wiley.com.

Wiley also publishes its books in a variety of electronic formats and by print-on-demand. Some content that appears in standard print versions of this book may not be available in other formats.

Library of Congress Cataloging-in-Publication Data

Names: Bhadouria, Rahul, 1982– editor. | Upadhyay, Shweta, 1987– editor. | Tripathi, Sachchidanand, editor. | Singh, Pardeep, editor.
Title: Urban ecology and global climate change / edited by Rahul Bhadouria, University of Delhi, Shweta Upadhyay, Banaras Hindu University, Sachchidanand Tripathi, University of Delhi, Pardeep Singh, PGDAV College, University of Delhi.
Description: First edition. | Hoboken, NJ : Wiley, 2022. | Includes bibliographical references and index.
Identifiers: LCCN 2022001032 (print) | LCCN 2022001033 (ebook) | ISBN 9781119807186 (cloth) | ISBN 9781119807193 (adobe pdf) | ISBN 9781119807209 (epub)
Subjects: LCSH: Urban ecology (Biology)--Environmental aspects. | Urban ecology (Sociology)--Environmental aspects. | Urbanization. | Overpopulation. | Climatic changes.
Classification: LCC QH541.5.C6 U735 2022 (print) | LCC QH541.5.C6 (ebook) | DDC 577.5/6--dc23/eng/20220112
LC record available at https://lccn.loc.gov/2022001032
LC ebook record available at https://lccn.loc.gov/2022001033

Cover Design: Wiley
Cover Images: © lightkey/Getty Images; erlucho/Getty Images; Mike_Kiev/Getty Images; B'Rob/Getty Images; Chones/Shutterstock; shomos uddin/Getty images; James O'Neil/Getty Images

Set in 9.5/12.5pt STIXTwoText by Straive, Pondicherry, India
Printed and bound by CPI Group (UK) Ltd, Croydon, CR0 4YY

C9781119807186_210322

Contents

List of Contributors

Rituparna Acharyya
Department of Geography
School of Earth Science
Central University of Karnataka
Karnataka
India

Saiema Ahmedi
Department of Biosciences
Jamia Millia Islamia
New Delhi
Delhi
India

Mohammad Ali
Institute of Forestry and Environmental Sciences
University of Chittagong
Chittagong
Bangladesh

Arisha Arora
Department of Biotechnology
Motilal Nehru National Institute of Technology Allahabad
Prayagraj
India

Department of Biosciences and Bioengineering
Indian Institute of Technology Guwahati
Assam
India

Subhasree Banerjee
School of Oceanographic Studies
Jadavpur University
Kolkata
West Bengal
India

Sarika Bano
Dr. B.R. Ambedkar Center for Biomedical Research
University of Delhi
New Delhi
Delhi
India

Sunny Bansal
RCG School of Infrastructure Design and Management
IIT Kharagpur
Kharagpur
West Bengal
India

VIT School of Planning and Architecture
VIT Vellore
Vellore
Tamil Nadu
India

Vidhu Bansal
VIT School of Planning and Architecture
Vellore Institute of Technology
Vellore
Tamil Nadu
India

Department of Architecture and Regional Planning
Indian Institute of Technology Kharagpur
West Bengal
India

Gunja Baretha
Department of Forestry
Wildlife & Environmental Sciences
Guru Ghasidas Vishwavidyalaya
(A Central University) Bilaspur
India

Jayprakash Chadchan
VIT School of Planning and Architecture
VIT Vellore
Vellore
Tamil Nadu
India

Abhra Chanda
School of Oceanographic Studies
Jadavpur University
Kolkata
West Bengal
India

Krishna Kumar Chandra
Department of Forestry
Wildlife & Environmental Sciences
Guru Ghasidas Vishwavidyalaya
(A Central University)
Bilaspur
India

Meenakshi Chaurasia
Department of Botany
University of Delhi
New Delhi
Delhi
India

Saptamita P. Choudhury
School of Biotechnology
Kalinga Institute of Industrial Technology
Bhubaneswar
Odisha
India

Dr. B.R. Ambedkar Center for Biomedical Research
University of Delhi
New Delhi
Delhi
India

Sanjay Kumar Dey
Dr. B.R. Ambedkar Center for Biomedical Research
University of Delhi
New Delhi
Delhi
India

Joy K. Dey
Documentation and Publication Section,
Central Council for Research in Homoeopathy
Ministry of AYUSH, Govt. of India
New Delhi
Delhi
India

Dey Health Care and Research Foundation
Nalikul
West Bengal
India

Govind Gupta
Environmental Science Discipline
Department of Chemistry
Manipal University Jaipur
Jaipur
Rajasthan
India

Tomáš Hák
Environment Centre
Charles University
Prague
Czech Republic

Faculty of Humanities
Charles University
Prague
Czech Republic

Eyadul Islam
School of Oceanographic Studies
Jadavpur University
Kolkata
West Bengal
India

Sharmila Jagadisan
VIT School of Planning and Architecture
Vellore Institute of Technology
Vellore
Tamil Nadu
India

Nishi Jain
Dr. B.R. Ambedkar Center for Biomedical Research
University of Delhi, New Delhi
Delhi
India

Department of Biotechnology
Amity University-Noida
Noida
Uttar Pradesh
India

Svatava Janoušková
Environment Centre
Charles University
Prague
Czech Republic

Faculty of Science
Charles University
Prague
Czech Republic

Anina James
Department of Zoology, Deen Dayal Upadhyaya College
University of Delhi
New Delhi
Delhi
India

Jabbar Khan
Environmental Science Discipline
Department of Chemistry
Manipal University Jaipur
Jaipur
Rajasthan
India

Arun Kumar
Bihar Agriculture University
Bhagalpur
Bihar
India

Rajesh Kumar
Department of Forestry
Wildlife & Environmental Sciences, Guru Ghasidas Vishwavidyalaya, (A Central University) Bilaspur
India

Arunashish Maity
School of Oceanographic Studies
Jadavpur University
Kolkata
West Bengal
India

Nikhat Manzoor
Department of Biosciences
Jamia Millia Islamia
New Delhi
Delhi
India

Rohit Mukherjee
School of Oceanographic Studies
Jadavpur University
Kolkata
West Bengal
India

Anirban Mukhopadhyay
Disaster Preparedness Mitigation, and Management (DPMM)
Asian Institute of Technology
Pathumthani
Bangkok
Thailand

Indrajit Pal
Disaster Preparedness, Mitigation, and Management (DPMM)
Asian Institute of Technology
Pathumthani
Bangkok
Thailand

Musarrat Parween
National Institute of Advanced Studies
Indian Institute of Science Campus
Bangalore
Karnataka
India

Kajal Patel
Department of Botany
University of Delhi
New Delhi
Delhi
India

Samin Poudel
UNIGIS Kathmandu, UNIGIS Dual Degree: Kathmandu Forestry College (Nepal) and Department of Geoinformatics-Z_GIS, University of Salzburg (Austria)

Niloy Pramanick
School of Oceanographic Studies
Jadavpur University
Kolkata
West Bengal
India

Kottapalli S. Rao
Department of Botany
University of Delhi
New Delhi
Delhi
India

Joy Sen
Department of Architecture and Regional Planning and Joint Faculty, RCG School of Infrastructure Design and Management, IIT Kharagpur
Kharagpur
West Bengal
India

Shahnawaz
Department of Geoinformatics - Z_GIS
University of Salzburg
Salzburg
Austria

Deepti Sharma
TerraNero Environmental Solutions Pvt. Ltd.
Varun Garden, Manpada
Thane (W)
Maharashtra
India

Him L. Shrestha
UNIGIS Kathmandu
Kathmandu Forestry College
Kathmandu
Nepal

Riddhi Shrivastava
Environmental Science Discipline,
Department of Chemistry
Manipal University Jaipur
Jaipur
Rajasthan
India

Poornima College of Engineering
Jaipur
Rajasthan
India

Naveen K. Singh
Environmental Science Discipline
Department of Chemistry
Manipal University Jaipur
Jaipur
Rajasthan
India

Ranjana Singh
Government Model Degree College, Arniya
Bulandshahar
Uttar Pradesh
India

Rishikesh Singh
Institute of Environment & Sustainable Development (IESD)
Banaras Hindu University
Varanasi
Uttar Pradesh
India

Department of Botany
Faculty of Science
Panjab University
Chandigarh
India

Vipin Kumar Singh
Department of Botany
Institute of Science
Banaras Hindu University
Varanasi
Uttar Pradesh
India

Prachi Sinha
Ministry of Home Affairs
Government of India
National Investigation Agency
New Delhi
Delhi
India

Sujata Sinha
Department of Botany
Deen Dayal Upadhyaya College
(University of Delhi)
New Delhi
Delhi
India

Pratap Srivastava
Department of Botany
Shyama Prasad Mukherjee Post-graduate College
University of Allahabad
Prayagraj
India

Gul-e-Noor Tanjina Hasnat
Institute of Forestry and Environmental Sciences
University of Chittagong
Chittagong
Bangladesh

Pramit Verma
Institute of Environment & Sustainable Development (IESD)
Banaras Hindu University
Varanasi
Uttar Pradesh
India

Foreword

R. K. Kohli

Nature produces enough for everyone's need. It also has in place the cyclic processes for the regeneration of nature's gifts. However, for his convenience man has introduced the concept of settlements in urban areas, beyond their carrying capacities. Hardly has any country evaluated the natural resource-carrying capacity of any area before the settlement of human population. This has not only disturbed the natural balance resulting in consequential impacts reflected in the form of pollution of air, water, soil, sound, and creation of perceptible urban heat island and thermal inversions but also resulted in social unrest, increased crimes, irritability, and impaction on human health.

Unfortunately, man evaluates every action in terms of economics without realising that the man-made currency does not follow the currency and principles of nature. The economists and ecologists unfortunately had been contradicting each other. The former had always been winning, and the latter cribbing. However, the last laugh is of the ecologist, especially when natural calamities and catastrophes occur. It is no exaggeration that the incidence and dimensions of such horrific events including floods, hurricanes, and pandemics have unfortunately increased lately, Undoubtedly, such events impact the future of earth's biosphere. For such reasons, the need for trees and green spaces, etc. for sustainable ecological balance is more in urban areas. Our actions need to match with the desired well-defined sustainable goals.

It is gratifying to note that the editors of the book '***Urban Ecology and Global Climate Change***' have chosen very well thought of five thematic sections focussing mainly on the interplay between cause and effect of urban pressure and climatic sustainability. These sections range from the perspective of (i) urban ecology and climate change, (ii) emerging technologies for designing urban landscape, (iii) urban biodiversity and natural resource exploitation, (iv) impact of fast urbanisation on the development sustainability, and (v) threat of climate change to ecological conservation. These themes have been very intelligently covered through 16 chapters. The choice of the editors for contributing authors of the 16 chapters is worth appreciation. It ranged from different parts of India besides scientists from Thailand, Czech Republic, Austria, Bangladesh, and Nepal.

In my opinion, the book will be very useful to the students and scientists not only in the field of environment and ecology but equally well for the town and country planning, policymakers, architects, managers of the river and coastal areas, and state and central government decision-makers, that too, not only in India, but the world over.

While congratulating the editors, authors, and the publishers for producing this title, I extend my best wishes for the success of the title which is very timely and relevant.

R. K. Kohli PhD, FNA, FASc, FNASc, FNAAS, FBS, FNESA
Vice Chancellor, Amity University Punjab, Mohali – 140306, India
Professor, JC Bose National Fellow, Certified Senior Emeritus Ecologist, ESA, USA
Former Vice chancellor, Central University of Punjab
2021

Foreword

P. K. Joshi

Global Climate Change largely impacts urban ecosystems and urban life. At the same time, the urban ecosystems are a key contributor to the global climate change. The recent reports on impacts of global climate change suggest that the impacts will be immense, enduring, and pernicious on entire human systems. Only scientific knowledge and actions at multiple scales can settle adaptation and mitigation to such impacts. However, to attempt this an interface between the urban ecosystems and the global climate change needs to be explored. The rate of urbanisation across the world has made it critical to keep urban ecosystems, their structural and functional components, and their dynamism central to the global environmental research and sustainability challenges. In fact, the science and policy communities have increasingly recognised this.

Urban Ecology and Global Climate Change is timely and a very welcome addition to the pertinent literature to unwrap the obstacles to adaptation and mitigation measures and sustainable development. It connects the relatively less known feedbacks between urban ecology and global environment change. The former is an eponymous discipline promoting resilient and sustainable urban spaces where human and nature coexist. The latter refers specifically to the rise in global temperatures since the mid-twentieth century to the present and its unprecedented impacts. Thus, by bridging these two important disciplines, the book fills an important niche which is no longer possible to support through a few research papers. The conditions are too complex in both the areas, urbanisation and climate change; thus, interlinkages need to be understood, documented, and reported for the wider audience. I am pleased to see that this volume is successful in achieving this.

In a way, the book examines the interacting forces of urbanisation and global climate change that are currently shaping the ecology of urban centres and the scope of future planning, design, and management strategies, while enhancing the environmental, social, and cultural values of these novel ecosystems. The book presents a wide range of contemporary aspects of urban ecology that are useful in ecological research, and that I hope others will find of benefit in design of research, or in reading and evaluating ongoing developments. I must congratulate the editors and authors to address intersections of queries of ecologist, environmentalist, geographers, urban planners, policymakers, and general public in this volume. It will serve as a wonderful guide to prompt these stakeholders to embrace the ecological concept of sustainability and understand the basics of urban ecology to deal with the questions of global climate change to create functional urban landscape in this challenging era.

P. K. Joshi
Professor – School of Environmental Sciences
Chairperson – Special Centre for Disaster Research
Jawaharlal Nehru University, New Delhi
2021

Section 1

Urban Ecology and Global Climate Change: Introduction

1

Urban Ecology and Climate Change: Challenges and Mitigation Strategies

Rishikesh Singh[1,2], *Pramit Verma*[1], *Vipin Kumar Singh*[3], *Pratap Srivastava*[4], *and Arun Kumar*[5]

[1] *Institute of Environment & Sustainable Development (IESD), Banaras Hindu University, Varanasi, India*
[2] *Department of Botany, Faculty of Science, Panjab University, Chandigarh, India*
[3] *Department of Botany, Institute of Science, Banaras Hindu University, Varanasi, India*
[4] *Department of Botany, Shyama Prasad Mukherjee Post-graduate College, University of Allahabad, Prayagraj, India*
[5] *Bihar Agriculture University, Bhagalpur, Bihar, India*

1.1 Introduction

Humankind is facing three major challenges viz. human overpopulation, urbanisation, and climate change with the onset of the twenty-first century (Steiner 2014). Presently, about seven billion (expected to reach 8.2 and 9 billion by 2025 and 2050, respectively) people are inhabiting the Earth which is more than any previous time. Urban areas and people living in the cities are increasing rapidly in size, globally (Mitchell et al. 2018). Over half (~54%) of the world's population is residing in the urban areas which is expected to grow to 60 and 80% by the year 2030 and 2050, respectively (Lee 2011; Vasishth 2015). Urbanisation phenomenon can be seen occurring on all the continents (except Antarctica); however, rapid urbanisation is happening, particularly in the Asia and Sub-Saharan Africa (Yu et al. 2017). Rapid urbanisation is putting severe stress on the planet Earth resulting in changes in the ecosystems from the landscape to the global scales (Steiner 2014; Colding and Barthel 2017). Urbanisation leads to the rapid conversion of natural pervious land surfaces to various impervious surfaces in the built forms like buildings and roads which resulted in changes in many ecosystem functions such as water infiltration and availability, species composition, soil properties, and thermal properties of the surfaces (Gaston et al. 2010; Seto et al. 2012; Yu et al. 2017). Urbanisation has not only affected the tangible features of the natural ecosystems but also resulted in the modifications of intangible aspects such as biogeochemical cycling and climate change (Kattel et al. 2013; Mitchell et al. 2018). Therefore, need for the proper planning and designing of the urban ecosystems has been arisen for reducing the ecological footprints (on per capita basis) of these ecosystems for managing the trio of challenges mentioned in the opening line of this chapter (Steiner 2014; Vasishth 2015).

Urban Ecology and Global Climate Change, First Edition. Edited by Rahul Bhadouria, Shweta Upadhyay, Sachchidanand Tripathi, and Pardeep Singh.

Urban ecosystems are not only the rich nodes of civilisation but also have become the engines of development and divers of various environmental changes even at fine spatial scales (Kattel et al. 2013; Verma et al. 2020a). Urban ecosystems are characterised by intensive human population and its supportive infrastructures such as built-up areas in the form of cities, towns, and megacities, developed by profound changes in the landscape structures and energy processes at the cost of natural ecosystems (Forman 2014; Jaganmohan et al. 2016; Yu et al. 2017; Mitchell et al. 2018; Verma et al. 2020a). The urban ecosystems are highly heterogeneous and fragile systems, which are facing many challenges due to massive human interferences (Ma et al. 2020). Since humans represent the core of urbanisation's structural processes, their activities related to aesthetic values, goods and services (food and water), energy and waste generation, and recycling are the driving factors of all the changes occurring in these ecosystems (Carpenter and Folke 2006; Chapin et al. 2011; Kattel et al. 2013). Thus, urban ecosystems are the complex ecosystems characterised by the interplays of socio-economic dimensions and biophysical (natural processes) interactions occurring at various spatio-temporal scales (Kattel et al. 2013; Ma et al. 2020). Most notably, the land-use change and the infrastructural development by the human decide the future of the urban ecosystems which are already facing various unprecedented social, demographic, technological, and environmental challenges (Niemelä 2014; Steiner 2014). Thus, there comes the need to understand the urban ecosystems as an ecological system which can play and sustain even under the human dominance (Grimm et al. 2008). In the next sub-section, the concept of urban ecology and its need for the sustainable urban development has been highlighted.

1.1.1 Urban Ecology

Nowadays most of the people live in the urban areas; however, they have limited understanding of the benefits derived from the interaction with the natural systems within the cities. Urban ecology is a scientific discipline which integrates a number of concepts from the natural and social sciences along with the landscape approach and ecosystems services at its core (Alberti 2008; Niemelä et al. 2011; Niemelä 2014) and represents the '*holistic ecology of urban areas*' (McDonnell et al. 2009; Jim 2011). Since this field is still emerging, its hypotheses, models, and theories are still in the process of testing and validation (Niemelä 2014). Since human activities are the dominant factors in shaping the urban ecosystems, the urban ecology unintentionally revolves around the processes and interactions attributed by the human actions (Verma et al. 2020a). For example, urban ecology helps in recognising the restorative behaviour of humans for the natural ecosystems and elucidating the mechanisms responsible for the structuring of urban communities during the process of urbanisation (Steiner 2014; Duffy and Chown 2016). However, poor understanding of the complex interactions between the socio-ecological and infrastructural developments in the urban ecosystems resulted in the social disharmony in the long term (Kattel et al. 2013). Therefore, Kattel et al. (2013) suggested a concept of complementary framework for urban ecology which represents the development of infrastructure and green spaces in an integrated manner to derive utmost ecosystem services. For instance, integrated development of flora and fauna along with the urban buildings, roads, and railway tracks for providing the utmost scope for the interactions between human inhabitants and wider communities through inter-habitat-community systems (Halpern et al. 2008).

Recent studies on urban ecology revealed that the field is now viewed as the 'ecology of the city' in addition to the 'ecology in the city' (Grimm et al. 2000; Pickett et al. 2001; Kattel et al. 2013; Childers et al. 2014). The 'ecology in the city' refers to the study of the natural ecological systems (fragments) within an urban ecosystem, i.e. different urban fragments as the analogues of their non-urban counterparts, whereas 'ecology of the city' represents a much wider context where urban ecosystem itself is studied as an ecological system, i.e. the study of the interactions of various biological, built, and social components of the ecosystems within a city (Vasishth 2015; Pickett et al. 2016; Verma et al. 2020b). Based on these diverse concepts of the urban ecology, following sections shall provide a brief understanding of the urban ecosystems. In the next sections, climate change as an emerging challenge to the urban ecosystems has been discussed, followed by the possible urban ecological approaches for mitigating the ill effects of climate change in these highly heterogeneous and fragile ecosystems.

1.2 Components of Urban Ecology

An urban ecosystem is composed of several tangible and intangible components. Tangible components include the physical structures which can be natural (such as flora and fauna, water bodies, mountains, urban agriculture, etc.) or human-made (such as built structures like buildings and building materials, roads, railways, health, and related infrastructural developments; energy sources like coal, liquified petroleum gas (LPG), wood; food supplies and waste generation, etc.). In addition, the intangible components include the ecosystems services derived from various natural systems, biogeochemical cycling, solar energy, and material flow in the urban areas (Verma et al. 2020a). These major components can be mainly divided into three sectors: (i) urban infrastructures associated with the urban heat island (UHI) effect; (ii) urban vegetation representing the green spaces and related ecosystem services; and (iii) urban metabolism which represents the flow of energy and materials within the urban ecosystems.

1.2.1 Urban (Built) Infrastructures

Urban infrastructures are built to provide the services and benefits to the urban inhabitants (Ma et al. 2020). Urban habitats are almost similar over large areas or human-managed cities and microclimatic regions (Savard et al. 2000) having a central square with paved areas, residential areas, urban parks, urban agriculture, and some disturbed/unmanaged plots (Lososová et al. 2018). Considerable impervious surfaces such as roofs, roads, and paving are the most common features of urban infrastructure which sets cities apart from adjacent rural areas (Vasishth 2015). Both the components (viz. natural and artificial) of urban infrastructure help in the maintenance of the ecosystem functioning and health (Kattel et al. 2013). For example, vegetation cover and water bodies in the urban and peri-urban areas provide habitats for biological diversity and help in maintaining the integrity of the natural/ecological and physical environments (Hofmann et al. 2012). In addition to the natural and artificial classification, urban infrastructures can also be classified into two major components viz. aboveground and belowground (Ma et al. 2020). Aboveground component comprises of the infrastructures at/on which citizens live in, walk on and ride on, such as buildings, parking lots, roads, sidewalks, green spaces, and public transportation (Andersson et al. 2014). Belowground component encompasses massive foundations of the buildings, subway lines, tunnels, gas lines, water and sewage pipes, stormwater management, electricity, and optical cables for providing various services to the citizens and ease in their livelihood (Sun and Cui 2018; Ma et al. 2020). Both the components of the urban infrastructures hold crucial significance and interact with each other in complex ways (Pandit et al. 2015). Aboveground infrastructures provide living spaces and comforts to the citizens, whereas belowground components play equally important services in the form of utilities, transportation, biomass, and structures which enable the urban areas for smooth functioning of the aboveground components (Ferrer et al. 2018; Ma et al. 2020). Moreover, natural components of the urban infrastructures such as plant roots and microbial communities also show strong competition for the space in the belowground components (Mullaney et al. 2015).

Land-use and land-cover changes are one of the major drivers of global change processes which should be taken into account considerably from ecological point of view (Vitousek 1994). Urban land cover has been projected to increase by 200% within the first three decades (2000–2030) of the twenty-first century (Elmqvist et al. 2013). These projections revealed that there has been and would be a massive investment in the development of the urban infrastructures at the cost of consumption of natural ecosystems/landscapes (Green et al. 2016). Increase in impervious surfaces and the materials used for their formation (dark asphalt and roofing materials) due to massive urbanisation have the ability to absorb the solar irradiance and influence the local climate and hydrological conditions (Vasishth 2015). UHI effect (described later) is one of the major outcomes of the increase in such urban built infrastructures (Jaganmohan et al. 2016). For managing the urban ecological components (e.g. biodiversity, nutrient cycling, etc.) influenced by the land-use change patterns, several conceptual frameworks, and models have been developed (Pickett et al. 2011). However, their proper implementation is lacking due to poor representation of the social components in these frameworks (Zipperer et al. 2011). Nowadays landscape urbanism is the emerging concept with non-hierarchical, flexible and strategic planning where landscapes in the urban areas are designed and managed as per the demand of the society (Kattel et al. 2013). Detailed elaboration of such strategies and frameworks has been given in the later sections of the chapter.

1.2.1.1 Urban Heat Islands

Modifications of the physical environment by the built structures during the process of urbanisation impede the energy distribution and composition of gases in the near-surface. It alters the microclimatic conditions by modifying the thermodynamics of the urban ecosystems which resulted in 2–5 °C higher ambient temperature than the surroundings (countryside/rural) areas (Phelan et al. 2015; Jaganmohan et al. 2016; Vasishth 2015; Duffy and Chown 2016). Such alterations in local and regional climatic conditions (temperature dynamics) by the urban infrastructures lead to the UHI effect (Dallimer et al. 2016; Zhou et al. 2017; Hu et al. 2019). The UHI effect is one of the most prominent human-made climatic phenomena in the urban ecosystems and has considerable ecological significance (Gaston et al. 2010; Akbari and Kolokotsa 2016; Yu et al. 2017). The UHI effect leads to the alteration of local and regional climatic conditions, impedes with the wind flows, and turbulence, related with shifts in cloud formation and precipitation, air pollution, and higher greenhouse gas (GHGs) emission (Seto and Shepherd 2009; Gaston et al. 2010). The increase in air pollution, heat stress, water quality, food security, and disparity in ecosystems services due to the UHI effects further affect the health and comfort of the urban inhabitants, thus, have adverse social, economic, and ecological impacts (Chang et al. 2016; Wong et al. 2016; Battles and Kolbe 2019). Moreover, UHI effect is further expected to contribute in the global climate change by increasing the GHGs (particularly CO_2) emission from the urban areas (Rosenzweig et al. 2010), and in response, the UHI effect's impact may further intensify in most of the cities (Chapman et al. 2017). However, the magnitude of the UHI effect depends on several local and regional factors such the latitude, weather and climatic conditions, diurnal conditions, rainfall, surrounding ecology, population and culture of the city, and the urban planning (Zhao et al. 2014; Vasishth 2015). For example, the increase in temperatures due to UHI effects during winter at higher altitudes, and during summer at lower altitudes may decrease and increase the costs of air-conditioning, respectively (Vasishth 2015). Thus, the UHI effects may have variable responses depending on the location of the urban area and it needs to be explored further under the changing environmental conditions.

1.2.2 Urban Vegetation

Urbanisation provides a unique habitat for vegetation growth. Urban vegetation (tree, shrub, or ornamental plants) is comprised of both native and exotic species (Cubino et al. 2021). Two major processes are involved in the growth of urban vegetation, viz. cultivated vegetation growth which represents the plants that are introduced and managed by the urban inhabitants, and spontaneous vegetation growth which have been established and colonised without the human-assistance (Cubino et al. 2021). Vegetation composition of the urban areas even differs at small scales depending on the several social factors (Čeplová et al. 2017). Introduction and cultivation of plants, especially ornamental or exotic plants for their aesthetic values by the urban inhabitants hold a crucial significance in current scenario. Exotic plants constitute a major portion of the urban vegetation; however, their contribution to the functioning and diversity at the socio-ecological scales have been less explored (Cook-Patton and Agrawal 2014; Pearse et al. 2018; Cubino et al. 2021). Most of the exotic species present in the urban areas are ornamental plants introduced intentionally by the humans (Čeplová et al. 2017; Lososová et al. 2018). Due to suitable environmental (heterogeneous) conditions or positive anthropogenic interferences (fertilisation and irrigation at regular intervals) in the urban areas, exotic plants have substantial scope to establish and colonise (Lososová et al. 2018). Human-induced dispersal of the plant species in the urban areas contributes to the plant distribution and community composition (Møller et al. 2012; Lososová et al. 2018). Moreover, local fauna and their preferences for the fruits/flowers also contribute to the biodiversity of the area. For example, the preference of frugivorous birds affects the dispersal of fleshy-fruited plant species in the urban areas (Møller et al. 2012). Moreover, birds help in seed dispersal of ornamental (exotic) trees from the gardens to the nearby natural ecosystems, thus, may facilitate plant invasion (Milton et al. 2007). Thus, exotic plants can be a major cause of ecosystem imbalances when the sufficient source of seeds/propagules and its dispersal agents are present in the nearby areas (Rai and Kim 2019).

Vegetation, particularly trees, plays a crucial role in maintaining the harmony in the urban ecosystems (Tigges et al. 2013). For example, trees store sufficient amount of carbon (C) and help in maintaining the overall C-pool of the urban ecosystems (Davies et al. 2011). Urban ecosystems have sufficient potential to store C in their above- and belowground components (Hutyra et al. 2011; Nowak et al. 2013), even in dense urban areas (Mitchell et al. 2018). Urban areas have abundant shade trees (recreational purpose), trees grown for hazard removal, or exotic trees, all have potential to store substantial amount of C in their vertical structures. However, C-density of the urban areas varies at spatio-temporal scales (Mitchell et al. 2018; Upadhyay et al. 2021). Detailed view on the urban C-stocks and their ecosystem services have been highlighted in the latter part of the chapter.

1.2.3 Urban Metabolism

The urban areas can be considered as an organism where consumption of materials, flow of energy and information, and waste generation (as end-products) are the common processes occurring at various spatio-temporal scales (Liu et al. 2013; Vasishth 2015; Verma et al. 2020a). These processes not only occur within a city but also affect the environment beyond the borders of the city, as like the natural organisms where different cells and tissues interact and involve in the metabolic processes and excrete the wastes outside the cell/body (Liu et al. 2013; Verma et al. 2020a). To understand the concept of material and energy supply for the functioning of the cities and the resultant waste (pollutants) generation in the urban ecosystems, the concept of urban metabolism has emerged (Restrepo and Morales-Pinzon 2018). The concept was first proposed by Wolman (1965), who believed that processes occurring in the urban systems are analogous to that occurring in the metabolic processes of the living organisms. This approach helps in quantifying and identifying the movement of energy and materials as well as management of the environmental problems in an urban ecosystem (Wang et al. 2021). Thus, the research focus has now been shifted from quantifying resource consumption and environmental impacts to identifying and analysing the internal processes and the mechanisms involved in the outcomes of end-products. Urban metabolism approach is getting wider attention of the urban ecology researchers as it helps in simulating the material flows and managing the environmental problems at different spatio-temporal scales (Wang et al. 2021). In a CiteSpace analysis, Wang et al. (2021) identified the research trends in the urban metabolism. They found that now research communities are focussing on different micro- and macro-scales in the urban areas such as by differentiating the central urban areas from the suburbs and rural areas to refine the results from the urban systems. Moreover, developing nations and the developing or less explored cities are being recognised as the new objects for the research, as several case studies are already available from the cities from the developed nations. In addition, future research should integrate the role of developing economies and the climate change phenomenon for exploring the urban metabolism at different scales (Wang et al. 2021). In the next sections, an insight has been given on the climate change and its impacts on the urban ecosystems. Moreover, the adaptation mechanisms of the urban ecology to the climate change has also been highlighted in the later sections.

1.3 Climate Change as Emerging Challenge for Urban Ecology

Climate change is the most challenging environmental change the whole humanity is facing nowadays (Niemelä 2014). It affects both the biotic and abiotic components of the urban ecosystems. The impact of climate change may become more severe with the UHI effects, particularly for the ageing and sensitive urban populations (UN 2011). Thus, climate change and rapid urbanisation are considered as the two major challenges the world is going through recently (Yu et al. 2017). Intensive land-use change and high consumption of fossil fuels for different purposes have resulted in the substantial GHGs emissions (~78%) from the urban ecosystems which further contribute to the global climate change (Kattel et al. 2013; Weissert et al. 2014; Mitchell et al. 2018). Sustainability of the urban development, and C and energy metabolism are emerging topics in the light of climate change scenario (Wang et al. 2021). Climate change is affecting the urban ecosystems in different ways such as by increasing the UHI effect, reducing the ecosystem services provided by the natural systems, occurrence of extreme events such as floods and droughts, wildfires, diseases, and health problems (Niemelä et al. 2010; Ma et al. 2020; Verma et al. 2020b). Thus, there is a need for proper land-use planning and improved infrastructural resilience for reducing the urban vulnerability to the extreme environmental events occurring (and expected to intensify) because of climate change-urbanisation nexus in the near future (Green et al. 2016; Ma et al. 2020). Moreover, there is a need of transdisciplinary research including both the natural and social sciences along with the major stakeholders for the climate change mitigation (Niemelä 2014). A brief insight on the impact of climate change on urban ecosystems has been given in the following sub-sections.

1.3.1 Urban Ecosystems as Indicators of Future Ecosystems

Urban ecosystems having comparatively higher temperature as compared to their surrounding (rural) areas are viewed as projections of the future ecosystems in the context of climate change (Grimm et al. 2008). These ecosystems represent the locations where human activities utilise higher proportion of primary productivity and produce comparatively higher amount of GHGs (CO_2 emission); thus, influencing the global C cycle (Gaston et al. 2013; Mitchell et al. 2018). Land-use change-related urban expansion has been reported as the cause of ~5% (1.38 PgC) emissions from the deforestation in the pan-tropics during the first three decades of the twenty-first century (Seto et al. 2012). Recent estimates on C-estimation revealed high potential of C-storage in the urban ecosystems, even at higher magnitude as proposed earlier (Davies et al. 2011). Therefore, researches on measuring the citywide C-stock and C-sequestration potential of different cities are the hot topics of research which will further help in developing strategies for climate change mitigation (Pedersen Zari 2019). Further, the development of novel plant communities and their formation processes under the combined impact of urbanisation and climate change may impact the health and livelihood of urban inhabitants (Knapp et al. 2017; Lososová et al. 2018). Healthy ecosystems have self-regulating capacity by regenerating the ecological and social health enabling humans to better adapt to the climate change, therefore, enabling the cities to evolve this ability should be the priority agenda of the future urban planning (Pedersen Zari 2019). A detailed insight on the impact of climate change on the urban flora has been given in the next sub-section.

1.3.2 Impact on Urban Flora

As mentioned earlier, urban plant communities are composed of both native and exotic origins with different traits and niches (Lososová et al. 2018). They hold crucial importance in the urban ecosystems by supporting massive biodiversity and providing ecosystems services such as temperature regulation, primary productivity, nutrient cycling, pollution reduction, carbon storage, and recreational opportunities to the human inhabitants (Dallimer et al. 2016; Lososová et al. 2018; Cubino et al. 2021). Studies revealed that climate change acts as a major driver for changes in the plant diversity, community structure, and composition (Lososová et al. 2018; Cubino et al. 2021). Plant species with different traits and origins have been supposed to respond differently to the future climate change. Based on the global climate model predictions for two representative concentration pathways (2.6 and 8.5), Lososová et al. (2018) reported that the perennial herbs and woody trees will respond more slowly to the climate change as compared to the fast-spreading annual herbs. However, studies revealed that the responses of both native and alien plant species will be similar under changing climate conditions (Lososová et al. 2018).

Different resource conditions such as soil water availability, precipitation, nitrogen deposition, photoperiod length, and CO_2 conditions also regulate the plant phenological events (Jeong et al. 2011; Cong et al. 2013). In temperate latitudes, climate change has played a major role in advancing and extending the growing seasons of plants (Menzel and Fabian 1999). Therefore, the projected climate change will alter the future species composition of the urban flora by modulating the temperature and precipitation conditions having direct impact on the plant phenological events (Neil et al. 2014; Lososová et al. 2018). Shifts in vegetation phenological events will have considerable impacts on different ecological functions which lead to the alteration in water, carbon and energy balances, and thus, primary productivity and interspecific interactions (Dallimer et al. 2016). Impact of urbanisation and climate change on the phenological events of birds (migration), amphibians (reproduction), plants (leafing and flowering), and arthropods (appearance and development) has been studied considerably (Grimm et al. 2008; Neil et al. 2010). Similarly, extensive studies have been done to measure the abundance and richness of bee species under changing climate and urbanisation conditions (Neil et al. 2014). A general decrease in bee species abundance and richness has been reported in the urban areas with respect to the climate change conditions. However, mechanistic understanding of these changes on flowering phenology and the pollinator communities is still needed to be explored (Neil et al. 2014). Therefore, for effective management of the urban green spaces, it is critically important to understand how the future plant and related pollinator communities will respond to the combined impacts of urbanisation and climate change conditions (Dallimer et al. 2016).

1.3.2.1 Invasive Species and Climate Change

As mentioned earlier, even under severe climate change conditions, the distribution and responses of both native and exotic plants will be similar. However, there will be greater chances of invasion of the alien species to the new sites which have become naturalised in the urban gardens/areas (Richardson et al. 2000; Lososová et al. 2018). Moreover, the risk of invasion will further intensify under the combination of warmer climate and urbanisation (with UHI effects), particularly in the parts of Europe (Central and Western) where native plants communities may not respond to these changes (Walther et al. 2009). In a study on the interactive effects of exotic plants and aridity, Cubino et al. (2021) observed that exotic plants showed better growth (height and leaf size) and responses under the warmer and drier regions, whereas poor responses under cooler and wetter regions as compared to the co-occurring native plants. This further indicates the chances of the spread of invasive species in the newly warmer and drier regions mediated by the changing climate and UHI effects. Thus, there is a need to consider the interactive effects of climatic variability and invasion to know the future plant community structure and composition in different regions of the world (Cubino et al. 2021).

In the next sections, results of bibliometric analysis performed for the urban ecology-climate change nexus have been presented, followed by the emerging features of urban ecology for the climate change mitigation.

1.4 Bibliometric Analysis for Urban Ecology and Climate Change Nexus

In order to observe the trends in research on the topics concerned with urban ecology and climate change during the past two decades (2001–2021), a bibliometric analysis was performed on 13 June 2021, using different search queries in the Scopus database (Singh et al. 2021). The search query was (TITLE-ABS-KEY ('urban ecology' AND 'climate change') AND PUBYEAR >1999. The query results in a total number of 137 documents published during the past two decades (from 2001 to the present, 13 June 2021). The year-wise growth of the research on the topic of urban ecology and climate change was considered along with country-specific research outputs. Further, institute-wise, research area-wise, author-wise, and publication type-wise distribution of research on the topic were also analysed. Though we searched from 2000 onwards, but we found that the researches are being published on this topic since 2004. Starting with merely one paper on the topic in the year 2004, the year 2020 observed a total number of 13 documents published within a single year (Figure 1.1a). Interestingly, a total of 13 papers have already been published on the topic in the year 2021 till 13 June only. This signifies the relevance of the topic and the increasing research attention in the current scenarios. The United States was the largest contributor of the papers under the country/territory-wise publication category with 47 documents. The United Kingdom and China ranked second and third by producing 17 and 16 papers each, while India stood at eighth rank with five papers published during the same period (Table 1.1). These observations revealed that there is an ample scope for research on urban ecology-climate change nexus in the developing countries which are expected to make a surge in urban population growth in the near future. Based on the Scopus data 2021, among the research institutes and universities globally, Arizona State University produced the maximum number of papers (8), followed by the Chinese Academy of Sciences (CAS) with six papers published during the last two decades (Table 1.1).

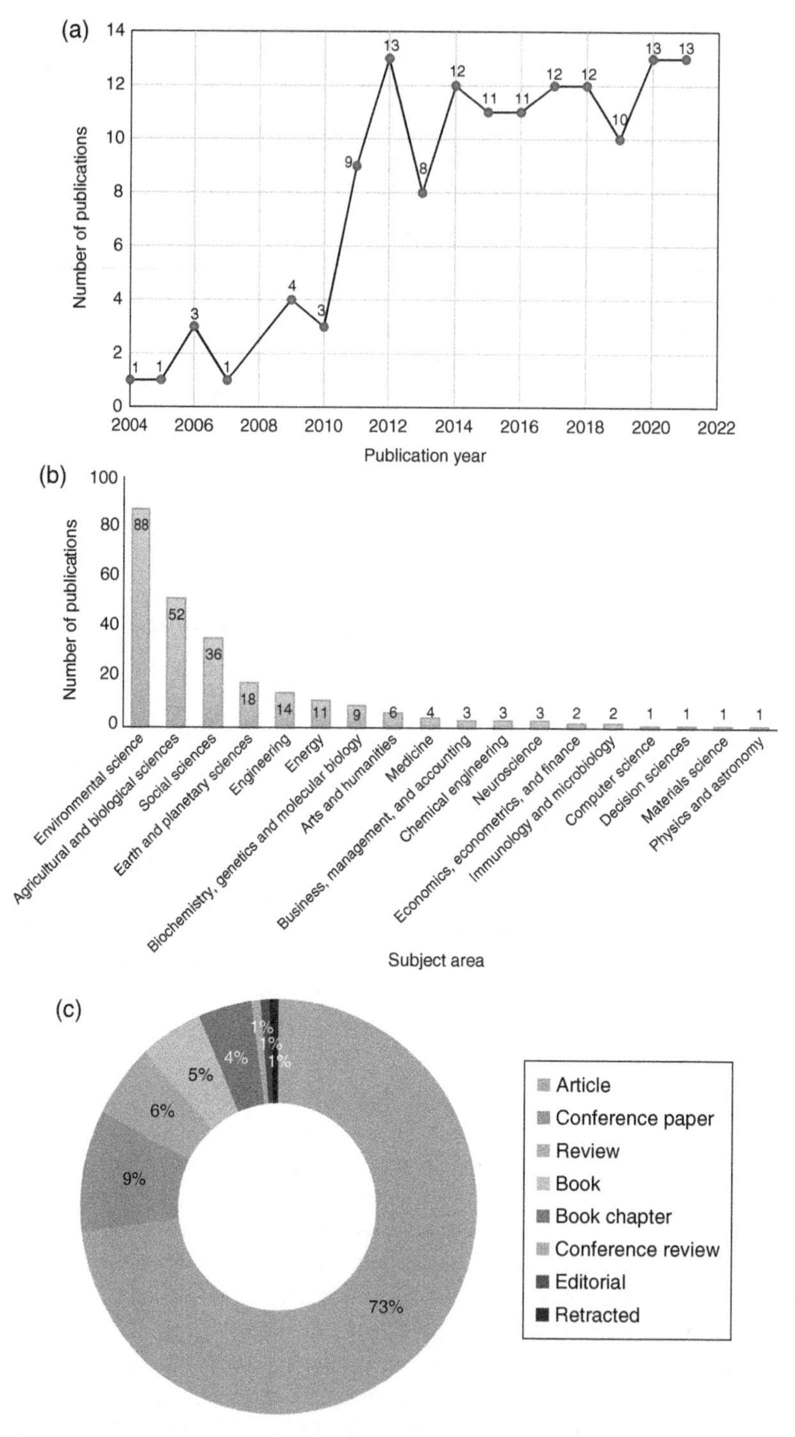

Figure 1.1 Bibliometric analysis results showing (a) year-wise publication growth, (b) subject area-wise distribution, and (c) article type-wise publications for the search query 'urban ecology AND climate change'. *Source:* Data from Scopus database (2021).

Table 1.1 Bibliometric analysis results showing top 20 countries, top 10 research institutes/ affiliations, top 20 authors, and top 10 research journal for the search query 'urban ecology AND climate change'.

Source category	No. of publications	Source category	No. of publications
Country-wise		*Country-wise*	
United States	47	Belgium	3
United Kingdom	17	Brazil	3
China	16	Chile	3
Australia	14	Colombia	3
Germany	11	Finland	3
New Zealand	6	Iran	3
Denmark	5	Sweden	3
India	5	Switzerland	3
Netherlands	5	Austria	2
Canada	4	Bulgaria	2
Affiliation-wise		*Affiliation-wise*	
Arizona State University	8	Research Centre for Eco-Environmental Sciences Chinese Academy of Sciences	4
Chinese Academy of Sciences	6	Helmholtz ZentrumfürUmweltforschung	4
KøbenhavnsUniversitet	5	University of Melbourne	4
NC State University	4	Humboldt-Universitätzu Berlin	3
University of Salford	4	Victoria University of Wellington	3
Author name-wise		*Author name-wise*	
Wu, J.	4	Gaston, K.J.	2
Dunn, R.R.	3	Goh, K.	2
Richter, M.	3	Kenawy, I.	2
Weiland, U.	3	Knapp, S.	2
Brans, K.I.	2	Livesley, S.J.	2
Davies, Z.G.	2	Lososová, Z.	2
Dobbs, C.	2	Neil, K.	2
Elkadi, H.	2	Qureshi, S.	2
Escobedo, F.J.	2	Steiner, F.	2
Frank, S.D.	2	Pedersen Zari, M.P.	2
Source title-wise		*Source title-wise*	
Landscape And Urban Planning	7	Urban Forestry And Urban Greening	3
Urban Ecosystems	6	Ecological Applications	2
Sustainability Switzerland	5	Forests	2
Ecosystems	3	IOP Conference Series Earth And Environmental Science	2
Acta Ecologica Sinica	3	Journal of Applied Ecology	2

Source: Data from Scopus database (2021).

Out of the total 137 publications, more than 80% of the total publications fall under the research areas of environmental sciences (88), agriculture and biological sciences (52), social sciences (36), earth and planetary sciences (18), and engineering (14) (Figure 1.1b). This signified that the interdisciplinary researches are increasing on the topic and focus on environmental impact assessment is given optimal importance. Author-wise analysis results revealed Wu J., Dunn R.R., Richter M., Weiland U., and Alavipanah S. as the five most productive authors publishing 4, 3, 3, 3, and 2 papers during the last two decades on the topic (Table 1.1). Landscape and Urban Planning (7), Urban Ecosystems (6), Sustainability (5), Ecosystems (3), and Urban Forestry and Urban Greening (3) were observed as the major sources which published researches on the topic in the recent decades (Table 1.1). These journals have been started with the aim of managing and improving the health and sustainability of the urban ecosystem. Among 137 papers, 73% were research articles whereas 9.5, 5.8, 5.1, and 4.4% were contributed by conference papers, review articles, books, and book chapters, respectively (Figure 1.1c). This signifies that the topic is getting considerable attention in different research publication types. Among 137 papers obtained for the search query, relevant articles were manually selected after going through the article in detail. Both original research and review articles written in the English language were considered for this purpose.

1.5 Emerging Features of Urban Ecology for Mitigating Climate Change

Urban ecosystems around the world are facing several major challenges among which increasing human population and climate change are the most prominent, making the sustainable urban development an elusive phenomenon (Childers et al. 2015). An increase in the frequency of extreme weather events induced by the climate change scenario is jeopardising the urban planners and designers to develop concrete adaptive measures for the urban settlements for the long term (Kattel et al. 2013; Steiner 2014). Thus, for sustainable cities, novel solutions and techniques for landscape planning having the potential of resilience and self-regulation from the extreme climatic conditions are needed (Childers et al. 2015; Yu et al. 2017). In this regard, urban vegetation can play a major role by managing the energy/radiation balance of the physical environment through the process of photosynthesis, evapotranspiration, and shadowing effects (Oke 2002; Yu et al. 2017). Moreover, Steiner (2014) highlighted four potential urban ecological design/planning measures, viz. application of ecosystem services, adaption of settlements for the natural disasters by managing green infrastructures, renewal of degraded landscapes, and improving the adaptive behaviour of people by linking knowledge about the surroundings to the actions.

The thematic evolution showed that there is a transitional shift in research focus during the last two decades (Figure 1.2). It can be seen that for the 2005–2014 period, social (climate change, urbanisation, and rural gradients) and ecological (phenology, conservation, vegetation, and UHI effect) aspects hold almost equal importance which get diversified in different emerging ecological aspects like ecology, ecosystem services, communities, resilience, and water during the 2015–2021 period. The 2015–2021 period showed the

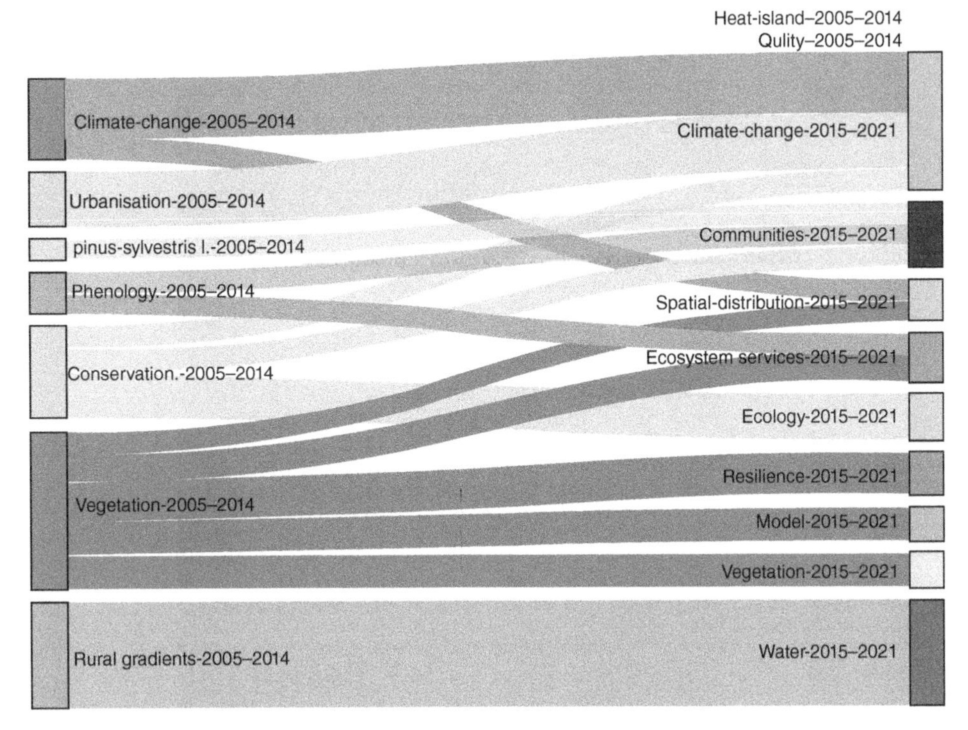

Figure 1.2 Thematic evolution of the urban ecology and climate change research areas during the last two decades. *Source:* Data from Web of Science Core Collection database (2021).

convergence of climate change and urbanisation, and conservation, whereas diversification of vegetation and conservation into communities, ecosystem services, ecology, and resilience showed that more research emphasis has now been given on ecosystem resilience and self-sustaining systems for the climate change mitigation. Moreover, the role of spatial modelling tools such as remote sensing and geographic information system (GIS) for developing sustainable urban ecosystems by vegetation mapping and risk analysis by the extreme weather events (climate change) has also been evolving considerably. Moreover, the role of vegetation phenology and its impact on various ecosystem services under the changing climatic conditions is also getting wider attention. The challenges to the urban ecosystems and adaptation strategies for the climate change have been briefly illustrated in Figure 1.3. In the next few sub-sections, ecosystem services and green infrastructure-related aspects have been elaborated while less emphasis has been given on later two mitigation measures in this chapter.

1.5.1 Ecosystem Services

Ecosystem health represents the system resilience, organisation, and vigour for the sustainable functioning (Costanza 1992). Ecosystem services are the major components of ecosystem health which not only provide benefits to the humankind but also help in

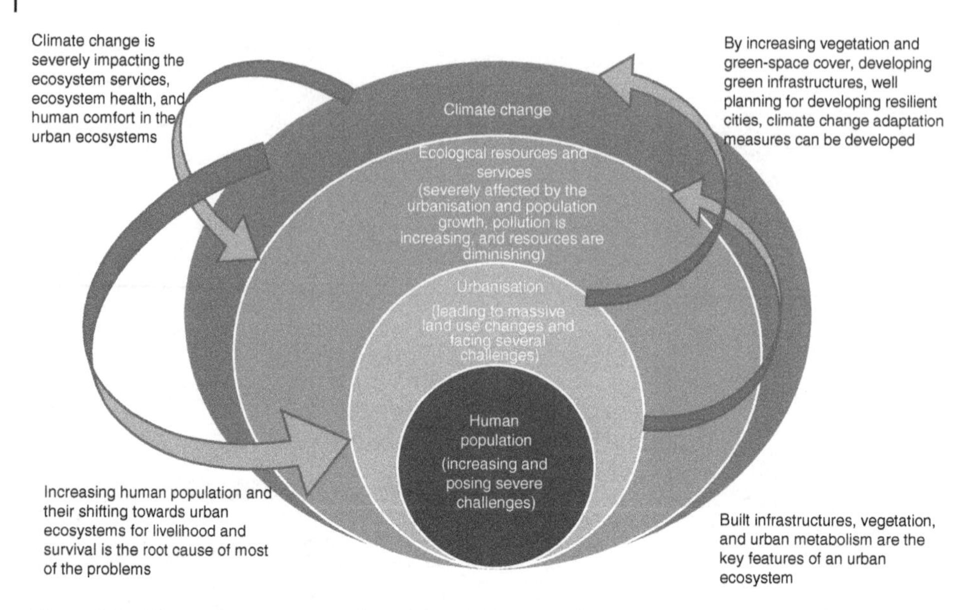

Figure 1.3 Illustrative representation of the challenges of the urban ecosystems and climate change adaptation strategies.

regulating the overall functioning of an ecosystem (Verma et al. 2020b). For example, urban and peri-urban forests help the urban society to tackle many of the issues such as protection from heat waves, flood/stormwater, pollution, etc. related to their livelihood (Green et al. 2016; Livesley et al. 2016). A few major ecosystem services (provisioning, supporting, cultural, and regulatory) provided by the urban vegetation include the temperature mitigation (urban cooling), noise level reduction, air purification, climate mitigation, nutrient cycling, C-sequestration, pollination, providing food (garden and farms), recreation opportunities, ecotourism, run-off mitigation, waste decomposition and detoxification, habitat for biodiversity, etc. (Steiner 2014; Dallimer et al. 2016; Aronson et al. 2017; Richards et al. 2019; Pedersen Zari 2019). Enumeration of ecosystem services of the urban ecosystems can be done by basic mapping to the valuation and management (McDonald and Marcotullio 2011). However, the extent of ecosystem services demanded by and provided to the urban dwellers varies from city to city throughout the world (Lososová et al. 2018; Richards et al. 2019). Moreover, an increase in extreme weather events has been reported to disrupt the provisioning of ecosystem services, causing an emerging challenge for finding appropriate adaptation measures (Wamsler et al. 2013). Site-specific enumeration of ecology and climate conditions may help in developing ecosystem services-based urban regeneration strategies (Pedersen Zari 2019). The ecosystem services are needed to be included in the major policy and decision-making agendas related to the urban planning and designing in the changing climate scenarios (Pedersen Zari 2019). Thus, with the growth of urban population and climate change, finding the appropriate methods for improving the provisioning of ecosystem services is needed (Richards et al. 2019). Moreover, research on evaluating ecosystem services in the urban ecosystems from the tropical regions is also needed to be explored.

1.5.2 Plant Adaptations

Not all the species are sensitive to the higher temperature and drought conditions, as some of the species may escape these extreme conditions by developing resilience and adaptations. In a study on species distribution under varying ranges of the European regions, Lososová et al. (2018) reported that about 45% of species do not show any direct relationship with the geographical distribution and climatic conditions. Thus, with the ongoing climate change, several species have been expected to decline, whereas some other species (particularly alien) which showed resilience to the climate change may spread, particularly in the European urban ecosystems. In other words, the space/niches created due to the decline of a species would be filled up by those species which have the tendency to cope-up with the increased temperature and drought conditions (Lososová et al. 2018). The fast-growing annuals (herbs) respond more quickly to the environmental changes as compared to the perennial herbs and woody vegetation (Grime 2001), thus these species may have wider distribution ranges in the future. The ruderal life strategy, production of large number of seeds/propagules and self-pollination traits help the annual herbs to track the environmental changes more quickly (Aarssen 1998; Lososová et al. 2018). Therefore, they are considered as the key indicators of the ongoing climate change. Further, the dominance of plants using the C_4 photosynthetic pathway has also been reported from the urban regions as compared to the non-urban regions of the European countries (Duffy and Chown 2016). It is expected that the C_4 plant species have more adaptive capacity to the localised warming conditions created by the UHI effects in the European regions which can be a strategy for the shift in future vegetation cover in these regions with the climate change (Duffy and Chown 2016). Such type of studies related to the species composition and climate change adaptation are needed from the tropical regions of the world as well.

1.5.3 Green Infrastructure

Green infrastructure is a strategic approach of landscape and environmental planning based on the principles of landscape and urban ecology (Niemelä 2014; Ramyar and Zarghami 2017). It represents the interconnected network of ecosystem structures such as green spaces (parks and gardens) and water bodies which are designed and managed for delivering numerous ecosystem services (European Commission 2013; Niemelä 2014; Green et al. 2016). These infrastructures are the critical components of the urban areas and provide several benefits to the humans in terms of ecological, environmental, and social challenges (Tzoulas and Greening 2011; Keniger et al. 2013; Ramyar and Zarghami 2017). Moreover, the concept of green infrastructure provides a common platform for scientists and researchers from ecology, social science, engineering, and landscape planners to interact and develop management systems for tackling different environmental issues (Pauleit et al. 2011). Green infrastructures can help in alleviating the impacts of UHI effect; however, UHI effect alternatively impacts the vegetation phenology, therefore, the role of these infrastructures in reducing UHI effect needs further research (Niemelä 2014; Dallimer et al. 2016). Therefore, increasing interests have been reported in investment in green infrastructure development and urban ecosystem regeneration in the recent years (Green et al. 2016). By having multiple benefits, green infrastructures are considered to play a major role in climate change adaptation (Ramyar and Zarghami 2017). Detailed viewpoint on different types of green infrastructures and their role in combating climate change has been given in the following sub-sections.

1.5.3.1 Green Space Development

Urban greening programmes are the leading features of the policies related to the climate change mitigation (Weissert et al. 2014). In view of ongoing climate change, potential research and management emphasis has been given to identifying the impacts of socio-demographic and environmental drivers on green spaces and the benefits derived from their conservation (Niemelä 2014; Verma et al. 2020c). The key tangible ecosystem services derived from the green spaces include mitigation of air pollution and UHI effect as well as physical and physiological health benefits to the residents (Verma et al. 2020b; Wang et al. 2020). Cooling effect provided by the green spaces is the most important ecosystem service which helps in mitigating UHI effect (Yu et al. 2017). The cooling effect of green spaces has been extensively explored by several researchers which involve two eco-physiological mechanisms viz. evapotranspiration and shadowing (Jiao et al. 2017; Wang et al. 2020). Size and characteristics (shape, structure, and composition/configuration) of green space is more important for cooling effect as it increases with increasing size of the green spaces (Jaganmohan et al. 2016; Yu et al. 2017); however, it is still controversial and several other factors come into the play (Monteiro et al. 2016). However, there is a threshold value of efficiency (TVoE) above which increase in vegetation cover may not lead to a consequent decrease in land surface temperature (Bao et al. 2016; Yu et al. 2017). Tree-based green spaces showed the highest cooling effect followed by bush and grassland (Kong et al. 2014). Since the canopy size and structure vary with the tree species, they provide different wind speed patterns which resulted in variable cooling effects (Armson et al. 2012). In addition, different trees have different eco-physiological mechanisms (e.g. evapotranspiration and leaf area index) which depend on the resource availability and management practices (Wang et al. 2020). Kuang et al. (2015) observed a positive correlation between the cooling effect of green space and normalised difference vegetation index (NDVI). In addition, the presence of water bodies along with the green spaces improve the cooling effect. For example, green spaces connected with water bodies showed higher cooling effect, whereas grassland-based green spaces showed weak cooling effect (Yang et al. 2020). Therefore, for climate change mitigation, interconnected green space and water body conservation and development are strongly suggested (Yu et al. 2017).

The magnitude of reduction of the temperature by one unit with the increase in vegetation (tree) cover is known as cooling efficiency (CE) which varies with the size and shape of the green spaces (Zhou et al. 2017; Wang et al. 2020). Compact green spaces with circular or square shapes provide more CE (Yu et al. 2017). Moreover, the regional climatic conditions also play a major role in shaping the vegetation structure and composition, thus, affecting CE (Richards et al. 2019). An extensive study was conducted by Wang et al. (2020) for 118 cities from 10 different biomes for observing the variation in CE. They used ordinary least squares (OLS) linear regression models with the percent of tree (Ptree) as independent variable and land surface temperature as dependent variable. They found the variation in CE from 0.04 to 0.57 °C with an average value of 0.17 °C where Ptree explained the 40.3% of the variation in the land surface temperature. Moreover, they observed that the biomes dominated by broadleaved trees reflect higher CE as compared to the biomes dominated by coniferous or sparse trees (e.g. savannas and shrublands). In addition to the inter-city variations, intra-city variations in CE were also observed, depending on the social and ecological contexts (Myint et al. 2015). Moreover, cities with hot and dry conditions (Mediterranean and desert biomes, e.g. Phoenix and Las Vegas) showed higher CE (Myint et al. 2015), whereas cities with hot and humid conditions (e.g. Nanjing, Beijing, Shenzhen, and Baltimore) reflected lower CE (Zhou et al. 2017). Exploring such scenarios for different cities from the developing and/or tropical nations could further help in mitigating UHI effects and adaptation to the climate change (Wang et al. 2020).

In addition to the direct reduction in land surface temperature, urban vegetation/tree shading also reduces the CO_2 emissions from the buildings by cutting air-conditioning costs during different weather conditions (Asgarian et al. 2015; Niemelä 2014). Moreover, visiting green space provides several benefits to the human in terms of health and well-being; however, there is no conclusive research on the mechanism underlying such results (Wu 2013). Therefore, there is a need for interdisciplinary researchers including the sociologists, psychologists, public health practitioners, anthropologists, and the ecologists to explore the mechanisms behind the linkages between human health and green spaces (Tzoulas and Greening 2011). Further, planning and management of green spaces of an area have been influenced by several factors including from the personal/resident/owner level to the government and political levels which determine the type and size of vegetation (Niemelä 2014). Urban green spaces in cities are needed to plan in such a way that they can contribute to the climate change mitigation, by integrating and planting trees having optimal evapotranspiration potential and physical shading (Niemelä 2014; Vasishth 2015).

1.5.3.2 Green-roofs

To mitigate the UHI effect, several measures have been applied and adapted in the urban areas. The examples include the use of heat-reflecting surfaces and materials like sunward-oriented roofs, roads and pavings, use of light-coloured materials, increase in the proportions of vegetation to the hard landscapes, etc. (Vasishth 2015). In addition to these measures, adoption of roof-top gardens or green-roofs is the emerging field of research for reducing the UHI effect and mitigating the climatic change (Niemelä 2014). As like the functions of green spaces, green roofs also help in cooling by evapotranspiration mechanism. Earlier, green roofs were thought as the burden to the buildings, but now they have been gaining wider attention by the researchers as well as the urban residents as the additional protective covering to the roofs for protection from the heat stress (Vasishth 2015). Studies revealed that well-designed and constructed green roofs may help in increasing the life of roofs and waterproofing structures even during the hot summer days (Vasishth 2015). However, research on exploring the ecosystem services to the humans (social and aesthetic benefits) provided by the green-roofs still needs attention of the scientific communities (Jungels et al. 2013; Niemelä 2014).

1.5.3.3 Green Building

The built infrastructures utilise massive amount of resources and energy and produce substantial amount of waste products and GHGs emission as the output (Singh and Raghubanshi 2020). In this view, the U.S. Green Building Council's (USGBC) Leadership in

Energy and Environmental Design (LEED) has launched a certification programme in 2000 with the aim '*to develop and encourage green building expertise across the entire building industry*' (www.usgbc.org). The LEED programme is one of the key certification programmes for developing site-specific building strategies and promoting the native plant diversity and water body conservation (Steiner 2014). This programme has four levels of certification, viz. certified, silver, gold, and platinum, depending on the credits received by a building project for six different categories which are: sustainable sites, water efficiency, indoor air quality, energy and atmosphere, innovation and the design process, and material and resources (Steiner 2014). Thus, the green buildings are the new initiatives launched by different agencies for the sustainable urban development and climate change adaptations. A few green building programmes viz. LEED India and the Green Rating for Integrated Habitat Assessment (GRIHA) in India are the two major certification programmes launched for regulating the infrastructure development and promoting sustainable building construction in India. Such programmes are working in congruence with the laws of the State and Central Governments of India (Singh and Raghubanshi 2020). Overall, researches on cost-effectiveness of green buildings and the benefits arising from them need to be scrutinised further in different regions of the world in view of climate change mitigation.

1.5.3.4 Urban Water Bodies

Most of the urbanisations have taken place at or around the bank of rivers/streams or water bodies, globally. Most of the water bodies flowing/situating along/around the cities have been overexploited and suffering from challenges like high pollution load, improper planning, and management to the extreme events (e.g. floods), etc. (Verma et al. 2020b). Similarly, most of the cities are characterised by receding groundwater levels and high water tables due to imbalances in the utilisation and recharge of water from the aquifers (de Graaf et al. 2019). However, the role of water bodies in mitigating the impact of climate change and improving urban health cannot be ignored or compromised. The hydrological cycle of the urban ecosystems determines the overall habitat and vegetation composition (Verma et al. 2020b). The water bodies act as urban cooling island (UCI) which played considerable role in mitigating UHI effect (Yu et al. 2017; Yang et al. 2020). For example, water bodies having square or circular shapes are more effective in providing the UCI effect as compared to water bodies with irregular or complicated shapes (Du et al. 2016). Thus, there is an urgent need to develop and design policies for urban water body management. This can be done by using a watershed management approach which will help in developing the water bodies along with their surrounding areas (and vegetation) by applying several modern tools like the use of remote sensing and GIS techniques (Ren et al. 2017).

1.5.4 Urban Vegetation and CO_2 Absorption

Vegetation stores a considerable amount of C in its different tissue components. Plants utilise the atmospheric CO_2 in the photosynthetic pathway to produce food and store C in their tissues (e.g. stem, branch, and roots) (Velasco and Roth 2010), thus, continuously help in mitigation of CO_2 emissions (Weissert et al. 2014). The C-sequestration potential of the urban vegetation holds a key motivation for their plantation as the climate change adaptation strategies (Schadler and Danks 2011). Studies suggested that during plantation drives, those areas which have limited vegetation should be planted first followed by areas having sufficient green cover for the effective and long-term understanding of the plant diversity, cover and health in relation to the surrounding conditions (Norton et al. 2015). Vegetation leads to reduction of atmospheric CO_2 considerably as compared to the other sectors of the urban areas, particularly during the growing seasons (Vesala et al. 2008). In a comprehensive review, Weissert et al. (2014) observed that dense vegetation may act as a potential local sink of atmospheric CO_2 in a city. They further reported that urban vegetation acts as a potential sink of atmospheric CO_2 during the growing season in the mid-latitude cities, whereas trees in tropical cities have the potential to absorb and sequester CO_2 from the residential areas throughout the year. However, comprehensive measurements and understanding of urban vegetation C-sequestration potential are still limited (Weissert et al. 2014). For example, the C-sink (or CO_2 uptake) potential of the urban vegetation may differ or considerably reduced when the overall emission of CO_2 from the urban areas are included in the overall C-budgeting (Weissert et al. 2014; Velasco et al. 2016; Zhao et al. 2016). Further, the C-sequestration potential of urban vegetation depends on climatic conditions, plant species, and management practices (Weissert et al. 2014). Temperature and precipitation are two major factors determining the plant establishment and growth attributes, thus, influencing the biomass accumulation and C-sequestration potential (Reich et al. 2014). Thus, changing climate scenario is also considerably impacting the C-sequestration potential of vegetation (Richards et al. 2019). Therefore, detailed spatio-temporal variation in urban vegetation C-storage behaviour in different regions of the world is needed to develop effective climate change mitigation strategies.

1.5.4.1 Urban Soils

The soils in the urban ecosystems have been extensively modified by different anthropogenic activities which include physical disturbances, waste deposition, filling materials, buildings, and management (irrigation and fertilisation) practices (Lorenz and Lal 2009; Raciti et al. 2012). As like urban vegetation, urban soils are the major storehouse of the C (Liu and Li 2012). Studies report that the urban soil C-stocks (organic and inorganic) are substantial as observed in the natural ecosystems (Vasenev and Kuzyakov 2018). Interestingly, urban soils contain a massive amount of C locked inside the impervious surfaces which have limited scope for decomposition, thus, act as a potential sink of C-stocks (Vasenev and Kuzyakov 2018). Moreover, several studies reported that the urban soils have significant potential of C-sequestration with proper management practices (Wang et al. 2019; Upadhyay et al. 2021). With the increase in atmospheric CO_2 concentration, temperature and growing seasons, and management practices, the inputs of C to the soil are also increasing which further improve the soil C-sequestration potential (Raciti et al. 2012). However, loss of soil organic C as soil CO_2 efflux with these changes in the surrounding conditions has also been observed in several studies (Raciti et al. 2012; Upadhyay et al. 2021). Therefore, there is a need to give more research attention on managing the urban soil C-stocks and improving the ways for more C-sequestration potential of soils, particularly in the tropical regions.

1.6 Conclusions and Future Research Directions

Humans are the major driving factors for most of the changes occurring in the urban ecosystems. Urban ecology provides a considerable understanding of the socio-ecological dimensions of urban ecosystems in relation to the human beings. For a more effective understanding of the processes and changes occurring in the urban ecosystems, concept of urban metabolism is getting wider attention. The climate change is affecting and expected to impact more severely the urban ecosystems in the near future. Urban green spaces and related water bodies provide several ecosystem services to the humans and can be developed as potential tools for the mitigation of climate change. In addition to green spaces, green roofs and green buildings are emerging as potential approaches for sustainable urban development. With proper planning and management strategies based on the integration of various emerging tools and techniques for developing resilient and self-regulating systems, urban ecology may help in mitigating the adverse effects of climate change on urban ecosystems.

Acknowledgements

Authors are thankful to University Grants Commission (UGC, Reference No. F.30-461/2019 (BSR)), and Science and Engineering Research Board (SERB, Reference No. PDF/2020/001607), New Delhi, India for financial support. RS is thankful to the Director, Institute of Environment & Sustainable Development (IESD), BHU, Varanasi; and Prof. Daizy R. Batish and Chairperson, Department of Botany, Panjab University, Chandigarh for providing necessary infrastructure for writing this chapter.

References

Aarssen, L.W. (1998). Why are most selfers annuals? A new hypothesis for the fitness benefit of selfing. *Oikos* 98: 606–612.

Akbari, H. and Kolokotsa, D. (2016). Three decades of urban heat islands and mitigation technologies research. *Energy and Buildings* 133: 834–842.

Alberti, M. (2008). *Advances in Urban Ecology: Integrating Humans and Ecological Processes in Urban Ecosystems*. New York: Springer-Verlag.

Andersson, E., Barthel, S., Borgström, S. et al. (2014). Reconnecting cities to the biosphere: stewardship of green infrastructure and urban ecosystem services. *Ambio* 43 (4): 445–453. https://doi.org/10.1007/s13280-014-0506-y.

Armson, D., Stringer, P., and Ennos, A. (2012). The effect of tree shade and grass on surface and globe temperatures in an urban area. *Urban Forestry & Urban Greening* 11: 245–255.

Aronson, M.F.J., Lepczyk, C.A., Evans, K.L. et al. (2017). Biodiversity in the city: key challenges for urban green space management. *Frontiers in Ecology and the Environment* 15: 189–196. https://doi.org/10.1002/fee.1480.

Asgarian, A., Amiri, B.J., and Sakieh, Y. (2015). Assessing the effect of green cover spatial patterns on urban land surface temperature using landscape metrics approach. *Urban Ecosystem* 18: 209–222.

Bao, T., Li, X., Zhang, J. et al. (2016). Assessing the distribution of urban green spaces and its anisotropic cooling distance on urban heat island pattern in Baotou, China. *ISPRS International Journal of Geo-Information* 5: 12.

Battles, A.C. and Kolbe, J.J. (2019). Miami heat: urban heat islands influence the thermal suitability of habitats for ectotherms. *Global Change Biology* 25: 562–576.

Carpenter, S.R. and Folke, C. (2006). Ecology for transformation. *Trends in Ecology and Evolution* 21: 309–315.

Čeplová, N., Kalusová, V., and Lososová, Z. (2017). Effects of settlement size, urban heat island and habitat type on urban plant biodiversity. *Landscape and Urban Planning* 159: 15–22.

Chang, C., Lee, X., Liu, S. et al. (2016). Urban heat islands in China enhanced by haze pollution. *Nature Communications* 7: 12509.

Chapin, F.S. III, Power, M.E., Pickett, S.T.A. et al. (2011). Earth stewardship: science for action to sustain the human–earthsystem. *Ecosphere* 2 (8): 1–20.

Chapman, S., Watson, J.E.M., Salazar, A. et al. (2017). The impact of urbanization and climate change on urban temperatures: a systematic review. *Landscape Ecology* 32: 1921–1935.

Childers, D.L., Pickett, S.T., Grove, J.M. et al. (2014). Advancing urban sustainability theory and action: challenges and opportunities. *Landscape and Urban Planning* 125: 320–328.

Childers, D.L., Cadenasso, M.L., Grove, J.M. et al. (2015). An ecology for cities: a transformational nexus of design and ecology to advance climate change resilience and urban sustainability. *Sustainability* 7 (4): 3774–3791.

Colding, J. and Barthel, S. (2017). An urban ecology critique on the "smart city" model. *Journal of Cleaner Production* 164: 95–101. https://doi.org/10.1016/j.jclepro.2017.06.191.

Cong, N., Wang, T., Nan, H. et al. (2013). Changes in satellite-derived spring vegetation green-up date and its linkage to climate in China from 1982 to 2010: a multimethod analysis. *Global Change Biology* 19: 881–891.

Cook-Patton, S.C. and Agrawal, A.A. (2014). Exotic plants contribute positively to biodiversity functions but reduce native seed production and arthropod richness. *Ecology* 95: 1642–1650.

Costanza, R. (1992). Toward an operational definition of ecosystem Health. In: *Ecosystem Health: New Goals for Environmental Management* (eds. R. Constanza, B.G. Norton and B.D. Haskell), 239–256. Washington, DC: Island Press.

Cubino, J.P., Borowy, D., Knapp, S. et al. (2021). Contrasting impacts of cultivated exotics on the functional diversity of domestic gardens in three regions with different aridity. *Ecosystems* 24: 1–16.

Dallimer, M., Tang, Z., Gaston, K.J., and Davies, Z.G. (2016). The extent of shifts in vegetation phenology between rural and urban areas within a human-dominated region. *Ecology and Evolution* 6 (7): 1942–1953.

Davies, Z.G., Edmondson, J., Heinemeyer, A. et al. (2011). Mapping an urban ecosystem service: quantifying above-ground carbon storage at a city-wide scale. *Journal of Applied Ecology* 48 (5): 1125–1134.

de Graaf, I.E., Gleeson, T., van Beek, L.R. et al. (2019). Environmental flow limits to global groundwater pumping. *Nature* 574 (7776): 90–94.

Du, H., Song, X., Jiang, H. et al. (2016). Research on the cooling island effects of water body: a case study of Shanghai, China. *Ecological Indicators* 67: 31–38.

Duffy, G.A. and Chown, S.L. (2016). Urban warming favours C_4 plants in temperate European cities. *Journal of Ecology* 104 (6): 1618–1626.

Elmqvist, T., Fragkias, M., Goodness, J. et al. (eds.) (2013). *Urbanization, Biodiversity and Ecosystem Services: Challenges and Opportunities*. Dordrecht: Springer.

European Commission (2013). *Green Infrastructure (GI): Enhancing Europe's Natural Capital*. Brussels: European Commission.

Ferrer, A.L.C., Thomé, A.M.T., and Scavarda, A.J. (2018). Sustainable urban infrastructure: a review. *Resources, Conservation and Recycling* 128: 360–372. https://doi.org/10.1016/j.resconrec.2016.07.017.

Forman, R.T. (2014). *Urban Ecology: Science of Cities*. Cambridge University Press.

Gaston, K.J., Davies, Z.G., and Edmondson, J.L. (2010). Urban environments and ecosystem functions. In: *Urban Ecology* (ed. K.J. Gaston), 35–52. Cambridge: Cambridge University Press.

Gaston, K.J., Ávila-Jiménez, M.L., and Edmondson, J. (2013). Managing urban ecosystems for goods and services. *Journal of Applied Ecology* 50 (4): 830–840.

Green, T.L., Kronenberg, J., Andersson, E. et al. (2016). Insurance value of green infrastructure in and around cities. *Ecosystems* 19 (6): 1051–1063.

Grime, J.P. (2001). *Plant Strategies, Vegetation Processes, and Ecosystem Properties*. Chichester, UK: Wiley.

Grimm, N.B., Grove, J.M., Pickett, S.T.A., and Redman, C.L. (2000). Integrated approaches to long-term studies of urban ecological systems. *Bioscience* 50: 571–584.

Grimm, N., Faeth, S., Golubiewski, N. et al. (2008). Global change and the ecology of cities. *Science* 319: 756–760.

Halpern, B.S., Walbridge, S., Selkoe, K.A. et al. (2008). A global map of human impact on marine ecosystems. *Science* 319: 948–952.

Hofmann, M., Westermann, J.R., Kowarik, I., and van der Meer, E. (2012). Perceptions of parks and urban derelict land by landscape planners and residents. *Urban Forestry & Urban Greening* 11: 303–312.

Hu, Y., Hou, M., Jia, G. et al. (2019). Comparison of surface and canopy urban heat islands within megacities of eastern China. *ISPRS Journal of Photogrammetry and Remote Sensing* 156: 160–168.

Hutyra, L.R., Yoon, B., and Alberti, M. (2011). Terrestrial carbon stocks across a gradient of urbanization: a study of the Seattle, WA region. *Global Change Biology* 17 (2): 783–797.

Jaganmohan, M., Knapp, S., Buchmann, C.M., and Schwarz, N. (2016). The bigger, the better? The influence of urban green space design on cooling effects for residential areas. *Journal of Environmental Quality* 45: 134–145.

Jeong, S.-J., Ho, C.-H., Gim, H.-J., and Brown, M.E. (2011). Phenology shifts at start vs. end of growing season in temperate vegetation over the Northern Hemisphere for the period 1982–2008. *Global Change Biology* 17: 2385–2399.

Jiao, M., Zhou, W., Zheng, Z. et al. (2017). Patch size of trees affects its cooling effectiveness: a perspective from shading and transpiration processes. *Agricultural and Forest Meteorology* 247: 293–299.

Jim, C.Y. (2011). Holistic research agenda for sustainable management and conservation of urban woodlands. *Landscape and Urban Planning* 100: 375–379.

Jungels, J., Rakow, D.A., Allred, S.B., and Skelly, S.M. (2013). Attitudes and aesthetic reactions toward green roofs in the Northeastern United States. *Landscape and Urban Planning* 117: 13–21.

Kattel, G.R., Elkadi, H., and Meikle, H. (2013). Developing a complementary framework for urban ecology. *Urban Forestry & Urban Greening* 12 (4): 498–508.

Keniger, L.E., Gaston, K.J., Irvine, K.N., and Fuller, R.A. (2013). What are the benefits of interacting with nature? *International Journal of Environmental Research and Public Health* 10: 913–935.

Knapp, S., Winter, M., and Klotz, S. (2017). Increasing species richness but decreasing phylogenetic richness and divergence over a 320-year period of urbanization. *Journal of Applied Ecology* 54: 1152–1160. https://doi.org/10.1111/1365 2664.12826.

Kong, F., Yin, H., James, P. et al. (2014). Effects of spatial pattern of greenspace on urban cooling in a large metropolitan area of eastern China. *Landscape and Urban Planning* 128: 35–47.

Kuang, W., Liu, Y., Dou, Y. et al. (2015). What are hot and what are not in an urban landscape: quantifying and explaining the land surface temperature pattern in Beijing, China. *Landscape Ecology* 30: 357–373.

Lee, R. (2011). The outlook for population growth. *Science* 333: 569–573.

Liu, C. and Li, X. (2012). Carbon storage and sequestration by urban forests in Shenyang, China. *Urban Forestry & Urban Greening* 11: 121–128.

Liu, J., Hull, V., Batistella, M. et al. (2013). Framing sustainability in a telecoupled world. *Ecology and Society* 18 (2): 26. https://doi.org/10.5751/ES-05873-180226.

Livesley, S.J., Escobedo, F.J., and Morgenroth, J. (2016). The biodiversity of urban and peri-urban forests and the diverse ecosystem services they provide as socio-ecological systems. *Forests* 7 (12): 291. https://doi.org/10.3390/f7120291.

Lorenz, K. and Lal, R. (2009). Biogeochemical C and N cycles in urban soils. *Environment International* 35: 1–8.

Lososová, Z., Tichý, L., Divíšek, J. et al. (2018). Projecting potential future shifts in species composition of European urban plant communities. *Diversity and Distributions* 24 (6): 765–775.

Ma, Y., Wright, J., Gopal, S., and Phillips, N. (2020). Seeing the invisible: from imagined to virtual urban landscapes. *Cities* 98: 102559.

McDonald, R. and Marcotullio, P. (2011). Global effects of urbanization on ecosystemservices. In: *Urban Ecology – Patterns, Processes, and Applications* (eds. J. Niemelä, J. Breuste, T. Elmqvist, et al.), 193–205. Oxford: Oxford University Press.

McDonnell, M., Breuste, J., and Hahs, A. (eds.) (2009). *Ecology of Cities and Towns: A Comparative Approach*. Cambridge: Cambridge University Press.

Menzel, A. and Fabian, P. (1999). Growing season extended in Europe. *Nature* 397: 659.

Milton, S.J., Wilson, J.R.U., Richardson, D.M. et al. (2007). Invasive alien plants infiltrate birdmediated shrub nucleation processes in arid savanna. *Journal of Ecology* 95: 648–661.

Mitchell, M.G., Johansen, K., Maron, M. et al. (2018). Identification of fine scale and landscape scale drivers of urban aboveground carbon stocks using high-resolution modeling and mapping. *Science of the Total Environment* 622: 57–70.

Møller, L.A., Skou, A.M.T., and Kollmann, J. (2012). Dispersal limitation at the expanding range margin of an evergreen tree in urban habitats? *Urban Forestry & Urban Greening* 11 (1): 59–64.

Monteiro, M.V., Doick, K.J., Handley, P., and Peace, A. (2016). The impact of greenspace size on the extent of local nocturnal air temperature cooling in London. *Urban Forestry & Urban Greening* 16: 160–169.

Mullaney, J., Lucke, T., and Trueman, S.J. (2015). A review of benefits and challenges in growing street trees in paved urban environments. *Landscape and Urban Planning* 134: 157–166. https://doi.org/10.1016/j.landurbplan.2014.10.013.

Myint, S.W., Zheng, B.J., Talen, E. et al. (2015). Does the spatial arrangement of urban landscape matter? Examples of urban warming and cooling in Phoenix and Las Vegas. *Ecosystem Health and Sustainability* 1: 1–15.

Neil, K., Landrum, L., and Wu, J. (2010). Effects of urbanization on flowering phenology in the metropolitan Phoenix region of USA: findings from herbarium records. *Journal of Arid Environments* 74: 440–444.

Neil, K., Wu, J., Bang, C., and Faeth, S. (2014). Urbanization affects plant flowering phenology and pollinator community: effects of water availability and land cover. *Ecological Processes* 3 (1): 1–12.

Niemelä, J., Saarela, S.-R., Söderman, T. et al. (2010). Using the ecosystem services approach for better planning and conservation of urban green spaces: a Finland case study. *Biodiversity and Conservation* 19: 3225–3243.

Niemelä, J., Breuste, J., Elmqvist, T. et al. (eds.) (2011). *Urban Ecology – Patterns, Processes, and Applications*. Oxford: Oxford University Press.

Niemelä, J. (2014). Ecology of urban green spaces: the way forward in answering major research questions. *Landscape and Urban Planning* 125: 298–303.

Norton, B.A., Coutts, A.M., Livesley, S.J. et al. (2015). Planning for cooler cities: a framework to prioritise green infrastructure to mitigate high temperatures in urban landscapes. *Landscape and Urban Planning* 134: 127–138.

Nowak, D.J., Greenfield, E.J., Hoehn, R.E., and Lapoint, E. (2013). Carbon storage and sequestration by trees in urban and community areas of the United States. *Environmental Pollution* 178: 229–236.

Oke, T.R. (2002). *Boundary Layer Climates*. London and New York: Routledge.

Pandit, A., Lu, Z., and Crittenden, J.C. (2015). Managing the complexity of urban systems. *Journal of Industrial Ecology* 19 (2): 201–204. https://doi.org/10.1111/jiec.12263.

Pauleit, S., Liu, L., Ahern, J., and Kazmierczak, A. (2011). Multifunctional green infrastructure planning to promote ecological services in the city. In: *Urbanecology – Patterns, Processes, and Applications* (eds. J. Niemelä, J. Breuste, T. Elmqvist, et al.), 272–285. Oxford: Oxford University Press.

Pearse, W.D., Cavender-Bares, J., Hobbie, S.E. et al. (2018). Homogenization of plant diversity, composition, and structure in North American urban yards. *Ecosphere* 9: e02105.

Pedersen Zari, M. (2019). Devising urban biodiversity habitat provision goals: ecosystem services analysis. *Forests* 10 (5): 391.

Phelan, P.E., Kaloush, K., Miner, M. et al. (2015). Urban heat island: mechanisms, implications, and possible remedies. *Annual Review of Environment and Resources* 40: 285–307.

Pickett, S.T.A., Cadenaso, M.L., Grove, J.M. et al. (2001). Urban ecological systems: linking terrestrial ecological, physical, and socioeconomic components of metropolitan areas. *Annual Review of Ecology and Systematics* 32: 127–157.

Pickett, S.T.A., Cadenasso, M.L., Grove, J.M. et al. (2011). Urban ecological systems: scientific foundations and a decade of progress. *Journal of Environmental Management* 92: 331–362.

Pickett, S.T., Cadenasso, M.L., Childers, D.L. et al. (2016). Evolution and future of urban ecological science: ecology in, of, and for the city. *Ecosystem Health and Sustainability* 2 (7): e01229.

Raciti, S.M., Hutyra, L.R., Rao, P., and Finzi, A.C. (2012). Inconsistent definitions of "urban" result in different conclusions about the size of urban carbon and nitrogen stocks. *Ecological Applications* 22: 1–58.

Rai, P.K. and Kim, K.H. (2019). Invasive alien plants and environmental remediation: a new paradigm for sustainable restoration ecology. *Restoration Ecology* 28 (1): 3–7.

Ramyar, R. and Zarghami, E. (2017). Green infrastructure contribution for climate change adaptation in urban landscape context. *Applied Ecology and Environmental Research* 15 (3): 1193–1209.

Reich, P.B., Luo, Y., Bradford, J.B. et al. (2014). Temperature drives global patterns in forest biomass distribution in leaves, stems, and roots. *Proceedings of the National Academy of Sciences of the United States of America* 111: 13721–13726.

Ren, N., Wang, Q., Wang, Q. et al. (2017). Upgrading to urban water system 3.0 through sponge city construction. *Frontiers of Environmental Science and Engineering* 11 (4): 9.

Restrepo, J.D.C. and Morales-Pinzon, T. (2018). Urban metabolism and sustainability: precedents, genesis and research perspectives. *Resources, Conservation and Recycling* 131: 216–224.

Richards, D., Masoudi, M., Oh, R.R. et al. (2019). Global variation in climate, human development, and population density has implications for urban ecosystem services. *Sustainability* 11 (22): 6200.

Richardson, D.M., Pyšek, P., Rejmánek, M. et al. (2000). Naturalization and invasion of alien plants: concepts and definitions. *Diversity and Distributions* 6: 93–107. https://doi.org/10.1046/j.1472-4642.2000.00083.x.

Rosenzweig, C., Solecki, W., Hammerm, S.A., and Mehrotra, S. (2010). Cities lead the way in climate-change action. *Nature* 467: 909–911.

Savard, J.P.L., Clergeau, P., and Mennechez, G. (2000). Biodiversity concepts and urban ecosystems. *Landscape and Urban Planning* 4: 131–142. https://doi.org/10.1016/S0169-2046(00)00037-2.

Schadler, E. and Danks, C. (2011). *Carbon Offsetting through Urban Tree Planting*. Burlington, VT: University of Vermont.

Seto, K.C. and Shepherd, J.M. (2009). Global urban land-usetrends and climate impacts. *Current Opinion in Environment Sustainability* 1: 89–95.

Seto, K.C., Güneralp, B., and Hutyra, L.R. (2012). Global forecasts of urban expansion to 2030 and direct impacts on biodiversity and carbon pools. *Proceedings of the National Academy of Sciences of the United States of America* 109 (40): 16083–16088.

Singh, R.P. and Raghubanshi, A.S. (2020). 'Green building' movement in India: study on institutional support and regulatory support. In: *Urban Ecology*, 435–455. Elsevier.

Singh, P., Borthakur, A., Singh, R. et al. (2021). A critical review on the research trends and emerging technologies for arsenic decontamination from water. *Groundwater for Sustainable Development* 14: 100607.

Steiner, F. (2014). Frontiers in urban ecological design and planning research. *Landscape and Urban Planning* 125: 304–311.

Sun, Y. and Cui, Y. (2018). Evaluating the coordinated development of economic, social and environmental benefits of urban public transportation infrastructure: case study of four

Chinese autonomous municipalities. *Transport Policy* 66: 116–126. https://doi.org/10.1016/j.tranpol.2018.02.006.

Tigges, J., Lakes, T., and Hostert, P. (2013). Urban vegetation classification: Benefits of multitemporal RapidEye satellite data. *Remote Sensing of Environment* 136: 66–75.

Tzoulas, K. and Greening, K. (2011). Urban ecology and human health. In: *Urban Ecology – Patterns, Processes, and Applications* (eds. J. Niemelä, J. Breuste, T. Elmqvist, et al.), 263–271. Oxford: Oxford University Press.

UN (2011). Department of Economic and Social Affairs; World Population Ageing 2009. *Population and Development Review* 37: 403.

Upadhyay, S., Singh, R., Verma, P., and Raghubanshi, A.S. (2021). Spatio-temporal variability in soil CO_2 efflux and regulatory physicochemical parameters from the tropical urban natural and anthropogenic land use classes. *Journal of Environmental Management* 295: 113141.

Vasenev, V. and Kuzyakov, Y. (2018). Urban soils as hot spots of anthropogenic carbon accumulation: review of stocks, mechanisms and driving factors. *Land Degradation & Development* 29 (6): 1607–1622.

Vasishth, A. (2015). Ecologizing our cities: a particular, process-function view of southern California, from within complexity. *Sustainability* 7 (9): 11756–11776.

Velasco, E. and Roth, M. (2010). Cities as net sources of CO_2: review of atmospheric CO_2 exchange in urban environments measured by Eddy covariance technique. *Geography Compass* 4: 1238–1259.

Velasco, E., Roth, M., Norford, L., and Molina, L.T. (2016). Does urban vegetation enhance carbon sequestration? *Landscape and Urban Planning* 148: 99–107.

Verma, P., Singh, R., Bryant, C., and Raghubanshi, A.S. (2020c). Green space indicators in a social-ecological system: a case study of Varanasi, India. *Sustainable Cities and Society* 60: 102261.

Verma, P., Singh, R., Singh, P., and Raghubanshi, A.S. (2020a). Urban ecology–current state of research and concepts. In: *Urban Ecology* (eds. P. Verma, P. Singh, R. Singh and A.S. Raghubanshi), 3–16. Elsevier.

Verma, P., Singh, R., Singh, P., and Raghubanshi, A.S. (2020b). Critical assessment and future dimensions for the urban ecological systems. In: *Urban Ecology* (eds. P. Verma, P. Singh, R. Singh and A.S. Raghubanshi), 479–497. Elsevier.

Vesala, T., Järvi, L., Launiainen, S. et al. (2008). Surface–atmosphere interactions over complex urban terrain in Helsinki, Finland. *Tellus B* 60: 188–199.

Vitousek, P.M. (1994). Beyond global warming: ecology and global change–Macarthur award lecture. *Ecology* 75: 1861–1876.

Walther, G.R., Roques, A., Hulme, P.E. et al. (2009). Alien species in a warmer world: risks and opportunities. *Trends in Ecology and Evolution* 24: 686–693. https://doi.org/10.1016/j.tree.2009.06.008.

Wamsler, C., Brink, E., and Rivera, C. (2013). Planning for climate change in urban areas: from theory to practice. *Journal of Cleaner Production* 50: 68–81.

Wang, C., Wang, Z.-H., Wang, C., and Myint, S.W. (2019). Environmental cooling provided by urban trees under extreme heat and cold waves in U.S. cities. *Remote Sensing of Environment* 227: 28–43.

Wang, J., Zhou, W., Jiao, M. et al. (2020). Significant effects of ecological context on urban trees' cooling efficiency. *ISPRS Journal of Photogrammetry and Remote Sensing* 159: 78–89.

Wang, X., Zhang, Y., Zhang, J. et al. (2021). Progress in urban metabolism research and hotspot analysis based on CiteSpace analysis. *Journal of Cleaner Production* 281: 125224.

Weissert, L.F., Salmond, J.A., and Schwendenmann, L. (2014). A review of the current progress in quantifying the potential of urban forests to mitigate urban CO_2 emissions. *Urban Climate* 8: 100–125.

Wolman, A. (1965). The metabolism of cities. *Scientific American* 213 (3): 179–190.

Wong, P.P.-Y., Lai, P.-C., Low, C.-T. et al. (2016). The impact of environmental and human factors on urban heat and microclimate variability. *Building and Environment* 95: 199–208.

Wu, J.G. (2013). The state-of-the-science in urban ecology and sustainability: a landscape perspective. *Landscape and Urban Planning* 125 (6): 298–303.

Yang, G., Yu, Z., Jørgensen, G., and Vejre, H. (2020). How can urban blue-green space be planned for climate adaption in high-latitude cities? A seasonal perspective. *Sustainable Cities and Society* 53: 101932.

Yu, Z., Guo, X., Jørgensen, G., and Vejre, H. (2017). How can urban green spaces be planned for climate adaptation in subtropical cities? *Ecological Indicators* 82: 152–162.

Zhao, L., Lee, X., Smith, R.B., and Oleson, K. (2014). Strong contributions of local background climate to urban heat islands. *Nature* 511: 216–219.

Zhao, S., Tang, Y., and Chen, A. (2016). Carbon storage and sequestration of urban street trees in Beijing, China. *Frontiers in Ecology and Evolution* 4: 1–8.

Zhou, W., Wang, J., and Cadenasso, M.L. (2017). Effects of the spatial configuration of trees on urban heat mitigation: a comparative study. *Remote Sensing of Environment* 195: 1–12.

Zipperer, W.C., Morse, W.C., and Gaither, J.G. (2011). Linking social and ecological systems. In: *Urban Ecology: Patterns, Processes, and Applications* (eds. J. Niemelä, J.H. Breuste, T. Elmqvist, et al.). New York: Oxford University Press.

Web links

Scopus database (2021). https://www.scopus.com/results/results.uri?sid=7bafb47309c29c7ec0444cd42ebf2afd&src=s&sot=b&sdt=b&origin=searchbasic&rr=&sl=51&s=TITLE-ABS-KEY(%22Urban%20ecology%22%20AND%20%22Climate%20change%22)&searchterm1=%-22Urban%20ecology%22%20AND%20%22Climate%20change%22&searchTerms=&connectors=&field1=TITLE_ABS_KEY&fields= (accessed 13 June 2021).

Web of Science Core Collection database (2021). https://www.webofscience.com/wos/woscc/summary/d7ee688e-5928-4013-8a00-75055184c8f8-0dcf753b/relevance/1 (accessed 13 June 2021).

2

Climate Change, Urbanisation, and Their Impact on Increased Occurrence of Cardiometabolic Syndrome*

Saptamita P. Choudhury[1,2], Arisha Arora[3,4], Nishi Jain[2,5], and Sanjay Kumar Dey[2]

[1] *School of Biotechnology, Kalinga Institute of Industrial Technology, Bhubaneswar, Odisha, India*
[2] *Dr. B.R. Ambedkar Center for Biomedical Research, University of Delhi, New Delhi, Delhi, India*
[3] *Department of Biotechnology, Motilal Nehru National Institute of Technology Allahabad, Prayagraj, Uttar Pradesh, India*
[4] *Department of Biosciences and Bioengineering, Indian Institute of Technology Guwahati, Assam, India*
[5] *Department of Biotechnology, Amity University-Noida, Uttar Pradesh, India*

2.1 Introduction

Environmental factors act as key facilitators for chronic non-communicable diseases. Similarly, urbanisation and climate change exaggerate the occurrences of such disorders. Over 60% of the world's population resides in cities and towns due to the huge rise in global development, and this proportion is going to increase up to 90% over the next few years (Nieuwenhuijsen 2018). Urban areas are the source of new discoveries and economic development but are also sources of pollution and diseases. With increasing population, human activities are also contributing to a major climate change. Climate changes are one of the defining issues of time and we are at the defining moment. The need for private vehicles is increasing day by day in order to have proper time management, to do work at a faster pace, and to maintain safety. According to the World Heart Federation report on urbanisation and cardiovascular diseases (CVD), human activities are changing with emerging urbanisation which results in pollution, loss of biodiversity, disturbance in ecosystem, and also Earth's temperature has elevated by the range of 0.85 °C approximately withinside the twentieth century and most of this warming came about by the year 1975 (De Blois et al. 2015). Over the last 30 years, the Global warming rate has increased by about 0.18 °C (De Blois et al. 2015). This was triggered by the poor development of infrastructures and less functional public transportation, lack of space, green areas, overuse of natural resources like coal, petroleum, global warming, etc. (Figure 2.1). All these are finally leading to subsequent higher rates of cardiovascular-related morbidity and mortality (Nieuwenhuijsen 2018). However, newer cardiometabolic treatments and

---

*Saptamita P. Choudhury, Arisha Arora, and Nishi Jain have equally contributed for this book chapter. Sanjay Kumar Dey is the corresponding author for the current book chapter.

Urban Ecology and Global Climate Change, First Edition. Edited by Rahul Bhadouria, Shweta Upadhyay, Sachchidanand Tripathi, and Pardeep Singh.

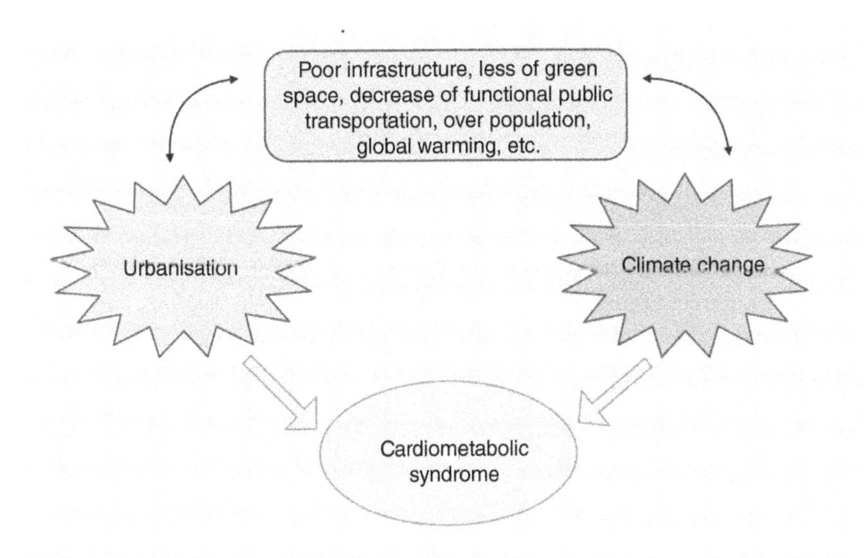

Figure 2.1 An overview of the various factors that are responsible for urbanisation and climate change finally leading to cardiometabolic syndrome.

therapeutic approaches can also help to reduce the burden of these syndromes (De Blois et al. 2015). To bring cardiometabolic syndrome-related risks under control, there are certain opportunities as well as challenges which have been described in this chapter.

2.2 Overview of Cardiometabolic Syndromes

Cardiometabolic syndrome is a cluster of insulin resistance and restricted high cholesterol responsiveness, elevated fasting, and epilepsy, which are all factors that influence glucose metabolism characterised by metabolic dysfunction. Often a list of cardiometabolic syndrome threats for individuals arises with glucose sensitivity. Individuals having cardiometabolic syndrome are substantially more likely to suffer from metabolic syndrome and twice as likely to have a sudden cardiac arrest (World Heart Federation 2015). The manufacturing and livestock revolutions in history produced more cholesterol-rich crops and carbs for intake than humanity needs (Miles et al. 2019). The addition of processed carbs, an abundance of saturated fats, and the shift from predator to civilised people have all led to the growth of obesity. According to the Global Burden of Disease report, India's adult cardiometabolic syndrome disease burden of 272 per 100 000 population was higher than the global average (Prabhakaran et al. 2016).

2.3 Pathophysiology of Cardiometabolic Syndromes

A multitude of pathophysiological cardiometabolic variables have been correlated with the severity of cardiovascular events. The variance of fat cells is a critical aspect of cardiometabolic risk. The cardiac syndrome's most prominent interpretation is abdominal obesity.

Adipose tissue is also an endocrine organ that expels adipokines which make a contribution to the atherogenic/diabetic physiological risk level attributed to hyperlipidaemia. A mismatch between energy consumption and expenditure contributes to abdominal fat. The pathogenicity of hyperglycaemia and hyperlipidaemia correlated with the cardiometabolic syndrome is probably triggered by variations in free fatty acid metabolism. It is a metabolically active tissue that metabolises a number of up-regulation and thrombogenic immune cells. Elevated plasma unsaturated fatty acid accumulation and inordinate release of lipids from adipocytes can hinder the ability of insulin to stimulate muscle glucose metabolism and suppress hepatic glucose production (Figure 2.2) (Kirk and Klein 2009). In juveniles, insulin tolerance tends to be related to a decline in the mitochondrial to nuclear DNA ratio. Elevated secretion of lipids from adipocytes is attributed to insulin sensitivity leading to a decline in glucose transmission through the muscles (Gill et al. 2005). Hyperinsulinemia has been reported as a significant prognostic factor in broad prospective epidemiologic

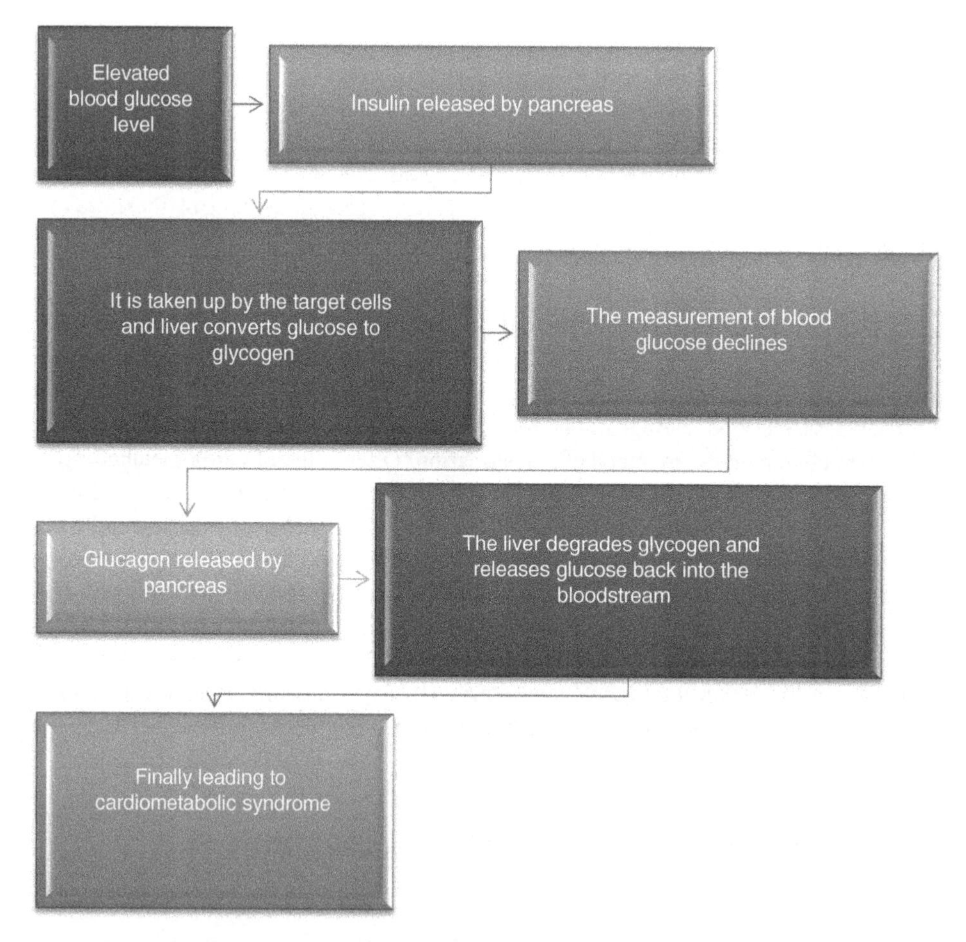

Figure 2.2 Dysregulation of sugar metabolism leads to cardiometabolic syndrome. *Source:* Based on Kirk and Klein (2009).

trials. In conclusion, there is considerable proof that the most prevalent forms of the metabolic syndrome are correlated to abdominal obesity, especially when it is followed by abdominal adipocytes accumulation.

2.4 Urbanisation as a Factor to Increase Cardiometabolic and Cardiovascular Disorders

2.4.1 The Driving Development of Urbanisation and Its Implications on Cardiovascular Syndrome in the Twenty-First Century

The birth of the metropolis has been a recent development in the latest generations. Backwoods regions have previously been urbanised; along with that many people have migrated to urban areas. In recent studies, it has been observed that the various risks of cardiometabolic and CVD begin in premature conditions of pregnant mothers, or at the time of birth. The risk increases more due to many factors like lack of physical activity, unhealthy diet, consumption of alcohol and tobacco, smoking, etc. Many of these exposures have increased due to the negative impacts of urbanisation. Although urbanisation has brought major improvements in the economy and many lifestyle opportunities like working variations, diversity to education, fast internet, and social world development and political mobilisation, but this poses a great obstruction towards maintaining a healthy lifestyle and behaviour (Figure 2.3). Taking into consideration man-made landscapes, often in certain cases, industrialisation has occurred way too fast, leading to many defects in the construction of living places. As a result, a person has started to live in insubstantial conditions starting from cardboard boxes to pavements, under bridges, near streets, slums, sidewalks tents, etc. Individuals living in such conditions have low nutritious diets and less

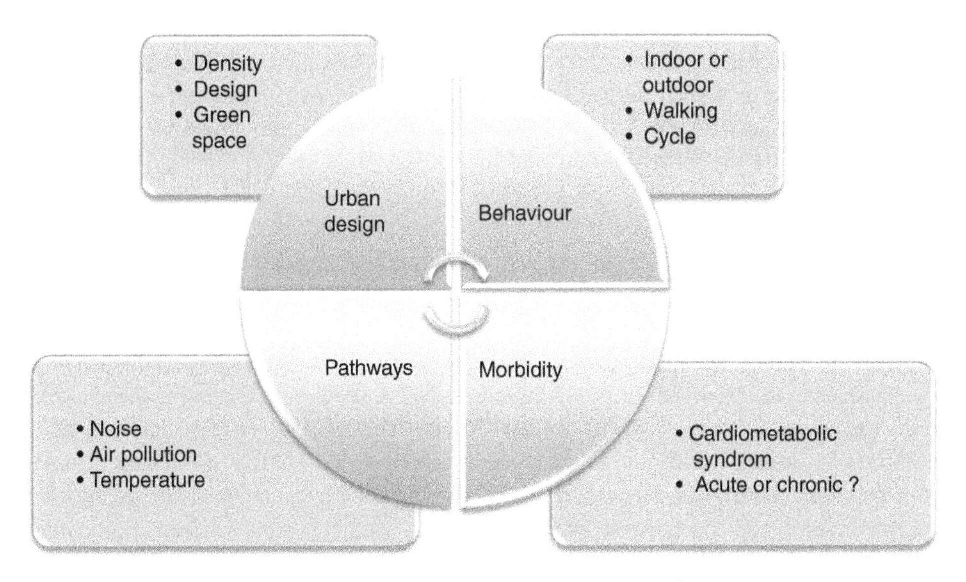

Figure 2.3 Complex urban planning and its impact on cardiometabolic syndrome.

opportunities for medication and other daily exercises. People living in extremely well-maintained houses are forced to sedentary behaviour due to lack of physical activity, less active in playing outside environment, not only that but much more addiction towards drinking and smoking. As a result, people residing in crowded places have a high risk to develop rheumatic heart disease (RHD), which causes damage to the heart muscle and heart valves. On the other hand, it has been said that children may suffer obesity due to maternal obesity during pregnancy; however, there is less proof to this statement (Castro et al. 2003). Individuals of low middle economic status are more open to street and cheap foods since they are easily available like open vendors as they have limited budget and less choice. Moreover, it has been noticed that people living in well-maintained conditions are also at threat of developing cardiometabolic syndrome often seen due to much addiction in the social world (Münzel et al. 2017a). In addition to that smoking rate, consumption of drugs, alcohol among youth has increased rapidly which is ultimately leading them to stress, depression, anxiety, obesity, and early-stage diabetes.

2.4.2 Mutualistic Relationship Between Urbanisation and Ecosystem

Industrialisation has been one of the most prominent causes of population changes in recent years, which is propelled by a multitude of societal, financial, and ecological mechanisms. Through the development of towns, communities, and infrastructure upgrades, urbanisation altered natural and previous rural habitats (Miller and Hutchins 2017). Smart employee population levels, expanded rough areas (e.g. roads and buildings), enhanced toxicity (e.g. air quality, light, soil), and high temperature are all characteristics of the novel, human urban setting. The urban sprawl resonance is a form in which cities are hotter than non-urban areas due to the increased impermeable surfaces (e.g. gravel and mortar) and significantly lower tree cover. The trend and intensity of the interaction between urbanisation and ecological consequences can differ and evolve with present, based on the geographic, societal, and financial factors as well as progress trajectories, according to emerging evidence (Bai et al. 2017). Poor air quality is a dynamic combination of airborne pollutants emitted by a wide range of sources, including factories, residential gasification heating systems, automobiles, and industrialisation. Domestic air emissions and urban chemical fumes are the third and ninth leading causes of death and disease, respectively. The latter two are contributing for 6.6 million deaths and 7.6% of global, with pollutants accounting for 3.5% of global disease burden (Münzel et al. 2017a).

CVD is also known to be the leading cause of mortality (Dey et al. 2020). Air contamination containing fine particulate matter with a diameter of less than 2.5 μm (PM2.5) is the world's leading cause of death rates. Rigorous disposal of pollutants from urban centres, combined with an increase in impermeable surfaces as a result of urbanisation, will lead to a steady decline in the health of urban aquatic habitats, defined as the urban stream syndrome (Bai et al. 2017). Pollution from faecal matter and microbial pathogens, as well as antimicrobial agents, is common in urban environments, particularly urban aquatic environments. Toxins from marine ecosystems can be exported to lands through irrigation of reclaimed water and urban-influenced river/stream water, posing a health risk (Miles et al. 2019). Changes in the global land use adversely affect the ecology and the atmosphere shape local and global climate by leading to thermal zone effect and disrupted biodiversity and propel multinational agricultural and forestry trading. Metropolitan composition can minimise power usage and fossil fuel used by transportation, but it also enhances the heat island effect and restricts groundwater infiltration. Abiotic stresses such as destruction and elimination of natural events, which inhibit ecosystem regeneration during advanced development stages, may be worsened by urban development (Johnson et al. 2019). While increased population and economic activities are often highlighted, studies indicate that the associated growth in environmental consequences, as manifested in property transition, generation of waste, air and water pollution, among other things, is much larger and faster. More analysis is required on such an intensifying and accelerating trend in urbanisation-environmental linkages (Li et al. 2019).

2.4.3 Why Is Urban Development a Challenge for Cardiometabolic Syndrome?

Heart disease has quite a significant global impact. The disease has a huge financial impact on healthcare services of any country. The persistent existence of the condition imposes a significant financial burden on patients and it has a negative impact on their quality of life. This strain has a negative impact on treatment compliance and adherence, which leads to further problems. The overall cost of medical care escalated by ~50% between patients aged 31–40 years and for those aged 61–70 years (World Heart Federation 2015). The expense of cardiometabolic syndrome in the society is indeed likely to rise as in the developed world, where cardiometabolic syndrome impacts a large percentage of working-age adults. In the long term, universal healthcare will face significant economic challenges as a result of rising general health costs, increasing maintenance costs, and a weak economy. If changes are not done, the pressure of cardiometabolic risk will fall on today's individuals and potential generations. Cardiometabolic syndrome harms people and children, inflicting physical signs of illness in low-income countries. Because of late diagnosis and/or a lack of access to adequate care, many of these children die young. Many who recover can face a lifetime of disabilities as a result of a grossly mistreated disease. An infant who loses a parent does not only deal with the traumatic aftermath of their loss but also with the financial challenges of growing up in a one-parent or no-parent household. Kids may indeed be proposed to assist with physical labour or domestic tasks, or they may be forced out of education to serve at a young age. They will have to be responsible for a member of the family who's been hospitalised with CVD. Discrimination towards girls and boys may develop in education, as well as later on in life in respect of employment and wages and benefits, and in many nations, these are the principal determinants of capital stability. This prejudice may deter some communities from disclosing their child's medical conditions, creating an obstacle to medical care and treatment (Figure 2.4).

2.4.4 Attempts to Combat Cardiometabolic Syndrome Risk Factors

Regular physical activities benefit an individual's well-being as well as his or her living standards. It encourages sustainable development, and also the prevention of obesity and excess weight gain, as well as social integration and civic health. Even after this counselling and the known benefits of fitness, global fitness levels are declining. According to

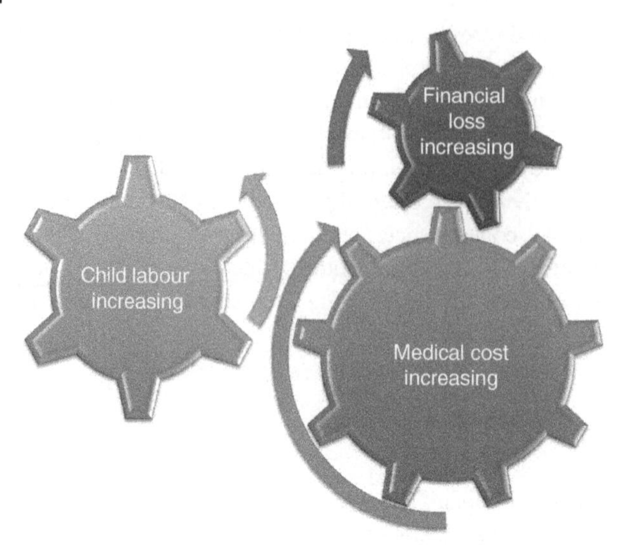

Figure 2.4 Societal stigma those are barriers to treat cardiometabolic syndrome.

recent global figures, inactivity exposes 60% of the world's population to health threats, resulting in numerous preventable deaths per year (Gupta et al. 2016). Inactivity is often dictated by the speed at which technology and modernisation are created, such as quicker home delivery of groceries, medications, and meals, which eventually leads to less moving and biking, electronic web browsing games, which attract kids to remain back at home and not actively play outdoors (Figure 2.5). Physical inactivity is tied to CVD or cardiometabolic risk factors such as hypertension, high blood sugar, and overweight (both

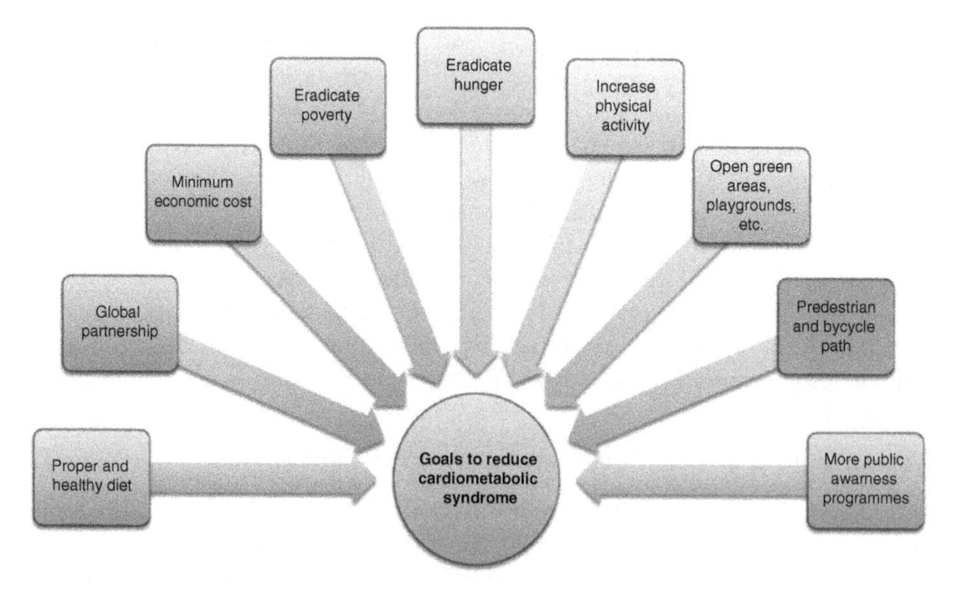

Figure 2.5 Plausible strategies to reduce cardiometabolic syndrome.

specifically and tangentially). Amounting to both the built urbanised area in which they live and the moderately active patterns and lifestyles that have emerged in this era, children living in cities may be especially restricted in their ability to partake in adequate physical activity. Physical activity such as walking and cycling to school is discouraged by development trends such as rush hour traffic, a lack of pedestrians, and overcrowded streets. Adolescents families may be much less willing to exercise as they simply couldn't afford or use fitness programmes, diet plans, and athletic centres. Individuals are getting more depressed as digital leisure activities become more prevalent. As a result, regular exercise promotional activities must overcome social, psychological, and financial obstacles. By leading the way in reducing levels of physical immobility, the medical care is better positioned to do so, promoting physical exercise for all individuals and proposing individual advice as part of broader wellness programmes. Merge partnerships between various governmental departments and rated, industries, community-based organisations, instructors, the internet, and the health service are more likely to result in strategic plans that reach a wide proportion of people and make suggestions, aid, and coordination for fitness exercises.

Dietary habits are changing across the world as a result of scale-up such as industrial prosperity, globalisation, and immigration. As a result of these habits, a trend of inadequacy and overnutrition has evolved, occasionally living side by side within the same nation, region, and even families. People in certain countries are growing up in a state of poverty and vulnerability, culminating in deficiency. The body triggers the storage and accumulation of fatty acids as a preventive mechanism of underfed people, increasing the risk of cardiometabolic syndrome and forming a tendency to obesity and diabetes. Saturated fat, processed food, glucose, and sodium diets have been attributed to four of the world's top leading causes of mortality: hypertension, diabetes, excess weight, and elevated cholesterol (World Heart Federation 2015). Unhealthy snacks are advertised in neon colours or sold with a game, ad campaigns with new figurative language; exposure of children to such commercials find it difficult to make a good decision and are therefore driven to ingest junk food. 'Eat for Goals!' was designed to encourage teenage individuals to accept a more balanced living and consume more healthy diet. 'Government School Feeding' and 'Nutrition Programmes' are undertaken to provide lunch for children who are in poverty. Such school-based feeding strategies have improved the predictive enrolment rates, decrease absences, and provided support for and perception of a healthier lifestyle for kids that will last into adult years (World Heart Federation 2015).

The relationship between urban growth and smokers is an experimentally proven fact. One instance, research over four decades ago reported that urban residents seem to be more smokers than rural residents (Maric et al. 2021). 'Smoking' is being considered as one of the nations critically significant public health challenges. The disadvantages of smoking are quite well acknowledged in current history. Kids are just unable to monitor their own surroundings due to dependence on adulthood, and they can be compelled to breathe smoke-filled air. Smoking may be conducted by adolescents for a variety of purposes. Other family and friends, such as siblings and parents, can have an influence on each other: whenever a relative smokes, the youngster is three times more likely to do the same. Girls are among the new targets of tobacco companies, as per the tobacco control collaboration, particularly in urban slums, where feminine smoker participation is still lower and the tobacco industry has recognised an economic boon to leverage. The implementation of the World Health Organisation (WHO's) 'CVD Prevention Programme' is among the main attractions in the battle against CVD (though it is a never-ending battle) (World Heart Federation 2015). In India, the 'Mobilising Youth for Tobacco-Related Initiatives' (MYTRI) are effective school-based tobacco control programmes that aim to expand tobacco control awareness and intervention in order to change behaviour patterns (Sidhu et al. 2018). Increased rates of CVD have been linked to industrial development, alterations in eating and exercise habits, cigarette smoking, and weight gain (Kumar et al. 2006). In Australia, a new law began to show up on 1 July 2012, authorising all smoking goods to be delivered in plain packaging. The aim of these new amendments is to minimise the use of such packages for marketing and promotions, enhance the functioning of safety warnings, discourage any use of deceitful wrapping to generate delusional ideas about smoking performance and quality, reduce tobacco consumption and absorption, and eradicate strong bonds with brand names (Maric et al. 2021). Commercials in the press and training programmes may assist in countering business tricks. Adults could be persuaded to quit smoking by advertising and awareness campaigns, thus having a significant impact on youth, or individuals could be alerted about the hazards of passive smoking, eventually trying to shield children from second-hand smoke. Anti-tobacco commercials and graphical packaging warnings – especially those with images – often lower the quantity of minors taking up smoking and expand the number of smokers dropping. Like many governments, the Indian government is also taking steps such as charging people who smoke in open areas. Despite the fact that there are many alternatives for adolescents to become more healthy, such as biking or commuting to work, this is not a possibility for them until congested road strategies or improvements to city infrastructures to include cycle tracks or large sidewalks are enacted (Figure 2.5). However, these are common practices in European and American nations.

The physical, financial, and social environments of many urban dwellers, especially young people, are driving them to lead pseudo lives in the light of drastic shifts in rapid and unplanned urban growth, as we are currently witnessing. Good infrastructure and working environments, food security and access to nutrient-dense foods, enhanced access to medical care, open environments for regular exercise, and education about physiological well-being and balanced living are only a handful of the many health determinants that should be resolved by government and societal action to encourage overall health. Groups of people and communities will profit massively from such actions: better family employment, reduced capital rates, and exacerbation of the emotional and behavioural problems associated with cardiometabolic syndrome are only a few of the advantages to be achieved. There is indeed a larger, right to life incentive for towns to act decisively against CVD. The 'Universal Declaration of Human Rights' (United Nations 1948) was published by the United Nations in 1948. Article 25 states that everyone has the right to a certain quality of living that is suitable for his or her health and well-being, including food, clothes, accommodation, medical treatment, and other required social services. Daily workout (3–5 days a week; 30–60 minutes a day), along with improved food safety and an intake of 500 calorie restriction a day, is the first phase of treatment. Weight loss by dietary restriction, independent of diet composition, is the most significant nutritional factor in lowering cardiometabolic rate (CMR) (Brunzell et al. 2008).

When communities increase in popularity, maintaining, and extending healthy and smoke-free open areas – such as playgrounds becomes extremely important to ensure suitable areas for the family-sports and physical activities. For the community, urban sprawl may either offer a benefit or a reward. The cost: cardiometabolic syndrome-related suffering which could have been eliminated, costing both social and political opportunity. The pledge: CVD mitigation, empowering urbanites to excel. Since the future has yet to be understood, we must work together to make sure that the societies of the long run are healthy environments for everyone.

2.5 Climate Change as a Risk Factor to Increase the Occurrence of Cardiometabolic Syndrome

2.5.1 Changing Climate Is One of the World's Principal Concerns

When the average weather, extreme weather, and its distribution across the globe see a statistically significant shift, it is termed as climate change (Wu et al. 2016). The main influencing causes are changes in one or more factors that decide the status of weather. These factors include temperature, precipitation, wind, and sunshine (Kim and Grafakos 2019). Climate change is our biggest concern nowadays as the human race, specifically impacted by non-negligible human physical activities. According to the European Environment Agency (EEA), the global average surface temperature has increased by 0.74 °C in the twentieth century, the global sea level has been rising 1.8 mm per year since 1961, and the Arctic sea ice has been shrinking by 2.7% per decade. Moreover, mountain glaciers are contracting, ocean water is becoming more acidic, and extreme weather events occur more often (Weng et al. 2014; Yang et al. 2013). This changing climate has created a great deal of concern for people prone to heart disease as the unpredictable weather has adverse effects on our physiology and blood flow rate such as oxidative stress, low-grade inflammation, vascular dysfunction, etc., which creates stress on the heart and increases the risk of cardiometabolic disease especially in aged people and people with a history in diseases (Münzel et al. 2017c).

2.5.2 Indicators That Have Been Predominant Contributors to Climate Change

The most significant factor for the drastic climate change that has been seen across the globe are primarily due to human activities. Greenhouse gas emission is the predominant reason for the global warming threat that we all are dealing with today among which CO_2 is the largest contributor. With existing climate change policies and the sustainable development practices, we follow global greenhouse gas emissions are likely to grow in the coming years (Metz et al. 2007). The emitted greenhouse gas is the prevalent cause of the changing climate especially the rise in temperature which is recognised to be a significant risk factor for people with diabetes and hypertension. By promoting the hypertension and diabetes, environmental factors may contribute substantially to the development and enhancement of cardiometabolic diseases (Münzel et al. 2017b, 2017c).

2.5.3 Health Impacts of Climate Change

Today, sustainable development is the biggest challenge to humans as a race. Unambiguous climatic changes have direct impacts on human health (Costello et al. 2009; Epstein 2001), especially when infectious, respiratory, and CVD including cardiometabolic syndrome are taken into account (Altizer et al. 2013; Bouzid et al. 2014). Most pathogens require vector carriers or hosts for their lifecycle. Drastic climate and weather conditions facilitate the survival, reproduction, distribution, and transmission of these and as a result, the disease. These co-morbid infectious diseases can certainly exaggerate the complications of cardiometabolic syndrome. Thus changing climate has a huge impact on these infectious and non-infectious diseases especially cardiometabolic diseases (Kim 2016). Patients that are at a risk for CVD are hit severely by fluctuating climates as well as the pollutants present in the air which directly impact the blood pressure and put these patients at a great risk (Laden et al. 2006).

In many parts of the world, the extreme weather conditions that are a product of the greenhouse gas emission has also created a barrier in people being able to go out and exercise which has a direct impact on the heart health as it increases obesity (Moellering and Smith 2012). High temperature also seems to have an impact on the blood pressure reading creating stress on the heart and increasing chances of cardiometabolic and CVD (Noordam et al. 2019).

2.5.4 Potential Method for Improvement of Cardiometabolic Disorder Conditions by Reducing Greenhouse Gases

Cardiometabolic syndrome and CVD are a huge concern for the human welfare sector of our society and represent the most threatening chunk of non-communicable disease-related deaths. It is obvious to us that the severely changing climate contribute drastically to global ischemic heart disease and stroke incidences, and is thought to have an even more significant impact worldwide. However, changes on our part have greatly contributed to contribute to the wellbeing of those who are greatly at risk. Strict air quality regulations have created a significant impact on increase in life expectancy, truly decreasing the mortality rate for sick patients in the last five decades. We should change our activity level if pollution levels are high by limiting our outdoor time.

Extremely hot and cold weather poses a tremendous amount of stress and risk on such patients National Institute of Environmental Health Sciences (NIEHS 2013). Air conditioning and other means to control the temperature for such patients can prove to be very helpful. Some other lifestyle changes that could drastically decrease greenhouse gas emissions are changes in our occupational style, cultural patterns, and consumer choice in buildings followed by reducing the usage of car, efficient driving style while use of public transport should be increased.

2.5.5 Introduction to Obesity and Its Associated Risk Factors Influencing Cardiometabolic Syndrome

Obesity is simply defined as the excess of body weight with respect to the height of the person. In adults, obesity and overweight are calculated by a simple index of weight-for-height that is the body mass index (BMI). It is calculated using a person's weight in kilogrammes divided by the square of his height in metres (kg m^{-2}). Obesity, a complex disease characterised by abnormal or excessive adiposity, fat accumulation or body fat (BF) leads to impairment of health as it affects not only in terms of the body size but also metabolically. Obesity greatly increases the risk of chronic disease morbidity including cardiometabolic syndrome, disability, depression, type 2 diabetes, CVD, and certain cancers leading to mortality. The challenge in this is that obesity and overweight initially were largely considered as a preventable disease but has evolved into a global burden. Since the 1960s, worldwide obesity has nearly tripled and reached epidemic proportions. Recently published data show that there is a gradual increase in the global prevalence of overweight adults (Figure 2.6) and that overweight and obesity, and its associated comorbidities cause over four million deaths yearly, worldwide (Grundy 2016; G. B. D. Obesity Collaborators et al. 2017).

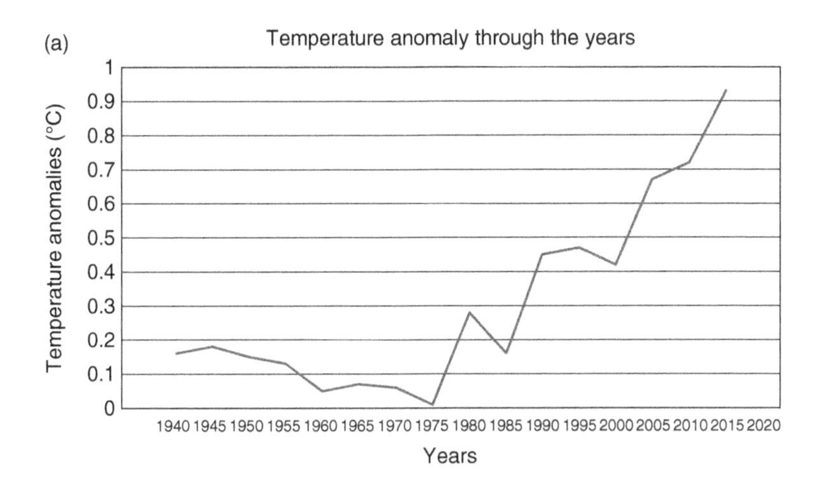

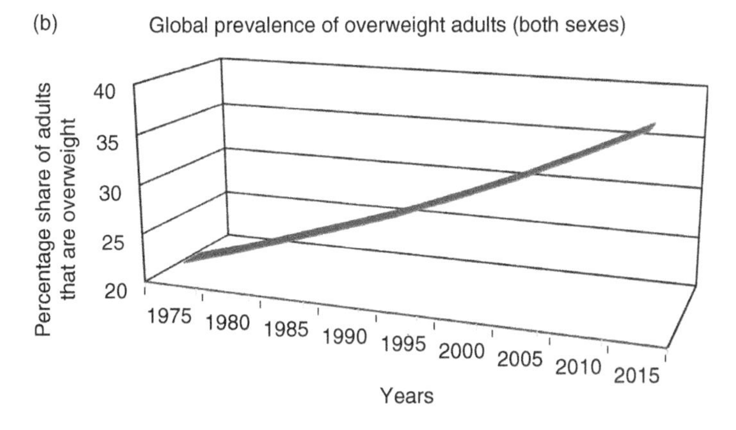

Figure 2.6 Graphical representations showing how ecology and urbanisation have impacted obesity and cardiometabolic syndrome with years (d) cardio-metabolic disease (CMD) per 100 000 population (WHO) is congruent with the data for (a) global temperature abnormalities (NOAA 2013), (b) global prevalence of overweight adults ('Our World in Data: Obesity' 2017), and (c) percentage of global urban population ('The World Bank' 2013) through a span of almost last 50 years.

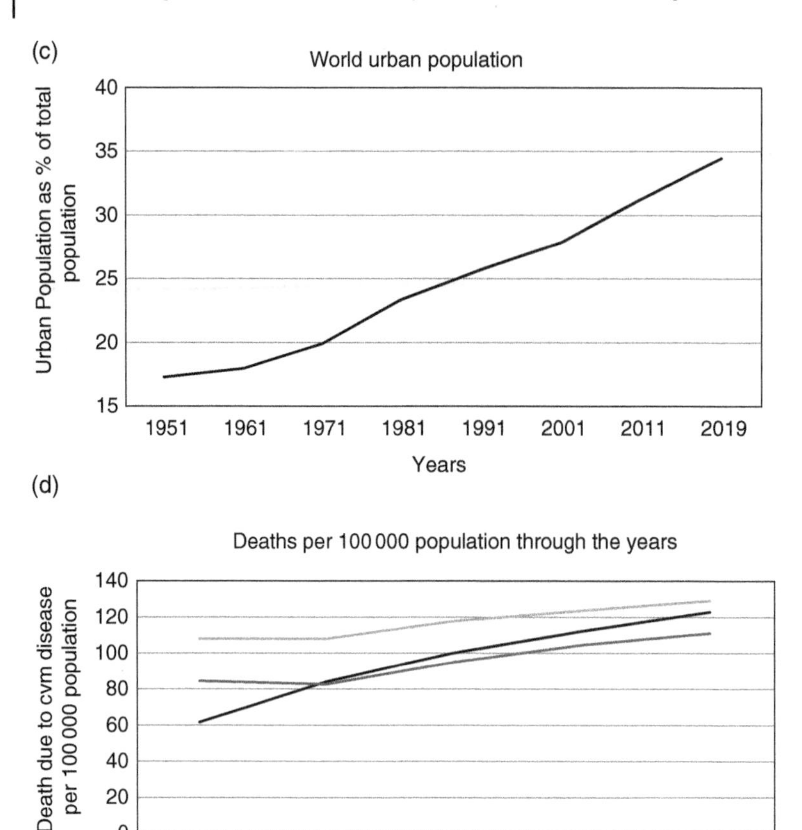

Figure 2.6 (Continued)

Obesity is fundamentally caused by an imbalance in dietary and nutritional intake (Hruby and Hu 2015). The disproportion of the calories consumed and calories expended, along with higher intake of energy-dense foods which are high in fat and sugars poses a major burden to chronic diseases and health complications such as insulin resistance, type 2 diabetes, inflammation, cardiometabolic syndrome, coronary heart disease (CHD), CVD, liver disease, cancer, and neurodegeneration (Figure 2.7) (Saltiel and Olefsky 2017).

It is also known that a high BMI is a major risk factor for non-communicable diseases such as CVD (mainly heart disease and stroke), which were the leading cause of death in 2012, diabetes, musculoskeletal disorders, especially osteoarthritis, and cancers including endometrial, breast, ovarian, prostate, liver, gallbladder, kidney, and colon. As the BMI increases, the risk of the cardiometabolic syndrome greatly increases. There is huge evidence of the sharp increase in the obesity-associated cardiometabolic risks and morbidities. These findings have led to an increase in studies that link and relate the fundamental reasons and risk factors in obesity and their mode of action which lead to these diseases (Table 2.1).

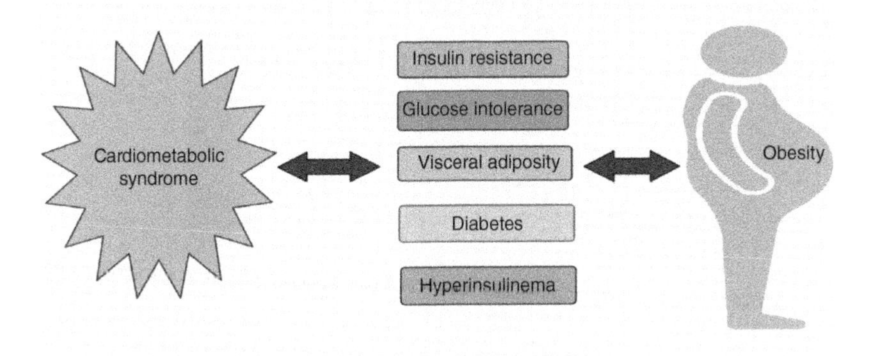

Figure 2.7 Individual suffering from obesity leads to insulin resistance, glucose intolerance, visceral adiposity, diabetes, hyperinsulinemia which are the major causes for cardiometabolic syndrome. *Source:* Based on Saltiel and Olefsky (2017).

Table 2.1 Obesity-induced risk factors and associated comorbidities.

Risk factors due to obesity	Mode of action	Associated morbidities	References
Chronic low-level inflammation and oxidative stress	Causes DNA damage, affects growth-promoting cytokines, and immune modulation	1) Chronic local inflammation which is a result of gastroesophageal reflux disease is a probable cause of oesophageal adenocarcinoma 2) Gallbladder inflammation is a strong risk factor for gallbladder cancer 3) Chronic ulcerative colitis, hepatitis are risk factors for different types of liver cancer 4) Atherosclerosis formation, impaired fibrinolysis, increased risk for CVD, including stroke and venous thromboembolism	Berger (2014); Fruh (2017); Gregor and Hotamisligil (2011); Bishayee (2014); Randi et al. (2006)
Increased blood levels of insulin and insulin-like growth factor-1 (IGF-1) (hyperinsulinemia or insulin resistance)	Cause cells to divide more than usual which increases the chances of cancer cells being made	Promotes the development of colon, kidney, prostate, and endometrial cancers, type 2 diabetes	Roberts et al. (2010); Gallagher and LeRoith (2015); Dey and Senapati (2021b)
Adipokines, hormones produced by fat cells	Stimulate or inhibit cell growth, induce cell proliferation	Example: Leptin, an adipokine is linked to the pathophysiology of breast cancer, obesity-*induced* hypertension, neurodegenerative diseases including Alzheimer's disease	Modzelewska et al. (2019); Ray (2018)
Fat cells may also have direct and indirect effects on other cell growth regulators	Dysregulation of rapamycin, mTOR pathway, and AMP-activated protein kinase	Cancer, diabetes, cardiovascular disease, neurodevelopmental, and neurodegenerative disorders	Saxton and Sabatini (2017); Takei and Nawa (2014)

2.5.6 The Impact of Urbanisation on Epidemiology of Obesity and Overweight in Relation to Cardiometabolic Syndrome

It is reported that excessive weight gain or obesity is not the result of a single cause, such as genetic tendency or the socioeconomic level of living as they do not justify the rapid increases in obesity in the nineteenth century. A number of other research suggested that urbanisation is one of the most important factors of an obesogenic environment as it entails an environment that promotes gaining weight and contributes to obesity (Swinburn et al. 1999) (Figure 2.6). There are various factors in the model of urbanisation which has led to the increased prevalence of overweight and obesity in the world over the last 50 years (Pinchoff et al. 2020). Several other factors include residential environment, cultural structure, social relationships, and extent of surrounding areas that changed during the phase of urbanisation and industrialisation. These are considered as some of the most important environmental factors impacting nutrition intake and physical activity which is a major cause of obesity. Industrialisation, a product of urbanisation in the UK and the United States of America (USA) from the start of the nineteenth century though made life easier but lead to a directly associated increase in the frequency of obesity in society (Congdon 2019).

After the association of the increasing trend of urban population and obese people in the developed countries, obesity epidemic spread in low- and middle-income economies were also linked to the lifestyle changes and a decrease in physical activity due to the sedentary nature of working, changing modes of transportation due to crunched time and ease of availability and few restrictions on access to or availability of junk food. Though urbanisation has led to supportive policies in almost every sector, such as health, agriculture, transport, urban planning, environment, food processing, distribution, and marketing; built environments lead to less physical activity, fewer open markets and farm stands and less outdoor recreational space. Also, technology advancements lead to lesser outdoor play, an increase in time spent watching television and playing computer games, more advertisements, and mass media marketing of junk food and carbonated beverages resulting in poorer diets and high-calorie food intake (Bahadoran et al. 2016). Even the job sector has seen a surge in jobs of a more sedentary nature (such as manufacturing and desk jobs) and fewer active jobs (such as farming). The effects of this environment have majorly impacted the health of individuals, not only physical health but also intellect, mental development, as well as on physical and emotional structure which has increased the incidence of chronic comorbidities, and become the leading cause of deaths worldwide.

Another major issue regarding urbanisation is associated with climate change. Urbanisation leads to the migration of people from rural places to the cities, which forces rapid, inadequate, and poorly planned expansion of cities. According to World Economic Forum Reports 2015, developing countries account for around two-thirds of annual greenhouse gas emissions, caused by their economic growth and rapid urbanisation, poor infrastructure, and activities of people (World Heart Federation 2015). These directly affect climate change and increase the potential of natural catastrophes to cause unprecedented damage. Climate change, increasing ambient temperature, carbon-extensive construction and large generation of greenhouse gases decreases the quality of life in the cities, reduces the conducive environment for physical activity which lays a huge burden on weight management.

Obesity research has highlighted that physical activity levels are declining, and sedentary behaviours increasing, not only in developed countries, such as the United States, but also in low- and middle-income countries, such as India, China, Bangladesh, and UAE (Radwan et al. 2018). The proportion of overweight individuals has seen a considerable spike between the years 1980 and 2020 from 40% to almost 75% in the United States, Canada, England, Spain, Austria, Italy, France, and Korea. According to the World Factbook-CIA 2016–2020, United States, Jordan, Saudi Arabia are among the top countries with the highest overweight people as a percentage of the total population. India though has a lower percentage of these people as compared to global ranking yet has an alarmingly high number of overweight people increasing every year.

Along with it, dietary changes and nutritional transition of the growth in fast food consumption have been significant worldwide, and affect both developed and less developed societies (Kallio et al. 2015). Lifestyle influenced excessive and life-threatening accumulation of BF, and the associated risk of obesity with a huge number of cardiometabolic disorders and cancer results in the notable increase in morbid obesity worldwide. Also, the rate of obesity-associated diseases is increasing in children at the same rate as the adult. If positive changes are not implemented to keep the ill effects of urbanisation in control, 38% of the world's adult population will be overweight and nearly 20% will be obese, by the year 2030 (Hruby and Hu 2015). In the developed countries, with prevalent increasing trends of established urbanisation and no control measures in check, it is estimated that 85% of adults being overweight or obese by 2030. Therefore, it has now become imperative that in urban planning, an environment should be provided taking public health and the risk of obesity into account. This is an urgent necessity considering that metabolic dysregulations, CVD, and cancer facilitated by obesity are the major contributors of mortality worldwide for the last half a century.

2.5.7 Obesity, a Major Risk Factor for Prevalent Cardiometabolic Syndrome

It has been reported that overweight and obesity are the causes of more number of deaths worldwide than underweight. Studies investigating the cause for the increase in frequency of obesity have identified a strong correlation between obesity and urbanisation and also between obesity and physical activity and chronic diseases (Dey and Senapati 2021a). This is so because obesity facilitates the incidence of cardiometabolic syndromes including insulin resistance, dyslipidaemia, and hypertension which together leads to chronic CVD, stroke, some types of cancer, and type II diabetes (Table 2.1). According to WHO, obesity and life-threatening excessive accumulation of BF is a major requisite risk factor and largest contributor in the aetiology of cardiometabolic syndrome and disorders.

The association of obesity and cardiometabolic syndrome is known to increase the risk of type 2 diabetes mellitus by fivefold and CVD by threefold. The impact of obesity-linked cardiometabolic risk is so adverse that various present studies aim to establish more specific and optimal cut-off values for anthropometric indicators of obesity to prevent the burden of healthcare on the economy (Figure 2.8) (Macek et al. 2020).

Major obesity complications relate to the pathology of disrupted metabolic homeostasis regulated primarily by several adipokines and cytokines such as leptin, adiponectin, tumour-necrosis-factor (TNF)-α, and interleukin (IL)-6. Obesity complications disrupt the

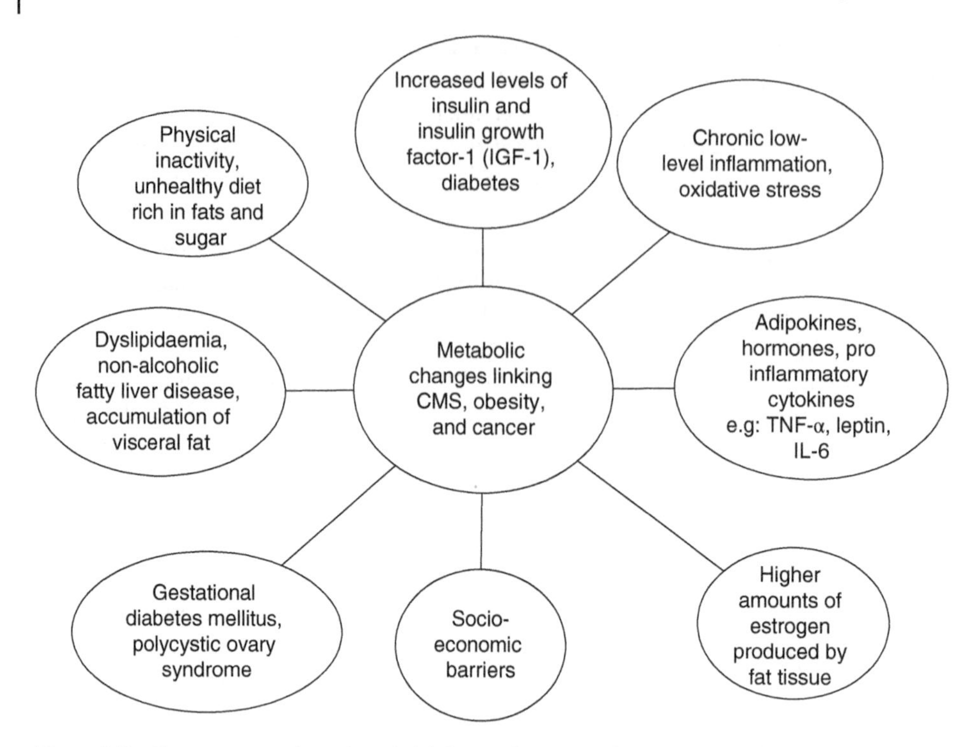

Figure 2.8 The most prevalent shared risk factors between obesity, cancer, and cardiometabolic syndrome. CMS, cardiometabolic syndrome.

normal adipose physiology in both men and women differently. It causes non-alcoholic fatty liver disease (NAFLD) which is the most common chronic liver disorder and many cardiometabolic disorders including high fasting triglycerides; low 'good' HDL cholesterol, high levels of C-reactive protein in the blood, a marker of chronic inflammation and elevated blood pressure. The coexistence of two or more of the cardiometabolic syndromes facilitated by obesity doubles the risk of CVD and heart attack and stroke and increases the risk of diabetes fivefold in a period of 5–10 years (Virani et al. 2020).

The rise of progressive urbanisation and related globalisation of unhealthy, sedentary, and stressful lifestyles result in overweight, which has potential adverse health consequences such as CVD including CHD, atrial fibrillation, venous thromboembolism, and congestive heart failure. The urban setting and fast-moving face of life in the cities have led to an increasing trend in fast food consumption and out-of-home eating lifestyle. It ultimately culminates in a lower diet quality with lower micronutrient balance, high calorie, and fat intake. These impair the fat metabolism, homeostasis of glucose and insulin, oxidative stress, sleep disorders, and lipid and lipoprotein disorders which are together also termed as cardiometabolic disorders (Bahadoran et al. 2016). The association of obesity and related cardiometabolic disorders was initially highly prevalent in the United States, Europe where the implementation of urbanisation and industrialisation first began. The trend of two- to threefold increase in the prevalence of overweight and obesity was reported in developing countries of the *Middle East* and southwest Asia between the late nineteenth century and continues till date. Incidence studies of CVD highlight that between the age of 20 and 79 years, overweight and obesity correlated with the early development of major cardiometabolic disorder stress followed by CVD and resulted in higher mortality with obesity. Half of the world's population with metabolic healthy obesity develop metabolic syndrome and poses increased threat of CVD compared with those with a healthy normal weight. Therefore, the presence of metabolically healthy obesity is also a high-risk state rendering greater cardiovascular risk age, BMI, and obesity are also risk factors for stroke, venous thromboembolism risk, and hypertension in patients. Metabolic syndrome related to obesity is also linked to other issues caused by obesity which include sleep issues, obstructive sleep apnea, and breathing troubles including asthma (Drager et al. 2013). In this regard, the latest research is focused on the development of compositions and compounds which are cardio-protective and anti-cardiac in nature (Kundu et al. 2019a). Research in this area is a key to lower the burden of CVD in the society (Kundu et al. 2019b).

2.5.8 Obesity, a Major Risk Factor for Prevalent Metabolic Syndrome in Women

Firstly, maternal obesity, overweight, and gestational weight gain in women aggravate many conditions such as PCOS (polycystic ovary syndrome) which associate with increased incidence of diseases falling under metabolic abnormalities including dyslipidaemia, hypertension, glucose intolerance, and also major reproductive complications including infertility, pregnancy complications, gestational diabetes mellitus (GDM), gestational hypertension, pre-eclampsia, and delivery of a preterm or growth-restricted baby (Osibogun et al. 2020).

Prior studies have found that women with increased pathogenesis of PCOS are more likely to have increased subclinical CVD markers and clinical outcomes of insulin resistance. This worsens the reproductive, metabolic, and psychological behaviour of women greatly including anxiety, depression, OCD (obsessive–compulsive disorder), and poor quality of life. PCOS and its strong link with infertility and obesity (~80%) could be managed by lifestyle management of proper nutritional intake and physical activity, which is always the advised first line of treatment. A small percentage (5–10%) of weight loss has led to improved PCOS conditions in women (Chella Krishnan et al. 2018).

Secondly, the pathology of the disruption of metabolic homeostasis between men and women is different due to differences in the normal adipose physiology. The location of the white adipose tissue could be under the skin as subcutaneous adipose tissue (SAT) or in the deep abdominal region as visceral adipose tissue (VAT). It is reported that VAT depots confers more cardiometabolic risk and are higher in adult men in comparison to premenopausal women who possess more SAT depots for the storage of fat (Chella Krishnan et al. 2018). On the other hand, increased risk of cardiometabolic syndrome in middle-aged women above 50 years of age as compared to men has been attributed to the loss of this cardiometabolic protection in postmenopausal women (Chella Krishnan et al. 2018).

Another major and most common cardiometabolic deregulation diagnosed frequently in pregnant women is GDM which involves a degree of glucose intolerance. GDM affects 5–25% of all pregnant women worldwide and is the largest group of cardiometabolic disorders in pregnant women followed by a higher risk of developing hypertension, hypercholesterolemia, and cancers of nasopharynx, lung kidney, breast, and thyroid glands (Peng et al. 2019; Han et al. 2018) (Figure 2.8). Thus, the dysregulations in cardiometabolic factors are the most common pathway for potential interrelatedness between preterm birth and CVD. It is crucial for women to control overweight through lifestyle changes to prevent the incidence of cardiometabolic disorder and potential multiple comorbidities during pregnancy and beyond (Grieger et al. 2021).

Also, cardiometabolic syndrome in women linked with obesity is due to social problems and socio-economic cultures in many underdeveloped and developing countries where there are security and safety issues that restrict the free movement, social interactions, and physical activity of children, particularly women leading to increased weight during childhood. All these varied reasons culminate in a nearly 15% higher rate of the risk of cardiometabolic disorders, mental and physical stress facilitated by obesity in women as compared to men.

2.5.9 Childhood Obesity, a Growing Concern

Another prevailing concern worldwide is childhood obesity in both developing and developed countries. The data from NHANES (National Health and Nutrition Examination Survey) in the year 2015 and 2016 highlighted that there was gradual increase in the overall prevalence of obesity from childhood, adolescents, and youth to adults from nearly 14 to 19% and 40%, respectively. Recently, childhood obesity has seen a considerable spike and the environment around the home of the child is the most prevalent cause. Along with the prevalence of obesity, 70% of the children had a minimum of one cardiometabolic risk factor (WHO 2019a). Insulin resistance due to obesity was more prevalent in young girls than boys. In terms of number, it is alarming that around 38 million children worldwide under the age of five years were reported to be overweight in 2019 (WHO 2019b). Childhood obesity leads to breathing difficulties, increased bone injuries, and psychological breakdown. Obese children have a predisposition for premature death, insulin resistance, adult obesity, and chronic inflammation. Various studies suggest the shared link between early markers of CVD, NAFLD with paediatric obesity. Therefore, paediatric obesity and its striking association of comorbidities such as asthma, fatty liver, sleep apnea, hypertension, orthopaedic problems, and type 2 diabetes in children bear economic and psychological burden on the children as well as their families along with the health issues (Corvalán et al. 2010).

2.5.10 Cardiometabolic Syndrome Associated Cancer Facilitated by Inflammation and Obesity

The relationship between metabolic syndrome and cancer is also one of growing concerns in the last few decades. Cardiometabolic syndrome and cancer are the major contributors to the burden of chronic disease and mortality worldwide (Koene et al. 2016). There is emerging evidence for possible interaction between the two with shared risk factors suggesting a similar underlying mechanism of the pathologies (Table 2.1) (O'Neill and O'Driscoll 2015). Additionally, each individual risk factor for metabolic syndrome has also an association with cancer. While inflammation is attributed to be the one of the major amalgamating factors in the onset and progression of the diseases, additional mechanisms or factors have also been identified, which mainly includes diabetes, insulin sensitivity, and obesity (Figure 2.8). It has been reported along with other factors sharing biological roles, phosphatase and tensin homolog (PTEN) is a crucial gene involved in regulating the Akt pathway (Arora et al. 2018) and PTEN mutations are linked to regulating insulin sensitivity and obesity (Mitchell 2012). Thereby, tracking back one of the major causes of cardiometabolic syndrome is obesity. Therefore, it becomes important to understand if cancer, CVD, and obesity have a shared biology. Studies in the United States in the later nineteenth century on cancer reported the association of BMI index and cancer deaths. It was observed that men with a BMI ≥ 40 had a 52% higher death rate from all cancer, while women with a BMI ≥ 40 had a 62% higher death rate from all cancers than men and women of normal weight, respectively. Thereby, if obesity is prevented in individuals, around 10–20% of cancer deaths in the United States could be prevented.

Scientific evidence shows that metabolic syndrome and obesity are contributors to an estimated 6% of all cancers (4% in men, 7% in women) diagnosed in 2007. Beyond being a major risk factor for diabetes, which itself is a risk factor for most cancers, obesity has long been understood to be associated with increased risk of oesophageal, colon, pancreatic, postmenopausal breast, endometrial, and renal cancers. More recently, evidence has accumulated that overweight and/or obesity raise the risk of cancers of the gallbladder, liver, ovaries (epithelial), and advanced prostate cancer, as well as leukaemia (Esposito et al. 2012). The complex relation between all three can be mediated by several factors including diet, physical activity and hormonal signalling (insulin-like growth factor signalling), oxidative stress (Dara Hope and Derek 2012). Pro-inflammatory cytokines predominantly secreted by immune cells and also other cell types like endothelial and adipocytes affect all tissues and organ systems including the vasculature system (Dinarello and Pomerantz 2001). The aetiology of cardiometabolic syndrome involves accumulation of visceral fat. This promotes the synthesis and release of pro-inflammatory cytokines leading to enhanced oxidative stress. Infiltration of immune cells particularly inflammatory macrophages in the adipose tissues further enhance the cytokine burden causing chronic inflammation (DeMarco et al. 2010). Pro-inflammatory cytokines produced within adipose tissues are also found to be elevated in the serum of obese people. Inflammatory cytokines such as IL-6, TNF-α, leptin have been extensively associated with obesity, CVD (upto 20%), and cancer (Grivennikov and Karin 2011; Bielecka-Dabrowa et al. 2007). Therefore, IL-6 and TNF-α and other potential pro-inflammatory cytokines are probably the factors linking cardiometabolic syndrome, obesity, and cancer together, and pose as potential targets for the treatment of these associated pathologies (Figure 2.8). Therefore, it has been established that a major cause of cancer is metabolic imbalance conditions including hypercholesterolemia, diabetes, hypertension, specific drug use. Effective interventions to reduce their prevalence could potentially reduce cancer risk

2.6 Conclusion

The book chapter covered the evidence that changes in the urban lifestyle is one of the major causes leading to obesity and associated chronic diseases including cardiometabolic syndrome, metabolic dysregulations, and also cancer. It is important to formulate strategies on public health based on bearable lifestyle changes and sustainable urbanisation to curb obesity in the population and risk of cardiometabolic syndrome and associated cancer.

Lifestyle interventions, weight management, and low-calorie food intake are crucial to mitigate increasing global burden of obesity, cardiovascular, and metabolic risk factors. Along with this improvement of medical and research facilities to develop anti-hypertensive and cardio-protective compounds is a crucial step towards treatment (Kundu et al. 2021a, 2021b). Individual sensitisation and choices are crucial in the case of nutritional intake and the promotion of healthy lifestyle. Government, private, technical, financial, and institutional capacities have to work together for the implementation of adequate land-use planning, well-governed cities, and more sustainable infrastructure and environment with open and recreational spaces to mitigate impacts of climate change and unhealthy lifestyles on health. The World Health Organisation has provided a global strategy and action plan to strengthen initiatives for the prevention and control of noncommunicable diseases. Similarly, to curb these life-threatening disorders individual alertness and population-based collective approach is the need of the hour.

Paths for bicycling, walking, etc. are also essential. Instead of junk foods, balanced diets can be promoted for regular intake by the children and individuals with genetic backgrounds of developing metabolic syndromes. More research should be commenced urgently to decipher the underlying molecular mechanisms connecting metabolic syndrome with other environmental and societal factors to generate sustainable strategies to develop a better world for the future.

Acknowledgements

All authors of the book chapter would like to acknowledge the Dr. B.R. Ambedkar Center for Biomedical Research (ACBR), University of Delhi, India for various help to complete the current work. Sanjay Kumar Dey acknowledges the University of Delhi, Institute of Eminence grant (IoE/2021/12/FRP).

Authors also acknowledge the use of the academic user license of BioRender (https://biorender.com) for preparing few of the figures shown in the current book chapter.

References

Altizer, S., Ostfeld, R.S., Johnson, P.T. et al. (2013). Climate change and infectious diseases: from evidence to a predictive framework. *Science 341* (6145): 514–519. https://doi.org/10.1126/science.1239401.

Arora, N., Gavya, S.L., and Ghosh, S.S. (2018). Multi-facet implications of PEGylated lysozyme stabilized-silver nanoclusters loaded recombinant PTEN cargo in cancer theranostics. *Biotechnology and Bioengineering 115* (5): 1116–1127. https://doi.org/10.1002/bit.26553.

Bahadoran, Z., Mirmiran, P., and Azizi, F. (2016). Fast food pattern and cardiometabolic disorders: a review of current studies. *Health Promotion Perspectives 5* (4): 231–240. https://doi.org/10.15171/hpp.2015.028.

Bai, X., McPhearson, T., Cleugh, H. et al. (2017). Linking urbanization and the environment: conceptual and empirical advances. *Annual Review of Environment and Resources 42* (1): 215–240. https://doi.org/10.1146/annurev-environ-102016-061128.

Berger, N.A. (2014). Obesity and cancer pathogenesis. *Annals of the New York Academy of Sciences 1311*: 57–76. https://doi.org/10.1111/nyas.12416.
Bielecka-Dabrowa, A., Wierzbicka, M., and Goch, J.H. (2007). Proinflammatory cytokines in cardiovascular diseases as potential therapeutic target. *Wiadomości Lekarskie 60* (9–10): 433–438. https://europepmc.org/article/med/18350717.
Bishayee, A. (2014). The role of inflammation and liver cancer. *Advances in Experimental Biology and Medicine 816*: 401–435. https://doi.org/10.1007/978-3-0348-0837-8_16.
Bouzid, M., Colón-González, F.J., Lung, T. et al. (2014). Climate change and the emergence of vector-borne diseases in Europe: case study of dengue fever. *BioMed Central Public Health 14* (1): 781. https://doi.org/10.1186/1471-2458-14-781.
Brunzell, J.D., Davidson, M., Furberg, C.D. et al. (2008). Lipoprotein management in patients with cardiometabolic risk: consensus conference report from the American Diabetes Association and the American College of Cardiology Foundation. *Journal of the American College of Cardiology 51* (15): 1512–1524. https://doi.org/10.2337/dc08-9018.
Castro, J.P., El-Atat, F.A., McFarlane, S.I. et al. (2003). Cardiometabolic syndrome: pathophysiology and treatment. *Current Hypertension Reports 5* (5): 393–401. https://doi.org/10.1007/s11906-003-0085-y.
Chella Krishnan, K., Mehrabian, M., and Lusis, A.J. (2018). Sex differences in metabolism and cardiometabolic disorders. *Current Opinion in Lipidology 29* (5): 404–410. https://doi.org/10.1097/mol.0000000000000536.
Congdon, P. (2019). Obesity and urban environments. *International Journal of Environmental Research and Public Health 16* (3): 464. https://doi.org/10.3390/ijerph16030464.
Corvalán, C., Uauy, R., Kain, J., and Martorell, R. (2010). Obesity indicators and cardiometabolic status in 4-y-old children. *American Journal of Clinical Nutrition 91* (1): 166–174. https://doi.org/10.3945/ajcn.2009.27547.
Costello, A., Abbas, M., Allen, A. et al. (2009). Managing the health effects of climate change: lancet and University College London Institute for Global Health Commission. *The Lancet 373* (9676): 1693–1733. https://doi.org/10.1016/s0140-6736(09)60935-1.
Dara Hope, C. and Derek, L. (2012). Obesity, type 2 diabetes, and cancer: the insulin and IGF connection. *Endocrine-Related Cancer 19* (5): F27–F45. https://doi.org/10.1530/ERC-11-0374.
De Blois, J., Kjellstrom, T., Agewall, S. et al. (2015). The effects of climate change on cardiac health. *Cardiology 131* (4): 209–217. https://doi.org/10.1159/000398787.
DeMarco, V.G., Johnson, M.S., Whaley-Connell, A.T., and Sowers, J.R. (2010). Cytokine abnormalities in the etiology of the cardiometabolic syndrome. *Current Hypertension Reports 12* (2): 93–98. https://doi.org/10.1007/s11906-010-0095-5.
Dey, S.K., Saini, M., Prabhakar, P., and Kundu, S. (2020). Dopamine β hydroxylase as a potential drug target to combat hypertension. *Expert Opinion on Investigational Drugs 29* (9): 1043–1057. https://doi.org/10.1080/13543784.2020.1795830.
Dey, S.K. and Senapati, S. (2021a). *in vivo* models for obesity and obesity related carcinogenesis. In: *Obesity and Cancer* (eds. S. Kumar and S. Gupta), 279–300. Singapore: Springer https://doi.org/10.1007/978-981-16-1846-8_14.
Dey, S.K. and Senapati, S. (2021b). Insulin and insulin-like growth factor-1 associated cancers. In: *Obesity and Cancer* (eds. S. Kumar and S. Gupta), 25–48. Singapore: Springer https://doi.org/10.1007/978-981-16-1846-8_3.

Dinarello, C.A. and Pomerantz, B.J. (2001). Proinflammatory cytokines in heart disease. *Blood Purification 19* (3): 314–321. https://doi.org/10.1159/000046960.

Drager, L.F., Togeiro, S.M., Polotsky, V.Y., and Lorenzi-Filho, G. (2013). Obstructive sleep apnea: a cardiometabolic risk in obesity and the metabolic syndrome. *Journal of the American College of Cardiology 62* (7): 569–576. https://doi.org/10.1016/j.jacc.2013.05.045.

Epstein, P.R. (2001). Climate change and emerging infectious diseases. *Microbes and Infection 3* (9): 747–754. https://doi.org/10.1016/s1286-4579(01)01429-0.

Esposito, K., Chiodini, P., Colao, A. et al. (2012). Metabolic syndrome and risk of cancer: a systematic review and meta-analysis. *Diabetes Care 35* (11): 2402–2411. https://doi.org/10.2337/dc12-0336.

Fruh, S.M. (2017). Obesity: risk factors, complications, and strategies for sustainable long-term weight management. *Journal of the American Association of Nurse Practitioners 29* (S1): S3–S14. https://doi.org/10.1002/2327-6924.12510.

G. B. D. Obesity Collaborators, Afshin, A., Forouzanfar, M.H. et al. (2017). Health effects of overweight and obesity in 195 countries over 25 years. *The New England Journal of Medicine 377* (1): 13–27. https://doi.org/10.1056/NEJMoa1614362.

Gallagher, E.J. and LeRoith, D. (2015). Obesity and diabetes: the increased risk of cancer and cancer-related mortality. *Physiology Reviews 95* (3): 727–748. https://doi.org/10.1152/physrev.00030.2014.

Gill, H., Mugo, M., Whaley-Connell, A. et al. (2005). The key role of insulin resistance in the cardiometabolic syndrome. *The American Journal of Medical Sciences 330* (6): 290–294. https://doi.org/10.1097/00000441-200512000-00006.

Gregor, M.F. and Hotamisligil, G.S. (2011). Inflammatory mechanisms in obesity. *Annual Review of Immunology 29*: 415–445. https://doi.org/10.1146/annurev-immunol-031210-101322.

Grieger, J.A., Hutchesson, M.J., Cooray, S.D. et al. (2021). A review of maternal overweight and obesity and its impact on cardiometabolic outcomes during pregnancy and postpartum. *Therapeutic Advances in Reproductive Health 15*: 2633494120986544. https://doi.org/10.1177/2633494120986544.

Grivennikov, S.I. and Karin, M. (2011). Inflammatory cytokines in cancer: tumour necrosis factor and interleukin 6 take the stage. *Annals of Rheumatic Diseases 70* (Suppl 1): i104–i108. https://doi.org/10.1136/ard.2010.140145.

Grundy, S.M. (2016). Metabolic syndrome update. *Trends in Cardiovascular Medicine Trends 26* (4): 364–373. https://doi.org/10.1016/j.tcm.2015.10.004.

Gupta, R., Mohan, I., and Narula, J. (2016). Trends in coronary heart disease epidemiology in India. *Annals of Global Health 82* (2): 307–315. https://doi.org/10.1016/j.aogh.2016.04.002.

Han, K.-T., Cho, G.J., and Kim, E.H. (2018). Evaluation of the association between gestational diabetes mellitus at first pregnancy and cancer within 10 years postpartum using National Health Insurance Data in South Korea. *International Journal of Environmental Research and Public Health 15* (12): 2646. https://doi.org/10.3390/ijerph15122646.

Hruby, A. and Hu, F.B. (2015). The epidemiology of obesity: a big picture. *PharmacoEconomics 33* (7): 673–689. https://doi.org/10.1007/s40273-014-0243-x.

Johnson, J.C., Urcuyo, J., Moen, C., and Stevens, D.R. 2nd. (2019). Urban heat island conditions experienced by the Western black widow spider (*Latrodectus hesperus*): extreme heat slows development but results in behavioral accommodations. *PLoS One 14* (9): –e0220153. https://doi.org/10.1371/journal.pone.0220153.

Kallio, K.A., Hätönen, K.A., Lehto, M. et al. (2015). Endotoxemia, nutrition, and cardiometabolic disorders. *Acta Diabetologica 52* (2): 395–404. https://doi.org/10.1007/s00592-014-0662-3.

Kim, E.J. (2016). U.S. Global Change Research Program: the impacts of climate change on human health in the United States: a scientific assessment. *Journal of the American Planning Association 82* (4): 418–419. https://doi.org/10.1080/01944363.2016.1218736.

Kim, H. and Grafakos, S. (2019). Which are the factors influencing the integration of mitigation and adaptation in climate change plans in Latin American cities? *Environmental Research Letters 14*: 105008. https://doi.org/10.1088/1748-9326/ab2f4c.

Kirk, E.P. and Klein, S. (2009). Pathogenesis and pathophysiology of the cardiometabolic syndrome. *The Journal of Clinical Hypertension (Greenwich) 11* (12): 761–765. https://doi.org/10.1111/j.1559-4572.2009.00054.x.

Koene, R.J., Prizment, A.E., Blaes, A., and Konety, S.H. (2016). Shared risk factors in cardiovascular disease and cancer. *Circulation 133* (11): 1104–1114. https://doi.org/10.1161/CIRCULATIONAHA.115.020406.

Kumar, R., Singh, M.C., Singh, M.C. et al. (2006). Urbanization and coronary heart disease: a study of urban-rural differences in northern India. *Indian Heart Journal 58* (2): 126–130. https://europepmc.org/article/med/18989056.

Kundu, S., Thelma, B.K., Maulik, S.K. et al. (2019a). Novel anti-hypertensive and anti-cardiac hypertrophic compounds. IN Patent 201711036983 A.

Kundu, S., Dey, S.K., Thelma, B.K. et al. (2019b). An anti-hypertensive cardio-protective composition. IN Patent 201811005899 A.

Kundu, K., Dey, S.K., Thelma, B.K., et al. (2021a). Quinolone-based anti-hypertensive cardio-protective composition. IN Patent 202111026777.

Kundu, S., Dey, S.K., Kumar, S. et al. (2021b). Novel anti-hypertensive cardioprotective composition comprising of DISPIRO[1H-PERIMIDINE-2(3H),2"(3"H)-[1H]PERIMIDINE. IN Patent 202111026998.

Laden, F., Schwartz, J., Speizer, F.E., and Dockery, D.W. (2006). Reduction in fine particulate air pollution and mortality: extended follow-up of the Harvard Six Cities study. *American Journal of Respiratory and Critical Care Medicine 173* (6): 667–672. https://doi.org/10.1164/rccm.200503-443OC.

Li, E., Parker, S.S., Pauly, G.B. et al. (2019). An urban biodiversity assessment framework that combines an urban habitat classification scheme and citizen science data [methods]. *Frontiers in Ecology and Evolution 7* (277) https://doi.org/10.3389/fevo.2019.00277.

Macek, P., Biskup, M., Terek-Derszniak, M. et al. (2020). Optimal cut-off values for anthropometric measures of obesity in screening for cardiometabolic disorders in adults. *Scientific Reports 10* (1): 11253. https://doi.org/10.1038/s41598-020-68265-y.

Maric, D., Bianco, A., Kvesic, I. et al. (2021). Analysis of the relationship between tobacco smoking and physical activity in adolescence: a gender specific study. *Medicina 57* (3): 214.

Metz, B., Davidson, O., Bosch, P. et al. (2007). *Climate Change 2007 - Mitigation of Climate Change*. AR4 Climate Change 2007: Mitigation of Climate Change. The Intergovernmental Panel on Climate Change (IPCC) https://doi.org/10.1017/CBO9780511546013.015, https://www.ipcc.ch/report/ar4/wg3/.

Miles, L.S., Breitbart, S.T., Wagner, H.H., and Johnson, M.T.J. (2019). Urbanization shapes the ecology and evolution of plant-arthropod herbivore interactions. *Frontiers in Ecology and Evolution 7* (310) https://doi.org/10.3389/fevo.2019.00310.

Miller, J.D. and Hutchins, M. (2017). The impacts of urbanisation and climate change on urban flooding and urban water quality: a review of the evidence concerning the United Kingdom. *Journal of Hydrology: Regional Studies 12*: 345–362. https://doi.org/10.1016/j.ejrh.2017.06.006.

Mitchell, F. (2012). Diabetes: PTEN mutations increase insulin sensitivity and obesity. *Nature Reviews Endocrinology 8* (12): 698. https://doi.org/10.1038/nrendo.2012.186.

Modzelewska, P., Chludzińska, S., Lewko, J., and Reszeć, J. (2019). The influence of leptin on the process of carcinogenesis. *Contemporary oncology (Poznan, Poland) 23* (2): 63–68. https://doi.org/10.5114/wo.2019.85877.

Moellering, D.R. and Smith, D.L. (2012). Ambient temperature and obesity. *Current Obesity Reports 1* (1): 26–34. https://doi.org/10.1007/s13679-011-0002-7.

Münzel, T., Camici, G.G., Maack, C. et al. (2017c). Impact of oxidative stress on the heart and vasculature: part 2 of a 3-part series. *Journal of the American College of Cardiology 70* (2): 212–229. https://www.jacc.org/doi/abs/10.1016/j.jacc.2017.05.035.

Münzel, T., Sørensen, M., Gori, T. et al. (2017a). Environmental stressors and cardio-metabolic disease: part II-mechanistic insights. *European Heart Journal 38* (8): 557–564. https://doi.org/10.1093/eurheartj/ehw294.

Münzel, T., Sørensen, M., Gori, T. et al. (2017b). Environmental stressors and cardio-metabolic disease: part I-epidemiologic evidence supporting a role for noise and air pollution and effects of mitigation strategies. *European Heart Journal 38* (8): 550–556. https://doi.org/10.1093/eurheartj/ehw269.

NIEHS (2013). Global environmental health newsletter. https://www.niehs.nih.gov/research/programs/geh/geh_newsletter/2013/7/index.cfm (accessed 10 March 2021).

Nieuwenhuijsen, M.J. (2018). Influence of urban and transport planning and the city environment on cardiovascular disease. *Nature Reviews Cardiology 15* (7): 432–438. https://doi.org/10.1038/s41569-018-0003-2.

NOAA (2013). Products by category. https://www.ncdc.noaa.gov/temp-and-precip/ (accessed 10 March 2021).

Noordam, R., Ramkisoensing, A., Loh, N.Y. et al. (2019). Associations of outdoor temperature, bright sunlight, and cardiometabolic traits in two European population-based cohorts. *The Journal of Clinical Endocrinology & Metabolism 104* (7): 2903–2910. https://doi.org/10.1210/jc.2018-02532.

O'Neill, S. and O'Driscoll, L. (2015). Metabolic syndrome: a closer look at the growing epidemic and its associated pathologies. *Obesity Reviews 16* (1): 1–12. https://doi.org/10.1111/obr.12229.

Osibogun, O., Ogunmoroti, O., and Michos, E.D. (2020). Polycystic ovary syndrome and cardiometabolic risk: opportunities for cardiovascular disease prevention. *Trends in Cardiovascular Medicine 30* (7): 399–404. https://doi.org/10.1016/j.tcm.2019.08.010.

Our World in Data (2017). Obesity. https://ourworldindata.org/obesity (accessed 10 March 2021).

Peng, Y.-S., Lin, J.-R., Cheng, B.-H. et al. (2019). Incidence and relative risk for developing cancers in women with gestational diabetes mellitus: a nationwide cohort study in Taiwan. *British Medical Journal Open 9* (2): e024583. https://doi.org/10.1136/bmjopen-2018-024583.

Pinchoff, J., Mills, C.W., and Balk, D. (2020). Urbanization and health: the effects of the built environment on chronic disease risk factors among women in Tanzania. *PLoS One 15* (11): e0241810. https://doi.org/10.1371/journal.pone.0241810.

Prabhakaran, D., Jeemon, P., and Roy, A. (2016). Cardiovascular diseases in India: current epidemiology and future directions. *Circulation 133* (16): 1605–1620. https://doi.org/10.1161/CIRCULATIONAHA.114.008729.

Radwan, H., Ballout, R.A., Hasan, H. et al. (2018). The epidemiology and economic burden of obesity and related cardiometabolic disorders in the United Arab Emirates: a systematic review and qualitative synthesis. *International Journal of Obesity 2018*: 2185942. https://doi.org/10.1155/2018/2185942.

Randi, G., Franceschi, S., and La Vecchia, C. (2006). Gallbladder cancer worldwide: geographical distribution and risk factors. *International Journal of Cancer 118* (7): 1591–1602. https://doi.org/10.1002/ijc.21683.

Ray, A. (2018). Cancer and comorbidity: the role of leptin in breast cancer and associated pathologies. *World journal of clinical cases* 6 (12): 483–492. https://doi.org/10.12998/wjcc.v6.i12.483.

Roberts, D.L., Dive, C., and Renehan, A.G. (2010). Biological mechanisms linking obesity and cancer risk: new perspectives. *Annual Review of Medicine 61*: 301–316. https://doi.org/10.1146/annurev.med.080708.082713.

Saltiel, A.R. and Olefsky, J.M. (2017). Inflammatory mechanisms linking obesity and metabolic disease. *Journal of Clinical Investigation 127* (1): 1–4. https://doi.org/10.1172/jci92035.

Saxton, R.A. and Sabatini, D.M. (2017). mTOR signaling in growth, metabolism, and disease. *Cell 168* (6): 960–976. https://doi.org/10.1016/j.cell.2017.02.004.

Sidhu, A.K., Kumar, S., Wipfli, H. et al. (2018). International approaches to tobacco prevention and cessation programming and policy among adolescents in India. *Current Addiction Reports 5* (1): 10–21. https://doi.org/10.1007/s40429-018-0185-z.

Swinburn, B., Egger, G., and Raza, F. (1999). Dissecting obesogenic environments: the development and application of a framework for identifying and prioritizing environmental interventions for obesity. *Preventive Medicine 29* (6 Pt 1): 563–570. https://doi.org/10.1006/pmed.1999.0585.

Takei, N. and Nawa, H. (2014). mTOR signaling and its roles in normal and abnormal brain development [Review]. *Frontiers in Molecular Neuroscience 7* (28) https://doi.org/10.3389/fnmol.2014.00028.

The World Bank (2013). Urban population (% of total population). https://data.worldbank.org/indicator/SP.URB.TOTL.IN.ZS (accessed 10 March 2021).

United Nations (2015). Universal declaration of human rights, United Nations, 2015. https://www.un.org/en/udhrbook/pdf/udhr_booklet_en_web.pdf (accessed on 10th March, 2020).

Virani, S.S., Alonso, A., Benjamin, E.J. et al. (2020). Heart disease and stroke statistics-2020 update: a report from the American Heart Association. *Circulation 141* (9): e139–e596. https://doi.org/10.1161/cir.0000000000000757.

Weng, Q., Xu, B., Hu, X., and Liu, H. (2014). Use of earth observation data for applications in public health. *Geocarto International 29* (1): 3–16. https://doi.org/10.1080/10106049.2013.838311.

WHO (2019a). Global health estimates: leading causes of death. https://www.who.int/data/gho/data/themes/mortality-and-global-health-estimates/ghe-leading-causes-of-death (accessed 10 March 2021).

WHO (2019b). Obesity and overweight. https://www.who.int/news-room/fact-sheets/detail/obesity-and-overweight (accessed 10 March 2021).

World Heart Federation (2015). WEF_Global_Risks_2015_Report. https://www.world-heart-federation.org/wp-content/uploads/2017/05/FinalWHFUrbanizationLoResWeb.pdf (accessed 17 March 2021).

Wu, X., Lu, Y., Zhou, S. et al. (2016). Impact of climate change on human infectious diseases: empirical evidence and human adaptation. *Environment International 86*: 14–23. https://doi.org/10.1016/j.envint.2015.09.007.

Yang, J., Gong, P., Fu, R. et al. (2013). The role of satellite remote sensing in climate change studies. *Nature Climate Change 3* (10): 875–883. https://doi.org/10.1038/nclimate1908.

Section 2

Urban Landscape Design Using Emerging Techniques

3

An Alternative Sustainable City Framework to Tackle Climate Change Issues in India

Sunny Bansal[1,2], *Jayprakash Chadchan*[2], *and Joy Sen*[3]

[1] *RCG School of Infrastructure Design and Management, IIT Kharagpur, Kharagpur, West Bengal, India*
[2] *VIT School of Planning and Architecture, VIT Vellore, Vellore, Tamil Nadu, India*
[3] *Department of Architecture and Regional Planning and Joint Faculty, RCG School of Infrastructure Design and Management, IIT Kharagpur, Kharagpur, West Bengal, India*

3.1 Introduction

Urbanization as a phenomenon has existed for long. Although it is an intricate socio-economic process, it broadly refers to the shift of population from rural hinterland to urban areas. World's urban population has rapidly grown from 751 million to 4.2 billion in the last seven decades (Ghosh 2019). As per the year 2018, 55% of the total population of the world resides in urban areas. This figure is expected to rise to 68% by the year 2050 (Phys.org 2018). The change in the world population (total, urban, and rural) from the year 1950 to 2050 is depicted in Figure 3.1. The World Urbanization Prospects – 2018 Revision Report projects that by the year 2050 urbanization wave could approximately add 2.5 billion more people to the urban areas of which roughly 90% addition will take place in the continents of Africa and Asia (UNDESA 2019). Between 2018 and 2050, developing nations like India, China, and Nigeria will contribute 37% of the world's urban population growth by each of them adding 416 million, 255 million, and 189 million urban dwellers respectively (UN News 2018).

By the year 2030, the number of world's megacities (population more than 10 million) is going to rise from 33 to 43 (Thornton 2019). Most of these megacities will be in the developing nations and hence, it becomes evident to focus on these nations. Taking the case of India, New Delhi urban agglomeration with 29 million population is the second largest urban agglomeration in the world closely followed by Mumbai with around 20 million population (The Economic Times 2018). Besides these two megacities, the country has three more megacities, namely, Kolkata, Chennai, and Bangalore. By the year 2030, the number of megacities in India will increase to seven by the addition of Ahmedabad and Hyderabad. In parallel, New Delhi is expected to grow and become the most populous urban agglomeration in the world by the year 2028 (Hindustan Times 2019). Urbanization and growth in GDP have become concurrent (Chen et al. 2014). These megacities are also the driving

Urban Ecology and Global Climate Change, First Edition. Edited by Rahul Bhadouria, Shweta Upadhyay, Sachchidanand Tripathi, and Pardeep Singh.

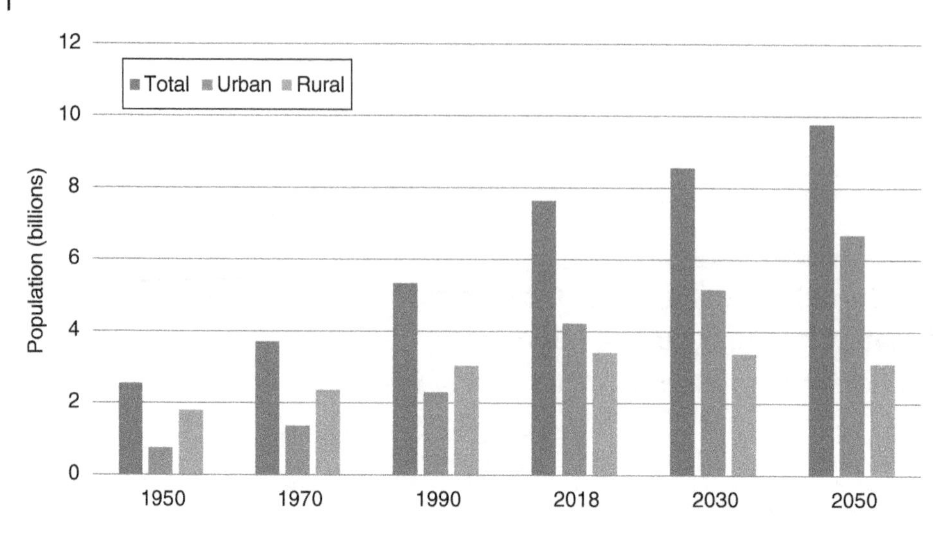

Figure 3.1 World population (Total, Urban, and Rural). *Source:* Data from UNDESA (2019).

engines of the economy around the globe and in India. However, with the growing urban population, there will be a rising need for housing, jobs, transportation facilities, energy systems, and other essential social infrastructures like healthcare, education, and recreation amenities.

The intensified and concentrated human activities in the megacities have led to a major contribution in the climate change (Krellenberg et al. 2014). Due to growing activities like construction, transportation, industrial production, land-use changes, and deforestation the urban ecological footprint of these megacities is escalating (Ramachandra et al. 2015). The megacities are trying hard for the improvements of their environment as they are the main source of greenhouse emissions. Megacities occupy around 2% of the total land on earth but contribute to about 70% of the greenhouse emissions (GHGs) (Bhushal 2019). There is an urgent need to find means to achieve socially and economically inclusive urban growth without hampering the environment.

Key solutions may be offered by the system of the city's planning, governance, and services provision. As the world continues to grow and urbanise, it is necessary to ensure sustainable urbanization especially in the developing nations (World Economic Forum 2018). Poorly managed urbanization is harmful to sustainable growth and development. The Sustainable Development Goals (SDGs) 2030 by the United Nations mention in the Goal 11 to 'Make cities and human settlements inclusive, safe, resilient and sustainable' (UNDP 2017). Thus, sustainable urbanization is crucial to universal sustainable development. Sustainable urbanization guarantees to address many crucial development challenges like disaster preparedness, energy usage, water availability, clean air, safe biodiversity, and adaptation to climate change. The book chapter has been divided into four sections. Section 3.1 (Introduction) presents the need and intent of the chapter. Section 3.2 portrays the consequences of urbanization in two categories – climate change and urban sprawl. Section 3.3 establishes the need for alternative sustainable development model and elaborates the various theories focussing on smart and sustainable growth. The concluding remarks have been depicted in the last section.

3.2 Urbanization and Its Consequences

Urbanization has major consequences besides the economic contribution. Two major impacts that lead to unsustainable urban development are climate change and urban sprawl. These impacts are discussed in detail in the upcoming sections.

3.2.1 Climate Change

As already discussed, India is the second biggest urban system in the world with over 5000 urban settlements. These urban settlements are totally dependent on energy for their existence and have a very high ecological footprint. As their population grows, the pressure on ecosystems to sustain these cities grows (Dovers and Butler 2015). A huge amount of water, food, and fuel is required to be brought to the cities and large quantities of sewage and garbage need to be transported out from the cities. This huge demand will noticeably bring massive environmental and climatic hazards along with it. Currently, India is the third largest contributor of GHG emissions of about three gigatons of CO_2 GHGs per year which is 7% of the global total (Ge and Friedrich 2020). Figure 3.2 shows the carbon footprints of 25 urban clusters in India in the year 2016.

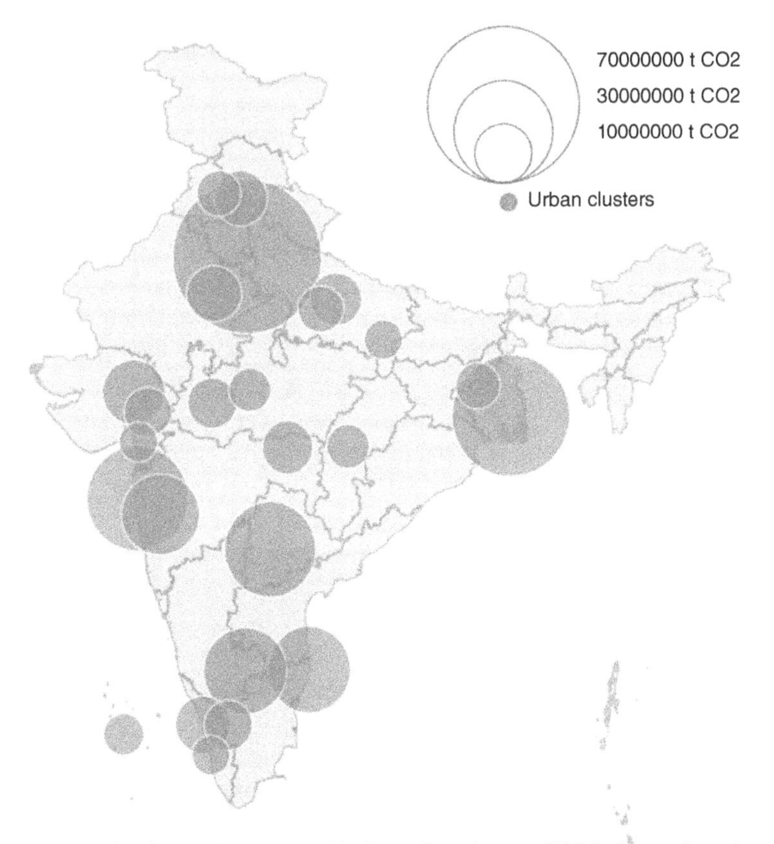

Figure 3.2 Carbon footprints of Indian urban clusters (2016). *Source:* Based on Moran et al. (2018).

The cities and megacities continue to build energy-intensive infrastructure, automobile-oriented road networks (without considering public transit, para-transit, and pedestrians), master plans that encourage long travels, insensitive waste disposal, and short-term local fugitive solutions. Urban sprawl and suburban living preference are proving to be negative for both urban and rural environments. One needs to understand that how different urban elements like transportation systems, industries, buildings, cooling systems, energy demand, etc. act as emitters and contribute to climate change directly. Globalisation and new technologies have led to the abandonment of traditional and vernacular construction techniques (Roy 2015). This has led to the use of cement, concrete, glass, and air conditioning systems to design homogenized buildings regardless of local conditions. Urbanization also has extreme impacts on groundwater tables, irrigation patterns, and land-use changes which indirectly contribute to climate change.

To date, the majority of investments and planning interventions are done in cities that do not consider the climatic effects. Even so, the effects of urbanization are converging with climate change in dangerous ways (Zhan et al. 2013). Many cities have already started experiencing the impact of natural and manmade disasters like urban floods, landslides, extreme weather, droughts, air pollution, water contamination, etc. A number of megacities and cities around the world are located along either the coastlines or rivers/floodplains and thus are quite vulnerable to such disasters. These disasters pose threat to millions of urban inhabitants especially to the most marginalized sections of the society. In the upcoming section, direct and indirect impacts of climate change in urban areas are discussed in specifics.

3.2.1.1 Direct Effects of Climate Change

a) *Sea Level Rise*: The most alarming effect of climate change is undoubtedly sea level rise. India has a coastline of about 7500 km including the island groups. In India, roughly 81 000 km^2 of land falls under the category of LECZ (Low Elevation Coastal Zone – less than 10 m of coastal elevation). This region houses over 60 million inhabitants of which more than 50% are urban residents including the Indian megacities like Mumbai, Chennai, and Kolkata. The Indian coast is projected to have a sea-level rise in between 30 and 80 cm in the next century. These coastal regions may suffer from key issues like salinization of water sources, scarcity of natural resources, destruction of ecosystems, and flooding especially during high tide (Panda 2020).
b) *Water and Sanitation*: Due to climate change, alterations in the water cycle and precipitation patterns are observed even now. This will further the already existing issues like water supply and water quality in urban areas, especially in drier regions like Delhi. Also, the poor age-old storm-water drainage systems, which are inadequate for even normal rains, may not withstand extreme rains. Urban floods in Mumbai have become an important concern. The city had a number of mangroves, wetlands, wastelands, and salt-pan lands acting as 'sponges'. But, slowly and systematically these spaces were reclaimed for the habitation of the rapidly expanding population which led to choking of the riverine and other natural drainage systems. The devastating flood of 26 July 2005 is an alarming example of the same when the city received a rainfall of 944 mm within 24 hours (Figure 3.3). The calamity cost more than 1000 lives and hundreds died

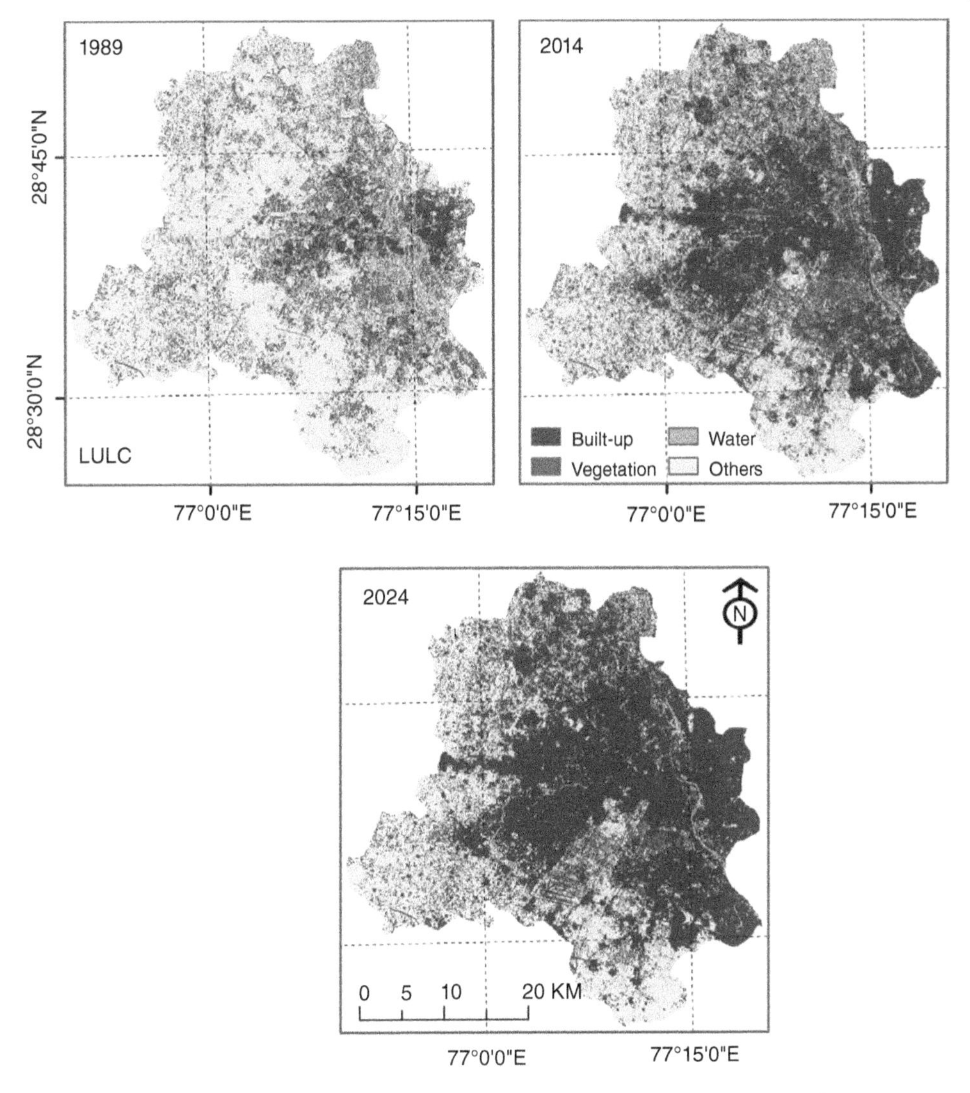

Figure 3.3 New Delhi's urban growth (1989–2024). *Source:* Tripathy and Kumar 2019/With permission of Elsevier.

in the aftermath due to various diseases. Economic losses were accounted for hundreds of crores of rupees. Similar urban floods took place in Chennai in 2015 where more than 400 people were reported dead and thousands were affected (The Hindu 2017).

c) *Health Issues*: Environment-related diseases are expected to increase due to climate change. For example, wetter or warmer breeding period, due to global warming, is ideal for mosquito-borne diseases like malaria and dengue. Lack of proper sanitation facilities and clean water will certainly upsurge food and water-borne diseases like hepatitis, diarrhoea, typhoid, cholera, etc. Also, due to increasing air pollution, warmer and drier cities are expected to see an escalation in respiratory diseases. The degrading air quality of Delhi is an example where the AQI (Air Quality Index) is being ranked as one of the worst in the world (Chatterji 2020).

d) *Heat Waves*: Urban heat island effect will make the cities and megacities warmer from their nearby areas. An average rise of 2–6 °C is expected due to waste heat produced by exhaustive usage of energy and land surface modification. Such extreme heatwaves can become more recurrent and intense and may kill thousands of people in urban areas (Beermann et al. 2016).
e) *Infrastructures*: All the aforementioned disasters like coastal and urban floods, cyclones, storms, etc. are a major risk to the already succumbing infrastructure. This includes transportation systems (roads, bridges, seaports, airports, and railway network), communication and telecommunication networks, power system, gas pipelines, water supply systems, drainage and sewage systems, industries, medical infrastructure, and housing (majorly informal and temporary housing) (Harvison et al. 2011).

3.2.1.2 Indirect Impact of Climate Change on Urban Areas

a) *Migration*: Disasters related to climate change like floods, droughts, cyclones, etc. are expected to further the migration from rural to urban (Podesta 2019). This shall overpopulate the cities and increase the share of poor and vulnerable urban residents. Approximately 500 million Indians are estimated to be affected by water issues due to global warming (Temple 2019).
b) *Economic Footprint*: Megacities and cities are the economic hotspots, and climate change is going to impact the urban economy. The World Bank estimates that there will be a loss of roughly 2% in nation's GDP for every meter rise in sea level (Ahmed and Suphachalasai 2014). A number of consequences of climate change will result in this decline, for example, increased expenses on healthcare, human resources deprecation, reduction in working hours/days at construction sites due to temperature rise, increase in food and fuel prices, expenditure on cooling systems as a response to global warming, etc.

3.2.2 Urban Sprawl

A major characteristic of urban growth in Indian megacities is sprawl. Urban sprawl is the unplanned and unrestricted spreading out or expansion of the urban areas into the rural or suburban regions. It is mostly dispersed, automobile-dependent, and low-density fragmented development. It generally increases the distance to be travelled for daily trips. It is a multidimensional phenomenon and difficult to define with a single definition. However, technically urban sprawl is 'a land use pattern with low levels of eight distinct dimensions – density, continuity, concentration, clustering, centrality, nuclearity, mixed uses, and proximity' (Galster et al. 2001).

Sprawl leads to consumption of excessive spaces leading to poor land management, high demand for transportation, and social segregation. It leads to higher energy and land consumption which threatens the urban and rural environments. This leads to a rise in GHG emissions and results in contributing to climate change (Bekele 2005). Figure 3.3 shows the urban growth for New Delhi between the years 1989 to 2024. Approximately, the urban land-use area increased from 160 to 450 km^2 between 1989 and 2014 and is expected to grow to 780 km^2 by 2024 (Tripathy and Kumar 2019). There are many impacts of urban sprawl which are discussed in the upcoming sections.

a) *Increased Air Pollution*: In many urban areas, vehicles are the major cause of air pollution and impact the overall health of the ecosystem. Longer commutes by private vehicles and traffic congestion lead to a drastic rise in GHG emissions, air pollution, and smog, and degrades the air quality of the region (Zhang and Batterman 2013).
b) *Water Pollution and Scarcity*: There is an increase in the non-point source contamination and pollution of water (runoff of oils, organic waste, metals, paints, construction site erosion, etc.) due to the presence of more roads and car parks in sprawled regions (Müller et al. 2020). Also, the pollutants in the air and soil erosion eventually add to water pollution. Most of the natural water filters like wetlands and water sources like lakes and ponds are reclaimed for urban sprawl. This leads to water scarcity, water pollution, and consequent water logging and urban floods. Several roads and highways increase the share of impervious cover and affect the groundwater too.
c) *Degraded Human Health*: Auto-dependent low-density development makes it hard for people to perform physical activity due to a sedentary lifestyle (Park et al. 2020). This further makes it difficult to maintain a healthy weight and higher BMI (Body Mass Index) level. This, in turn, leads to issues like hypertension, diabetes, high cholesterol, cardiovascular diseases, and risks overall mortality. Air pollution also leads to respiratory ailments like bronchitis, asthma, and even cancer. The urban heat island effect also leads to issues like heat strokes, heat cramps, general discomfort, and exhaustion.
d) *Costly Infrastructure*: Living in larger and dispersed settlements makes the public utilities and infrastructure more costly (Seto et al. 2014). Water supply, sewage, power, etc., become very expensive per household due to low density of population and longer lengths of pipes and powerlines mean more maintenance costs. As the development becomes private vehicle-oriented in the absence of transit facilities, more extensive infrastructure including highways, flyovers, bridges, and parking lots are constructed.
e) *Consumption of Energy*: Travel induced energy demand rises as the densities fall. Sprawled developments are generally less energy efficient than the compact urban developments. This leads to higher GHG (greenhouse gas) emissions and eventually climate change (International Energy Agency 2021).
f) *Natural and Protected Areas*: Impacts of urban sprawl on natural and protected areas are quite devastating. Sprawl leads to accessibility and proximity of urban activities to protected and natural areas. This impacts the ecosystems by the way of air, water, and noise pollution. Loss of wetlands, mangroves, meadows, natural and agricultural lands, fragmentation of forests and other habitats, disturbance wildlife species' migration corridors. Even a large number of parks and open spaces are converted to roads, parking lots, and malls. The degradation of East Kolkata Wetlands is one such example of sprawl eating up natural areas (East Kolkata Wetlands Management Authority 2021).
g) *Soil*: Urban sprawl deteriorates soil's quality by lowering its water permeability, soil biodiversity loss, reducing soil's capacity to work as a carbon sink (Galina 2016).

h) *Social Capital*: Urban sprawl is partly accountable for waning social capital. Compact neighbourhoods facilitate social interaction amongst the neighbours and promote public spaces while sprawl forms barriers and encourages private spaces (Civelli et al. 2021).
i) *Increased Traffic Congestion and Traffic-Related Fatalities*: Automobile-oriented development leads to traffic congestion as sprawled city residents spend three to four times more time driving than the compact and planned city residents (Wen et al. 2019). Increasing roads length and widening the roads leads to more sprawl and traffic. Heavy traffic leads to increased air pollution and more traffic crashes of which may prove to be fatal. Also, the residents spend more potion of their income on transportation compared to high-density area residents.

3.3 Need for Alternative Sustainable Urban Development Model

It is evident from the above discussion that the present trends of urban growth witnessed in megacities of India especially New Delhi, Mumbai, Chennai, and Kolkata are undesirable and unsustainable on many counts. The development trend has resulted in serious implications in terms of losing fertile agricultural land, increase in private vehicles leading to greenhouse gas emissions, increased consumption of energy resources, and also cities becoming more vulnerable to frequent disasters in terms of flooding, high air pollution and smog, water pollution and scarcity, etc.

Keeping in mind the current challenges faced by Indian megacities and also considering the possible implications to be faced by future upcoming cities, it is need of the hour to review and explore various alternative approaches to sustainable urban development as followed in other countries and develop a suitable framework that suits to the Indian context. To make the cities sustainable, the need is for a fair and equitable urbanization. In the upcoming sections, four alternative sustainable urban development theories are discussed in detail.

3.3.1 New Urbanism

New Urbanism is a strategy that supports sustainable development by promoting compact, mixed land use, interconnected streets and pedestrian qualities (Bohl 2000). New urbanism combines elements of the eighteenth and nineteenth-century European and American towns to provide a neighbourhood feel where everyone knows you personally and have easy access to transit, bicycle, and pedestrian pathways. New urbanism is also known as 'traditional neighbourhood design' and it is the answer to suburban sprawl and it is an effective way to counter communities who are dependent on individual private cars for making every trip (Hikichi 2003).

3.3.2 Transit-oriented Development (TOD)

Transit-oriented development has various definitions, but it is basically a mixed-use community that encourages people to live near transit services and to decrease their dependence on driving (Van Gieson 2006). In other words, TOD is the creation of compact and walkable communities centred on high-quality transit systems. TOD helps in solving the major global level concerns of peak oil and global warming. The creation of dense and walkable communities connected to a train line greatly helps in reducing the need for driving and the burning of fossil fuels. TODs are carefully designed to balance relatively high-density housing with shops, workplaces and transit access.

3.3.3 Smart Growth

Smart growth is primarily a reaction to the sprawling form of urbanization, characterized by low overall densities, a rigid specialization of land uses and a near total dependence on the automobile (Filion 2003). Smart Growth Network is the development that helps in serving the economy, community and the environment. Smart Growth attempts to provide a solution for the wide range of issues faced by communities especially the impact of sprawled development patterns, travel costs and time, conservation of prime agricultural land, etc. Smart Growth principles promote mixed land uses, compact development, providing a range of housing opportunities and choices, pedestrian-friendly communities, preserving open space, farmland and protecting environmental areas, infill development, multi-modal transport choices and community and stakeholder collaboration in development decisions (Smart Growth Network 2002). Smart Growth helps in significantly reducing public infrastructure and service costs, resulting in savings on roads, water, sewage, garbage collection, transportation and parking facilities compared with more scattered, automobile-dependent land-use patterns called 'urban sprawl' (Litman 2004).

3.3.4 Smart Cities and Sustainable Development

The initiative of 'Smart Cities and Sustainable Development' program in Europe by the European Commission (EC) and the European Investment Bank (EIB) aimed to secure the European Union's 2020 objectives by developing/ redeveloping smart, sustainable, and inclusive cities and communities in Europe. The key projected objective concerns were that 75% of the Europeans spend their lives in towns and cities, with 85% of GDP, 80% of all energy consumption and 75% GHG emissions are created in urban areas, while facing increasing economic, social, and environmental challenges (Russo et al. 2014). A number of factors like sustainability, high quality of life, and economic development define and characterize smart cities. These factors may be attained by the enhancement and augmentation of physical, human, and social capital and Information and Communication Technology (ICT) (European Commission 2011). Smart City is a technologically advanced city connecting people, city elements, and information for the creation of a sustainable greener city, economic innovations, and growth in quality of life with good governance (Ferrer 2017). A 'smart sustainable city' is a smart city that takes care of the needs of the current generation without compromising on the needs of the future generations (ITU-T. 2014). Smart cities are inclusive urban settlements which bring together formal and informal sectors, connect city cores with fringe areas, and distribute equal services to all. 'Promoting smart cities is about rethinking cities as inclusive, integrated, and livable' (World Bank 2012).

From the discussion above, it is evident that there is no single universally acceptable definition of 'Smart City'. Moreover, it is clear that the smart city approach complements sustainable development in one or the other way. Smart city concept aims for achieving sustainable development by means of balancing social, economic, and environmental factors by including Information Technology (IT), good governance, and citizen participation. Even the Smart Cities Mission by the Government of India believes that no sole description is apt to define the term. It proposes that smart city is quite contextual and varies among people, cities and countries. It depends on the development level, willingness to change and transform, and the resources and aspiration of the city and its people (MoUD 2015). The objective of the Mission is to endorse cities that provide basic infrastructure and a decent quality of life to their citizens, a clean and sustainable environment, and application of 'smart' solutions. Figure 3.4 depicts the five axes of smart cities and the relevance of sustainability is clearly visible (Mattoni et al. 2015).

3.4 Conclusion

Megacities and cities are the future of human habitation. These cities are to be planned and developed to reduce their negative impacts on the environment and at the same time be resilient to the consequences of climate change. Without proper planning and investment in the cities, people will keep on facing the unprecedented negative impacts of climate change. Not only that but issues like economic decline reduced quality of life and escalated social instability will also rise. Lack of appropriate knowledge and information sharing at

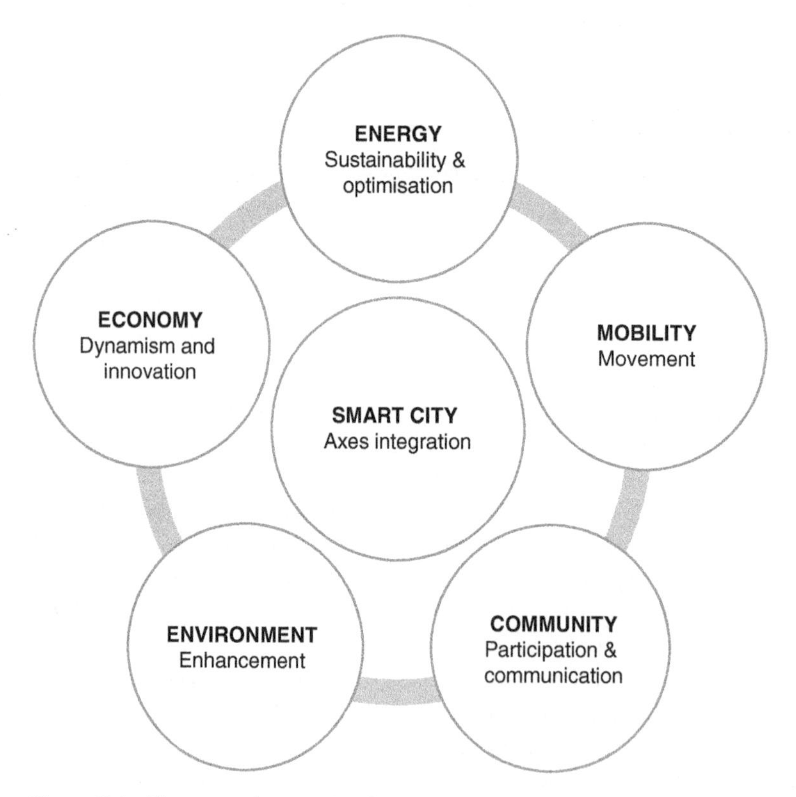

Figure 3.4 Five axes of smart city. *Source:* Mattoni et al. 2015/With permission of Elsevier.

various levels, be it local, regional, or national is a big concern. Even now, climate change is not taken as an emergency. One must not only depend on technological miracles to solve these issues but rather think of adapting to the situation better. For example, technology may not solve the issue of air pollution but re-defining mobility will. Instead of focusing on mitigation, attention should be paid to adaptation. To be actually fruitful, these strategies should be blended into urban planning frameworks. These frameworks to manage urban growth will ensure equal access to infrastructural facilities.

The urban development concepts discussed in Section 3.3, principally focus on attaining and encouraging sustainable development. While new urbanism and TOD are universal concepts, the approach of smart city mission is quite contextual. Its scope and complexity will vary between the developing and developed world. Nevertheless, smart city movement has interdependent hierarchies with contemporary sustainable urban growth theories. Thus, the actual application and effectiveness of smart cities cannot be realized without implementing concepts from the contemporary sustainable urban growth theories. Few important characteristics of smart city movement in accordance with contemporary sustainable urban growth theories are:

- Smart cities are dynamic processes.
- Smart cities ought to be livable cities.
- Smart cities are not limited to the ICT application.
- Concept of smart cities is advanced or extended version of contemporary sustainable urban growth theories.

Achieving sustainable growth in countries like India involves an integrated and inclusive method to tackle the issues at grass root level. Currently, the implementation of smart cities policies is not taking place in a holistic way but in a piecemeal manner. Contemporary urban growth theories possess immense potential and applicability to address the Indian complexities and diversities. But India has ignored and heeded little attention to these principles comprehensively. Therefore, few recommendations on the basis of alternative sustainable urban development model are listed below:

- Primary objective for the nation is to initiate measures and policies that take the developments on the path of sustainability.
- New international models or concepts shall not be blindly followed without proper analysis of their feasibility and application in the Indian context.
- Relationships and hierarchies of the various models and concepts with respect to sustainable urban development shall be studied and evaluated systematically.
- The focus shall be on addressing the core developmental and infrastructural issues rather than advanced IT applications.
- The measures taken shall be examined that they enhance the economic prosperity and quality of life in the long run.

For addressing and solving the existing key urban development issues of urban sprawl and climate change caused and faced by the Indian megacities, it is a call for rethinking, adopting, and executing the abovementioned principles in a cohesive way. The theories discussed in the chapter are workable concepts of alternative sustainable urban development to plan and design resource-efficient, people-centric, sustainable, and resilient cities.

References

Ahmed, M. and Suphachalasai, S. (2014). *Assessing the Costs of Climate Change and Adaptation in South Asia*. Metro Manila: Asian Development Bank.

Beermann, J., Damodaran, A., Jörgensen, K., and Schreurs, M.A. (2016). Climate action in Indian cities: an emerging new research area. *Journal of Integrative Environmental Sciences*: 55–66.

Bekele, H. (2005). *Urbanization and Urban Sprawl*. Stockholm: Kungliga Tekniska Högskolan.

Bhushal, R. (2019, October 3). Can Cities Save the World? Retrieved from The Third Pole: https://www.thethirdpole.net/en/energy/can-cities-save-the-world/.

Bohl, C.C. (2000). New urbanism and the city: potential applications and implications for distressed inner-city neighborhoods. *Housing Policy Debate* 11: 761–801.

Chatterji, A. (2020). *Air Pollution in Delhi: Filling the Policy Gaps*. New Delhi: Observer Research Foundation.

Chen, M., Zhang, H., Liu, W., and Zhang, W. (2014). The global pattern of urbanization and economic growth: evidence from the last three decades. *PLoS ONE 9* (8): e103799.

Civelli, A., Gaduh, A., Rothenberg, A.D. et al. (2021). Urban Sprawl and Social Capital: Evidence from Indonesian Cities. *2020 Virtual Meeting of the Urban Economics Association* (pp. 1-83). Syracuse: Syracuse University.

Dovers, S., and Butler, C. (2015, July 24). Population and Environment: A Global Challenge. Retrieved from Australian Academy of Science: https://www.science.org.au/curious/earth-environment/population-environment.

East Kolkata Wetlands Management Authority (2021). *East Kolkata Wetlands Management Action Plan 2021-2026*. Kolkata: Wetlands International South Asia.

European Commission (2011). *A European Strategy for smart, sustainable and inclusive growth*. Brussels: European Commission.

Ferrer, J.-R. (2017). Barcelona's Smart City vision: an opportunity for transformation. *Field Actions Science Reports*: 70–75.

Filion, P. (2003). Towards smart growth? The difficult implementation of alternatives to urban dispersion. *Canadian Journal of Urban Research* 12: 48–70.

Galina, C. (2016). The role of urbanization in the global carbon cycle. *Frontiers in Ecology and Evolution* 3, 144.

Galster, G., Hanson, R., Ratcliffe, M. et al. (2001). Wrestling sprawl to the ground: defining and measuring an elusive concept. *Housing Policy Debate* 12: 681–717.

Ge, M., and Friedrich, J. (2020, February 26). 4 Charts Explain Greenhouse Gas Emissions by Countries and Sectors. Retrieved from World Research Institute: https://www.wri.org/blog/2020/02/greenhouse-gas-emissions-by-country-sector.

Ghosh, I. (2019). 70 Years of Urban Growth in 1 Infographic. Retrieved February 8, 2021, from Weforum: https://www.weforum.org/agenda/2019/09/mapped-the-dramatic-global-rise-of-urbanization-1950-2020/.

Harvison, T., Newman, R., and Judd, B. (2011). *Ageing, the Built Environment and Adaptation to Climat Change*. Sydney: City Futures Research Centre.

Hikichi, L. (2003). *New Urbanism and Transportation*. Milwaukee: University of Wisconsin.

Hindustan Times (2019, February 6). India's Urbanisation not a Problem, but an Opportunity to Grow Sustainably. Retrieved from Gurugram News: https://www.hindustantimes.com/

gurgaon/india-s-urbanisation-not-a-problem-but-an-opportunity-to-grow-sustainably/story-diZ9K2dUadomJdYBB6Sm4M.html.

International Energy Agency (2021). *Net Zero by 2050: A Roadmap for the Global Energy Sector.* Paris: International Energy Agency.

ITU-T. (2014). *Smart Sustainable Cities: An Analysis of Definitions.* Geneva: International Telecommunication Union.

Krellenberg, K., Jordán, R., Rehner, J. et al. (2014). *Adaptation to Climate Change in Megacities of Latin America.* Santiago: United Nations.

Litman, T. (2004). *Understanding Smart Growth savings.* Victoria: Victoria Transport Policy Policy Institute (VTPI).

Mattoni, B., Gugliermetti, F., and Bisegna, F. (2015). A multilevel method to assess and design the renovation and integration of Smart Cities. *Sustainable Cities and Society* 15: 105–119.

Moran, D., Kanemoto, K., Jiborn, M. et al. (2018). Carbon footprints of 13 000 cities. *Environmental Research Letters*: 1–9.

MoUD (2015). *Smart City: Mission Statement and Guidelines.* New Delhi: Ministry of Urban Development.

Müller, A., Österlund, H., Marsalek, J., and Viklander, M. (2020). The pollution conveyed by urban runoff: A review of sources. *Science of The Total Environment 709*: 136125.

Panda, A. (2020, May 22). Climate Change, Displacement, and Managed Retreat in Coastal India. Retrieved from Migration Policy Institute: https://www.migrationpolicy.org/article/climate-change-displacement-managed-retreat-india.

Park, J.H., Moon, J.H., Kim, H.J. et al. (2020). Sedentary lifestyle: overview of updated evidence of potential health risks. *Korean Journal of Family Medicine* 41: 365–373.

Phys.org. (2018, May 16). UN: 68 Percent of World Population will Live in Urban Areas by 2050. Retrieved February 7, 2021, from Phys.org: https://phys.org/news/2018-05-percent-world-population-urban-areas.html.

Podesta, J. (2019, July 25). Brookings. Retrieved from The climate crisis, migration, and refugees: https://www.brookings.edu/research/the-climate-crisis-migration-and-refugees/.

Ramachandra, T.V., Aithal, B.H., and Sowmyashree, M.V. (2015). Monitoring urbanization and its implications in a mega city from space: Spatiotemporal patterns and its indicators. *Journal of Environmental Management* 148: 67–91. Retrieved from eSS Current Affairs.

Roy, S. (2015, September 28). Has Globalization and Technological Advancement Made Traditional Knowledge Irrelevant? Retrieved from Zingy Homes: https://www.zingyhomes.com/latest-trends/effect-of-globalization-technology-on-traditional-architecture/.

Russo, F., Rindone, C., and Panuccio, P. (2014). *The Process Of Smart City Definition At An EU Level. SUSTAINABLE CITY*, 979–989. Ashurst: WIT Press.

Seto, K., Dhakal, S., Bigio, A. et al. (2014). Human settlements, infrastructure and spatial planning. In: *Climate Change 2014: Mitigation of Climate Change. Contribution of Working Group III to the Fifth Assessment Report* (eds. O. Edenhofer, R. Pichs-Madruga, Y. Sokona, et al.), 923–1000. Cambridge: Cambridge University Press.

Smart Growth Network (2002). *Getting to Smart Growth: 100 Policies for Implementation.* New York: International City/County Management Association.

Temple, J. (2019, April 24). India's Water Crisis Is Already Here. Climate Change Will Compound it. Retrieved from MIT Technology Review: https://www.technologyreview.com/2019/04/24/135916/indias-water-crisis-is-already-here-climate-change-will-compound-it/.

The Economic Times (2018, May 16). Delhi Projected to become World's most Populous City around 2028: UN Report. Retrieved from The Economic Times|Politics: https://economictimes.indiatimes.com/news/politics-and-nation/delhi-projected-to-become-worlds-most-populous-city-around-2028-un-report/articleshow/64194461.cms.

The Hindu (2017, December 2). The Hindu. Retrieved from Revisiting the 2015 floods in Chennai: images from The Hindu: https://www.thehindu.com/news/cities/chennai/revisiting-the-2015-floods-in-chennai-images-from-the-hindu/article21248895.ece.

Thornton, A. (2019, Februrary 6). 10 Cities Are Predicted to Gain Megacity Status by 2030. Retrieved Februrary 6, 2021, from Weforum: https://www.weforum.org/agenda/2019/02/10-cities-are-predicted-to-gain-megacity-status-by-2030/.

Tripathy, P. and Kumar, A. (2019). Monitoring and modelling spatio-temporal urban growth of Delhi using Cellular Automata and geoinformatics. *Cities* 90: 52–63.

UN News (2018, May 16). Around 2.5 billion more people will be living in cities by 2050, projects new UN report. Retrieved from Department of Economic and Social Affairs: https://www.un.org/development/desa/en/news/population/2018-world-urbanization-prospects.html.

UNDESA (2019). *World Urbanization Prospects: The 2018 Revision (ST/ESA/SER.A/420).* New York: United Nations.

UNDP (2017, July 6). United Nations Development Programme. Retrieved from Sustainable Development Goals: https://www.undp.org/content/undp/en/home/sustainable-development-goals.html.

Van Gieson, J. (2006). *NOW IS THE TIME FOR TODS: Homebuyers Are Waiting in Line for Transit-oriented Development*. Washington DC: National Association of Realtors.

Wen, L., Kenworthy, J., Guo, X., and Marinova, D. (2019). Solving traffic congestion through street renaissance: a perspective from dense Asian cities. *Urban Science*: 1–21.

World Bank (2012, March 20). Who Needs Smart Cities for Sustainable Development? Retrieved from The World Bank: https://www.worldbank.org/en/news/feature/2012/03/20/who-needs-smart-cities-for-sustainable-development.

World Economic Forum (2018). *The Global Risks Report 2018*. Geneva: World Economic Forum.

Zhan, J., Huang, J., Zhao, T. et al. (2013). Modeling the impacts of urbanization on regional climate change: a case study in the Beijing-Tianjin-Tangshan metropolitan area. *Advances in Meteorology*: 1–8.

Zhang, K. and Batterman, S. (2013). Air pollution and health risks due to vehicle traffic. *Science of The Total Environment*: 307–316.

4

Integrated Water Resource Management for Future Water Security

Musarrat Parween

National Institute of Advanced Studies, Indian Institute of Science Campus, Bangalore, Karnataka, India

4.1 Introduction

With around 2.54% of total land area, India owns 4% of the world's water resources (ADRI 2021). Extensive population growth and industrialisation along with rising food security concerns have led to severe depletion of the world water resources. The per-capita availability of water which was 5177 m^3 per year in 1951 supporting a population of 361 million has fallen down abruptly to 1820 m^3 per year in 2001 for serving a population load of 1027 million (Govt of India 2009). This is further expected to fall to 1341 m^3 by 2025 and to 1140 m^3 by 2050. The blame could be put onto the extensive lapses in maintaining the quantity as well as quality of surface water resulting in overexploitation of groundwater for the growing agricultural, industrial, and domestic needs in India. Around two-thirds of the total freshwater in the world is used for irrigation (Water Aid India 2019). Groundwater development exceeds 100% in most of the states of India, most remarkably in Delhi, Haryana, Punjab, and Rajasthan. Deeper drilling for water has rendered contamination of aquifers with arsenic, fluoride, and nitrate and has led to high salinity of groundwater. As of 2010, chemical contamination continued to be affecting 9% of the Indian population (DDWS 2010).

The lack of strategies, planning and governance concerning the extraction and use of water, growing food demand, intensifying industrialisation, inappropriate discharge of wastewater, and lack of proper and sufficient means and mechanisms of wastewater treatment has put under scrutiny the future water security, not only in India but also worldwide (Parween and Ramanathan 2019). There is an urgent need to manage water resources in a sustainable manner. An integrated water resource management is the need of the current situation. This calls for a life cycle-based approach towards water. The management of rainwater, surface water, and groundwater resources in a coherent manner is essential. Rainwater harvesting and groundwater recharge could be viable and important aspects of water conservation. Recycling and reuse of water should be

Urban Ecology and Global Climate Change, First Edition. Edited by Rahul Bhadouria, Shweta Upadhyay, Sachchidanand Tripathi, and Pardeep Singh.

undertaken on priority basis, while introducing mechanisms to check overextraction of water through water metering and pricing.

Six hundred million people in India go through acute water shortage while the domestic water demand rose up by around 20% since COVID-19, owing to enhanced awareness about hygiene (Nikore and Mittal 2021). Wastewater has enormous potential of being used in several developmental processes such as the use of grey water in toilet flushes and untreated domestic wastewater for floriculture, public parks, and lawns (Parween and Ramanathan 2019; Water Aid India 2019). The use of particular class or type of wastewater can be selected or customised for the intended purpose. Wastewater is drenched with innumerable resources such as water, energy, organic matter, and nutrients such as nitrogen and phosphorus, which hold specific value and significance in several other economic, social, and environmental activities (Parween et al. 2016, 2017; Parween and Ramanathan 2019), and can therefore be diverted to be used as raw materials. However, inadequacy and unavailability of data with respect to flow of wastewater as well as their qualities have by and large hindered such progresses in the country.

Institutional inputs and corporate social responsibility have significant roles to play in this direction. Even the common man can join hands to this cause. Sensitisation and awareness among the common mass is expected to make a large difference. It is important to educate and empower communities facilitating a participatory approach in this field. Several sectors as well as disciplines need to be amalgamated in the water sector which itself is a multidisciplinary and trans-sectoral subject. An inter-sectoral and -disciplinary effort is indispensable in ensuring public participation and sustainability (Mohan et al. 2017).

4.2 Significance of the Study

4.2.1 Water Resources and Rising Water Insecurity

Out of the total water withdrawn worldwide, only 44%, i.e. $1716\,km^3\,yr^{-1}$, is actually consumed. The rest is lost in various forms. The largest share lost is mainly in agriculture, which is through evaporation in irrigated crop fields. The remaining 56% ($2212\,km^3\,yr^{-1}$) gets converted to wastewater and is eventually discharged in the form of effluent from the municipal and industrial sectors and as agricultural drainage (Figure 4.1).

The term 'water security' means to ensure that every person has easy and equitable access to adequate as well as safe water availability for basic everyday activities, such as drinking, cooking, and other domestic needs throughout on a sustainable basis.

The National Water Policy has kept provision of safe water for drinking and sanitation on top priority followed by domestic usage inclusive of livestock, food security, agricultural, and minimum ecosystem needs. The allocation of water promoting conservation and efficiency should be met only after the above needs are fulfilled. However, 86% of the available water resources in India was used in agriculture, 6% by industries, and 8% for domestic purposes (DDWS 2010). Water consumed in the agriculture sector has further reached a level as high as 90% in India, as against the figure of 69% in the world (AQUASTAT, FAO 2016; Parween and Ramanathan 2019). The sector-wise consumption pattern of freshwater within various sectors in India as well as the world has been depicted in Figure 4.2.

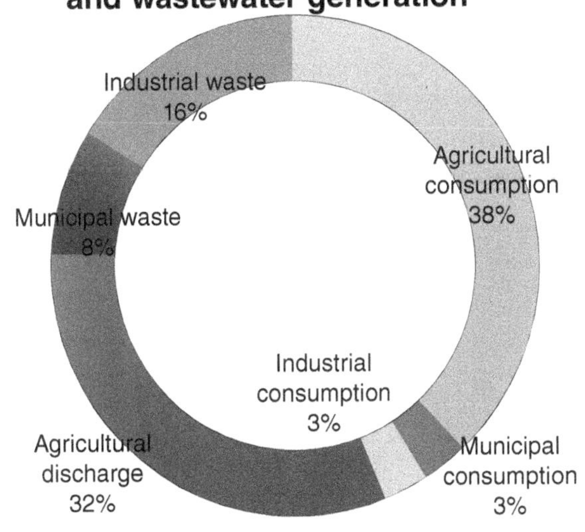

Figure 4.1 Global water consumption and wastewater generation in various sectors. *Source:* AQUASTAT; Mateo-Sagasta et al. (2015).

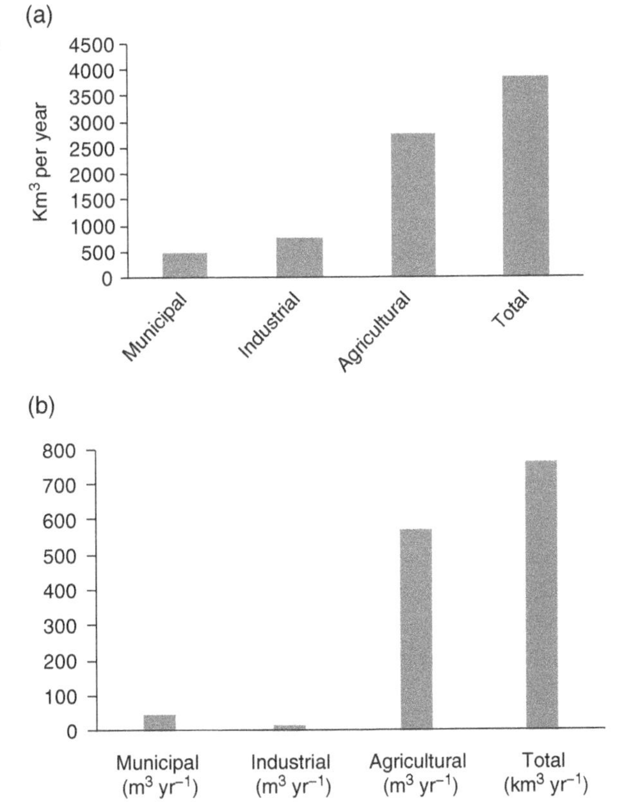

Figure 4.2 Sector-wise water consumption in India and the world. (a) World water consumption; (b) water consumption in India. *Source:* Gleick et al. (2014); AQUASTAT, FAO (2016).

4.2.2 Deteriorated Surface Water Bodies

The dependence of the developing countries including India on water bodies for numerous developmental activities (domestic, agricultural, and industrial) as well as for cleansing and disposal of wastes has become a grave threat to the existence of rivers and lakes. Most of the rivers in urban centres are ill, while many have already disappeared. Around 1.8 million people in the developing countries, mostly children, die each year from diseases related to contaminated water (WHO 2006). About 70–80% of the total domestic water supplied in India gets transformed to wastewater according to the Central Public Health and Environmental Engineering Organisation, Govt. of India (EY-ASSOCHAM 2019).

4.2.3 Overexploited Groundwater Resources

Groundwater has been serving more than 85% of the country's rural domestic water requirements, 50% of its urban water needs, and greater than 50% of its irrigation supplies (CGWB 2014; Dutta 2019). As revealed by the World Bank in 2010, around 20 million wells in India are used for the withdrawal of groundwater resources, mostly through private initiatives. Despite being a replenishable resource, accessibility of groundwater cannot be ensured everywhere and every time owing to its non-uniform availability with respect to space and time.

Most of the areas overexploited for groundwater extraction in the country are concentrated within three regions: (i) North-western covering parts of Punjab, Haryana, Delhi, and Western Uttar Pradesh. Despite abundance and replenishability of the resource in this region, indiscriminate withdrawals have led to overexploitation. (ii) Western region: Arid parts of Rajasthan and Gujarat with little scope of groundwater recharge. (iii) Peninsular region: including Andhra Pradesh, Tamil Nadu, and Karnataka with little water availability due to poor aquifer characteristics (CGWB 2014).

According to the Central Groundwater Board, India, an estimated 433 billion m^3 (bcm) of groundwater is annually replenishable, 58% of which is contributed alone by the monsoon rainfall recharge. While rainfall remains the major source of the country's annual replenishable groundwater resource with an input of 68% recharge, other sources such as canal seepage, return flow from irrigation, recharge from tanks, ponds, and water conservations structures too play a substantial role in the same, thereby facilitating 32% annual recharge taken together.

After allocating to natural discharge, the net annual groundwater availability remains 398 bcm out of the replenishable 433 bmc of the resource. The annual groundwater draft in 2011 was 245 bcm. The overall groundwater development in the country was about 62%. Out of the 6607 units (blocks/mandals/talukas/firkas) assessed, 1071 have been categorised as 'overexploited', i.e. the annual groundwater extraction exceeds the net annual groundwater availability, showing significant decline in long-term groundwater level trend during pre-monsoon and/or post-monsoon. Two hundred and seventeen units are 'critical' with a groundwater development of more than 90%, but within 100% of the net annual groundwater availability and significant decline in the abovementioned trend, 697 units are semi-critical, with development between 70 and 100%, and 4530 units are safe showing no decline in long-term groundwater level trend. Moreover, there exist 92 blocks/firkas which are completely underlain by saline groundwater (CGWB 2014).

4.2.4 Water Quality

As mentioned in Section 4.1, about 144 064, i.e. 9% of the habitations, are bound to face water quality issues in the form of chemical contamination. Among these, 6548 habitations spread over 8 states suffer from arsenic contamination, 26 131 habitations of 19 states from fluoride contamination, and 28 398 habitations of 15 states (inland and coastal) have saline groundwater. Iron contamination has been affecting 79 955 habitations of 21 states, and nitrate contamination is reported in 3032 habitations of 12 states. All of these contaminations are either natural or a repercussion of overexploitation of groundwater (DDWS 2010).

Several sources have reported bacteriological contamination mostly due to waterborne diseases, especially during the monsoon season. Spread of such diseases is witnessed every year due to poor sanitation and impact severely both maternal and child morbidity as well as mortality.

4.2.5 Abrupt Rainfall Pattern

The average rainfall that India receives per year is about 120 cm, which varies drastically with space and time. Approximately 75% rainfall is received during monsoon months (June to September), leaving behind eight relatively dry months. While the regions of coastal Karnataka, Konkan, Goa and north east India are drenched with more than 250 cm of rainfall annually, western Rajasthan receives merely about 30 cm. There are 559 departmental and part-time observatories of the India Meteorological Department and 8579 non-departmental rain gauge stations, out of which 5039 are non-reporting (CGWB 2014).

Good quantity and continuous rainfall as well as planned management practices such as groundwater augmentation and conservation with the help of government and private initiatives have led to improved groundwater situation, as reflected by the shifting of the stages of groundwater development towards better categories (CGWB 2014).

4.2.6 Government-led Initiatives

The Department of Drinking Water and Sanitation, Ministry of Rural Development, Govt. of India, claims to have accomplished a remarkable status in terms of provision of safe drinking water, with more than 90% of the rural households 'covered' as revealed by the National Sample Survey Office (NSSO) 65th round survey 2008–2009. Around 30% of the rural households in India have tapped water supply for the purpose of drinking. However, there is large variation in the figures among the different states ranging from less than 5% in Uttar Pradesh and Bihar to around 80% in Tamil Nadu and Himachal Pradesh (DDWS 2010). Economic status too plays a role in determining access to water also as depicted by the figures revealed by the UNICEF. While ~32% of the rich people have piped water connections on their premises, only around 1% of the poor enjoy this basic facility (DDWS 2010).

The rural water supply schemes have seen numerous programmes and missions running since several decades, such as the Accelerated Rural Water Supply Programme (ARWSP) in 1972, Rajiv Gandhi National Drinking Water Mission in 1986, and National Rural Drinking Water Programme in 2009. Nevertheless, the concerns relating to water security do not seem to diminish. There are still miles ahead to go in the fields of improved levels of water availability, quality, and sustainability. As reported, the ARWSP seems to have achieved substantially the target of providing basic water supply facilities in rural India. Figures showing coverage is overwhelming. However, reports obtained from the states have indicated that 494 610, i.e. 30% of the total 1 661 058 habitations, have slipped back from the category of covered to partially covered as in April 2010 (DDWS 2010). This could be attributed to reasons such as subsequent inclusion of newly formed habitations, better knowledge about areas facing water quality issues, more habitations falling under quality affected due to overextraction of groundwater, poor operation and maintenance, and sources turning non-functional.

4.2.7 Urban Water Crisis and Poor Management

Around 160 000 people are migrating from rural to urban regions every day. The current urban population in India needs 740 billion cubic meters water per year, which is expected to grow exponentially (EY-ASSOCHAM 2019). All major cities of India are facing extreme water crisis, as economic development and monetary profits have overshadowed the urban water resource management, be it Bengaluru, Chennai, Delhi, Kolkata, or Mumbai. Though the water issues faced by each are diverse, it finally results in inadequate water available for urban populations. During 2019, while on one hand Chennai was drenched down to the deepest water crisis it ever faced, Bengaluru had been averting water day zero with water tankers from the Cauvery reaching out to half the urban population.

Chennai, the automobile hub (60% of the country's automobile exporter) as well as a leading IT hub of the country, has witnessed dark days in terms of its water needs. In the absence of a perennial water source, the city is bound to depend on reservoirs fed by rain (Puzhal lake, Poondi Reservoir, and Cholavaram Lake). Just the urban centre contains 66% households with private wells. Such extensive overexploitation has led to extraction twice as that of the annual recharge (Kubernein Initiative 2020). The city is the first one to go dry (India today 2019). Studies indicate widening of the gap between demand and supply of water for Chennai in 2030, which is estimated to reach up to 400 million litres per day (MLD), in the absence of urban water management and conservation practices (Paul and Elango 2018). This could compromise on the water security and health of the urban population along with the economic share contributing to the country's development.

The Bengaluru Water Supply and Sewerage Board mainly supplies water obtained from the Arkavathi as well as the Cauvery rivers. This covers just 55% of the population, rendering the burden on groundwater resources. Twenty percent of the water supplied by the BWSSB remains unaccounted owing to leakage and theft (Kubernein Initiative 2020). Bengaluru is left with just around 80 lakes (that numbered 260 in the 1960s), and that too severely degraded/polluted. Rejuvenation and management of these lakes have immense potential in meeting the water requirements of the urban centre, thereby reducing the load on groundwater. While water is overexploited in these cities, efforts towards conservation and replenishment/rejuvenation are seldom made. The exponentially growing built-in area has covered several water recharge zones, thereby obstructing natural water recharge. Bengaluru, the IT hub, alone holds 40% of this industry in India (electronics, telecommunications, aerospace, biotechnology, artificial intelligence, etc.). The second and third largest industries are biotechnology and textile industries, respectively. The latter is water intensive. Agriculture and industries consume a great share of water in the city which is certain to overshadow domestic supply of water and lead the city to reach its water day zero sooner.

This is also the case with Delhi, which has little water resources allotted by nature, leaving it dependent on other states for its water supplies. Eighty two percent of the urban centre's water needs are met by the Delhi Jal Board piped water supply (60% from Yamuna, 34% from Ganga, and the rest from Bhakra storage and underground water). However, it is a pity that around 40% of the supplied water is left unaccounted and is unable to generate revenue (Safe Water Network 2016b). This water is lost to theft, leakages, unmetered usage, etc. Politically driven free/subsidised water as well as an over-surplus supply has led to indiscriminate use of water in the city. Delhi is home to 29 industrial areas and 20 large, 25 medium, and around 93 000 small-scale industries. According to the Central Pollution Control Board, the large- and medium-scale industries (mostly engineering, textile, chemicals, electronic, and electric goods) contribute to 50% of the total waste generated. Around 40% of sewage go into river Yamuna untreated. The river water is relentlessly polluted with several kinds of wastes, effluents, and discharges from domestic, industrial, and agricultural sources (Parween et al. 2014, 2021).

The water crisis in Kolkata and Mumbai is rather area specific and confined to certain zones. Uneven distribution of water resources in the city results in such disparities. The community deprived of adequate water are largely slum dwellers or the economically backward section of society. Mumbai, the commercial capital of India, has several challenges to face, such as uneven water resources in the city, unplanned urbanisation and settlements, and poor infrastructure and access to water resources. Water supplied to the city is obtained mainly from the lakes and reservoirs located along the rivers Amba, Patalganga, Ulhas, and Vaitarna, all of which depend on precipitation. The inconsistency in rainfall and exponentially growing population can widen the supply and demand gap to 1100 MLD, as per estimates (Safe Water Network 2016a). The economically weaker and slum regions, which make 40% of the population (Economic Survey 2011), face the crisis more severely. This has resulted in overexploitation of groundwater in the city. Effluents from refineries, fertiliser industries, and sewage have immensely polluted surface as well as groundwater. Despite bans on groundwater for drinking, there are 3950 dug wells and 2514 bore wells in the city (Safe Water Network 2016a). Urban planning and water management has a significant role to play in this case. Wastewater treatment plants have the potential to supply additional 496.8 MLD water, while a desalination plant could further add up to 200 MLD according to the Ministry of Water Resources, Central Ground Water Board (2021).

The main source of water in Kolkata is the Hooghly River. Eleven percent of the water needs are served by groundwater, 55% of which is contaminated with arsenic and therefore non-usable. This is a situation that arises due to overexploitation of groundwater. As the water tax was abolished in 2011 (Singh 2019), the situation became grim, leading to further exploitation by builders and private tankers. Unaccounted loss of water supplied is 35% in this city (Kolkata Municipal Corporation 2021). The major industries in the city (such as manufacturing and textile industries and paper and paper product industries) account for water-intensive activities. The Kolkata Municipal Corporation has estimated the potential for acquisition of 247 billion litres water through rainwater harvesting along with other urban water management and conservation practices. This could, however, be a big challenge for the city and would need several levels of planning and management in an integrated manner and with a holistic approach.

The concept of smart cities is the focus in government-led initiatives. A smart city is the one equipped with core infrastructural facilities along with digital transformation to create a clean and sustainable environment (EY-ASSOCHAM 2019). Water management is certainly the most essential foundation of the smart city mission. Smart cities are therefore expected to be aided with smart meters and metering management, leakage identification tools, pollution prevention, water flow maintenance and water quality monitoring, etc. along with systematic water recharge and rejuvenation schemes.

4.3 Methodology

4.3.1 Integrated Water Management

Water management with an integrated approach towards securing the present and future requirements of safe and adequate water should be aimed essentially at undertaking an integrated and life cycle- and ecosystem-based management model operable across all the three dimensions of sustainable development (social, economic, and environmental), irrespective of the geographical boundaries, local barriers, and localised sociopolitical motives. The keys for such an integrated management approach lie on the efforts to reduce, reuse, recycle, and recover. Therefore, the process must begin with the judicious use of water and lies upon conservation and management at every single stage by switching to products and processes that are water efficient. It also includes several other components such as treatment, recycling, reuse, recovery of essential components, wise and efficient disposal of non-utilisable or residual wastes, facilitation of aquifer recharge, tapping of rainwater, prevention of pollution and overexploitation, promotion of social awareness, and community participation and their empowerment. People can play a vital role by owning, managing, and maintaining water resources that they are directly associated with. An inclusive approach with multiple dimensions and multidisciplinary and multistakeholder approaches would be best suited to address issues related to future water security. The following could be some of the initiatives that could help formulate strategies to achieve an integrated water resource management.

4.4 Recommendations for an Integrated Management of Water Resources

4.4.1 Regulate Water Extraction

Several government schemes and programmes have facilitated the easy extraction of groundwater, especially through borewells/tube wells. Moreover, many of these are unauthorised and a result of private drilling. Strict regulations and parameters allowing such extractions need to be laid and implemented for all uses.

4.4.2 Water Quality

Restoring water quality is one of the greatest challenges which can be addressed with strong legislation and enforcement of laws pertaining to quality standards and testing protocols, overexploitation, poor operation and maintenance of facilities and structures, etc. Spreading awareness among communities about water quality, health and hygiene, and promotion of public participation in management and maintenance of water resources are essential.

4.4.3 Reduce, Reuse, and Recycle Water

Restricted allocation of water in industries, commercial institutions, and corporate houses which consume a lot of water in production or in maintaining the ambience through parks, gardens, fountains, swimming pools, etc. would help in checking the indiscriminate use of water. Water treatment, recycling, and reuse should be made mandatory by law and the use of treated/grey water be imposed wherever applicable. Annual rewards for best water conservation efforts can be attractive too.

4.4.4 Rainwater Harvesting

Rainwater harvesting, storage, or recharge should be imposed in commercial, institutional, and community buildings and also promoted elsewhere through incentives/subsidies.

4.4.5 Agricultural Reforms

Education and awareness of farmers is the key to the judicious use of water. Promotion of modern methods of irrigation such as sprinklers and drip irrigation is needed, especially in water-stressed areas. Farmers should also be educated about cropping patterns, crop rotation, water-intensive crops and varieties, impacts of commercial farming and overirrigation and methods of sewage water application, postharvest washing, etc. in order to facilitate water conservation as well as crop hygiene and protection.

Upscaling of farmer-managed ground and surface water resources model is also deeply required. Water metering/pricing could be helpful in regulating indiscriminate use.

4.4.6 Reusability of Wastewater

Wastewater carries immense potential of being used in several developmental processes determined carefully by the type of wastewater and use (Nikore and Mittal 2021). Wastewater contains numerous valuable resources such as water, energy, organic matter, and nutrients such as nitrogen and phosphorus, which can be used in other economic, social, and environmental activities and/or processes (Parween et al. 2017; Parween and Ramanathan 2019).

4.4.7 Conjunctive Use of All Resources

Conjunctive use of surface water, groundwater, and rainwater resources is beneficial in ensuring economic, consistent, and adequate supply of water throughout the year.

4.4.8 Sustainability of the Source

While ensuring a conjunctive use, it is important to ensure sustainability of the existing water source. All the resources as well as structures seeking storage, recharge, and water harvesting should be constructed in a scientific manner so as to ensure their functionality on a long-term basis. Lessons from past failures as well as improvisation could provide a viable, economic, and sustainable source of water.

4.4.9 Participatory Mechanism

The emphasis on participatory integrated water resource management at village, district, and state levels could serve best the strategy of a conjunctive use of rainwater, groundwater, and surface water in order to sustain the fulfilment of the country's water requirements.

4.4.10 Decentralisation of Action Plans

Decentralisation of institutional roles and responsibilities to support water security planning and implementation are the best ways to deal with the diverse problems of affected areas which are specific and unique to the location.

Several programmes run by the central and state governments have their own fragmented and specified agendas, some of which are overlapping, while others may be conflicting in nature. It is therefore pertinent to converge different development programmes and a holistic approach be acquired towards issues that are closely linked in contributing towards the future water security.

4.5 Conclusion

The rising concern over water crisis demands immediate action to ensure water security. Rivers, lakes, and ponds are deteriorated to a critical extent, rendering turning back a challenge and an expensive affair. Indiscriminate extraction of groundwater, especially in the urban centres of the country, has left the aquifers overexploited and severely contaminated, leading to poor quality of water which is even used for drinking. Abrupt rainfall and frequent draughts in several parts of the country have been adding up to the complexity of the problem. This is a matter of grave concern. Several missions and programmes initiated by government along with private efforts have been ineffective in improving the situation. It is therefore pertinent to start working towards an integrated water resource management and ascertain the sustainability of the sources. Systematic planning and management is required right from withdrawal to use as well as discharge of water. A full life cycle and ecosystem-based sustainable approach towards water which is the most indispensable need of mankind is therefore obligatory in urgency.

References

ADRI (2021). India Water Facts. Asian Development Research Institute. Retrieved from https://www.adriindia.org/adri/india_water_facts.

AQUASTAT, FAO (2016). AQUASTAT Database. Retrieved from http://www.fao.org/nr/water/aquastat/data/query/index.html?lang=en

CGWB (2014). *Dynamic Ground Water Resources of India*. Central Ground Water Board, Ministry of Water Resources, River Development & Ganga Rejuvenation, Government of India.

DDWS (2010). *Strategic Plan – 2011-2022, Department of Drinking Water and Sanitation – Rural Drinking Water, "Ensuring Drinking Water Security In Rural India"*. Ministry of Rural Development, Government of India.

Dutta, P. K. (2019). Why India Does Not Have Enough Water to Drink. India Today. Retrieved from https://www.indiatoday.in/india/story/why-india-does-not-have-enough-water to-drink-1557669-2019-06-28.

Economic Survey (2011). Union Budget and Economic Survey, Govt. of India. Retrieved from https://web.archive.org/web/20111008034945/http://indiabudget.nic.in/survey.asp.

EY-ASSOCHAM (2019). "Think Blue" Effective Water Management: Integrating Innovation and Technology, ASSOCHAM EY. Retrieved from http://www.spml.co.in/Download/Reports/Think-Blue-Effective-Water-Management-June-2019-EY-Assocham-Report.pdf.

Gleick, P., Pacific Institute, and Ajami, N. (2014). Data Table 2. Freshwater withdrawal by country and sector, 2013 update. In: *The World's Water Volume 8: The Biennial Report on Freshwater Resources* (ed. P. Gleick), 227–235. ISBN: 161091483X, 9781610914833.

Govt. of India (2009). Background note for consultation meeting with Policy makers on review of National Water Policy. Ministry of Water Resources.

India Today (2019). Chennai Water Crisis: Govt Promises Clean Drinking Water to Every Household by 2024. Retrieved from https://www.indiatoday.in/india/story/chennai-water-crisis-govt-promises-clean-drinking-water-to-every-household-by-2024-1557652-2019-06-28.

Kolkata Municipal Corporation (2021). Ground Water Information Booklet. Retrieved from: http://cgwb.gov.in/District_Profile/WestBangal/Kolkata%20Municipal%20Corporation.pdf.

Kubernein Initiative (2020). The Challenges of Urban Water Security and Growth. Working paper 1. Retrieved from https://kuberneininitiative.com.

Mateo-Sagasta, J., Raschid-Sally, L., and Thebo, A. (2015). Global Wastewater and Sludge Production, Treatment and Use. In: *Wastewater* (eds. P. Drechsel, M. Qadir and D. Wichelns). Dordrecht: Springer.

Ministry of Water Resources, Central Ground Water Board (2021). Ground Water Information Greater Mumbai District Maharashtra. Retrieved from http://cgwb.gov.in/District_Profile/Maharashtra/Greater%20Mumbai.pdf

Mohan, N.S., Parween, M., and Raj, B. (2017). Water challenges in India: seeking solutions with an integrated approach. *Current Science* 113 (11): 2074–2076.

Nikore, M. and Mittal, M (2021). "Arresting India's Water Crisis: The Economic Case for Wastewater Use," ORF Issue Brief No. 453, March 2021, Observer Research Foundation.

Parween, M., Ramanathan, A.L., Raju, N.J., and Khillare, P.S. (2014). Persistence, variance and toxic levels of organochlorine pesticides in fluvial sediments and the role of black carbon in their retention. *Environmental Science and Pollution Research* 21 (10): 6525–6546. https://doi.org/10.1007/s11356-014-2531-6.

Parween, M., Ramanathan, A.L., Raju, N.J., and Khillare, P.S. (2016). Persistent pesticides in fluvial sediment and their relationship with black carbon. In: *Geostatistical and Geospatial Approaches for the Characterization of Natural Resources in the Environment* (ed. N. Raju). Cham: Springer https://doi.org/10.1007/978-3-319-18663-4_54.

Parween, M., Ramanathan, A.L., and Raju, N.J. (2017). Waste water management and water quality of river Yamuna in the megacity of Delhi. *International Journal of Environmental Science and Technology* 14 (10): 2109–2124. https://doi.org/10.1007/s13762-017-1280-8.

Parween, M. and Ramanathan, A.L. (2019). Wastewater management to environmental materials management. In: *Handbook of Environmental Materials Management 9* (ed. C.M. Hussain), 1–24. Springer International Publishing https://doi.org/10.1007/978-3-319-73645-7_72.

Parween, M., Ramanathan, A.L., and Raju, N.J. (2021). Assessment of toxicity and potential health risk from persistent pesticides and heavy metals along the Delhi stretch of river Yamuna. *Environmental Research* 202: 111780. https://doi.org/10.1016/j.envres.2021.111780.

Paul, N. and Elango, L. (2018). Predicting future water supply-demand gap with a new reservoir, desalination plant and waste water reuse by water evaluation and planning model for Chennai megacity, India. *Groundwater for Sustainable Development* 7: 8–19, ISSN 2352-801X. doi:https://doi.org/10.1016/j.gsd.2018.02.005.

Safe Water Network (2016a). Drinking Water Supply for Urban Poor: City of Mumbai. Retrieved from https://www.safewaternetwork.org/sites/defaultfiles/Safe%20Water%20Network_Mumbai%20City%20Report.pdf.

Safe Water Network (2016b). Drinking Water Supply for Urban Poor: City of New Delhi. Retrieved from https://www.safewaternetwork.org/sites/default/files/Safe%20Water%20Network_Delhi%20City%20Report.PDF.

Singh, G. (2019) As Kolkata's Groundwater Level Depletes, Consequences Go Beyond Water Shortage, Mongabay. Retrieved from https://india.mongabay.com/2019/06/as-kolkatas-groundwater-level-depletes-consequences-go-beyond-water-shortage/.

Water Aid India (2019) India Water Factsheet. Retrieved from https://www.wateraidindia.in/sites/g/files/jkxoof336/files/india-water-fact-sheet-2019.pdf.

WHO (2006). *Guidelines for Drinking-water Quality: Incorporating First Addendum*, Recommendations. – 3rde, vol. 1. World Health Organization.

5

Water Urbanism and Multifunctional Landscapes: Case of Adyar River, Chennai, and Ganga River, Varanasi, India

Vidhu Bansal[1,2], *Sharmila Jagadisan*[1], *and Joy Sen*[2]

[1] *VIT School of Planning and Architecture, Vellore Institute of Technology, Vellore, Tamil Nadu, India*
[2] *Department of Architecture and Regional Planning, Indian Institute of Technology Kharagpur, Kharagpur, West Bengal, India*

5.1 Introduction

Our lives are enriched and surrounded by natural assets. One important asset is water, which is critical for socioeconomic development, food security and healthy ecosystems, and very essential for improving the living standards of the people (Ahern 2013). The rapidly increasing population, growing needs, and demands for economic development have put tremendous pressure on natural resources including water. The United Nation's sustainable development goals (SDGs) goal **6** states that 'Water sustains life, but safe clean drinking water defines civilization' and access to safe drinking water is a fundamental right of every individual (SDG 6). But unfortunately, climate change intensifies the water challenges by deficit, quality deterioration, altered watersheds, flooding, and changes in the course of the river system. Hence, addressing water management is often a key entry point to restore degraded lands and to enhance landscape resilience for the benefit of the people, economy, and environment. Our core ecological economies are embedded in society and culture that are themselves embedded in an ecological life support system (Costanza et al. 2012). Up until recently, these assets have been thought in terms of separate entities without understanding the profound reality of their interconnectedness. A city or region cannot be sustainable without full and functional consideration of ecological processes. In fact, 'multifunctionality' is a popular term in sustainable landscape systems, and in recent times, it has been imperceptibly influencing and strongly resonating within landscape planning and design. Multifunctionality is increasingly used to improve the functional performance of an area as it leverages efficient use of land, delivers more functionality, and services for a wider user base (public

Urban Ecology and Global Climate Change, First Edition. Edited by Rahul Bhadouria, Shweta Upadhyay, Sachchidanand Tripathi, and Pardeep Singh.

benefit), leading to better stewardship towards environmental, economic, and sociocultural aspects (Benedict and McMahon 2006). However, there is still a widespread lack of knowledge and global awareness to recognise the symbiosis between the quality and connectivity of natural assets with local environmental and economic performance. The strength of multifunctional landscapes is their ability to afford the needs of multiple stakeholders across broad spatial scales and sectors, thus appealing to diverse constituents with various social, economic, recreation, and ecological objectives (Benne and Mang 2015).

From the multifunctional perspective, the key to a sustainable future is to restore the biodiversity, make effort to adapt and build more resilience to a changing climate, and re-establish an ecologically healthy relationship between nature and culture (Borja et al. 2010). There is a need for integrated land use planning as it encourages synergies between different uses. Sustainable landscape planning has been strongly supported through major international policy agreements and can be generally defined as 'a condition of stability in physical and social systems achieved by accommodating the needs of the present without compromising the ability of future generations to meet their needs' (IUCN 1980; WCED 1987). The goal is to improve the 'functional performance' of current urban facilities besides generating new one and making changes in order to showcase new ideas for boosting urban sustainability.

5.2 Definitions and Perspectives in the Spectrum of Multifunctional Landscapes

Does this bring to the most pertinent question as to how a multifunctional landscape can be exactly defined? Are they green forest covers? Would it be limited to the urban landscape and food-producing patches of land? Are they water bodies and landscape in general? The present study examines the multifunctional landscapes in the context of physical, social, and cultural contexts.

Multifunctional landscape is such landscapes which provide multiple benefits ranging from various physical and cultural resources (Bollinger et al. 2011). Multifunctional landscapes can also be defined as the capacity of a particular landscape to supply multiple ecosystem services at the landscape level (Mastrangelo et al. 2013). The multifunctionality of the landscape arises because of its oversimplification of land use for food production. The oversimplification leads to stripping of the environment of all its natural benefits. Multifunctional landscapes, by definition, are designed for multidimensional benefits. Landscape architects, architects, and planners are charged with designing landscapes that meet diverse human needs, while also facilitating ecosystem functions (Yang et al. 2013). Multifunctional landscapes play a critical role in climate regulation, ecosystem services, and the resilience of the community in general (Fagerholm et al. 2020).

It requires deep system changes which focus on 'local landscapes' and a transition towards more sustainable and management practices. Multifunctional landscapes are composed of different types of ecosystems that represent comprehensively dynamic and heterogeneous activities. Multifunctional landscape design and planning imply definitions of diverse targeted and composite performance standards (Young 2016). Some of these standards relate to the delivery of supporting ecosystem services (e.g. protecting and enhancing biodiversity, soil formation, and nutrient cycling), provisioning services (e.g. production of energy and other utilitarian resources), and regulating services (e.g. air and water quality and climate moderation). Others relate to cultural and social services (e.g. visual quality, human health, and recreational opportunity) (Lovell and Johnston 2009).

Multifunctionality and landscapes have been a subject of interdisciplinary research for long. Some associate multifunctionality with a supply of a variety of ecosystem services to some as a climate control regulatory concept, for some it alleviates poverty, and for others it makes the community more resilient (Fagerholm et al. 2020). In many instances, the nature and its linkages to human well-being have been studied, and the results are always on the positive side. The multifunctionality of a landscape is of utmost importance when seeking an ecological restoration. It is also important if the cultural context of the community is to be understood. The study aims to understand the multifunctional landscapes in the context of India and how it transforms the physical, social, and cultural aspects of the environment. The objective of this study is to understand the concept of multifunctional landscapes and its significance in the Indian context through the following case studies:

- Adyar Ecological Restoration Project (Chennai) in terms of physical and social impacts on the surroundings
- Ganga river and the *Kunds* of Varanasi in terms of its social and cultural impacts on the surroundings

The two case studies are set in a different geographical context and hence would help in understanding the aspects of multifunctional landscapes in the context of India as perceived in its different parts.

5.3 Case Studies

The following case studies would examine the aspects of water urbanism and multifunctional landscapes in the context of India. The Adyar Ecological Restoration Project would look into the physical dimensions and the social impact of multifunctional landscapes on the community. The second case study of Ganga and the *Kunds* (ponds) of Varanasi would look into the water-based landscape of the city and its social and cultural implications for the community.

5.3.1 Case Study 1: Adyar Ecological Restoration Project

One of the case studies which focused on developing multifunctional landscapes and demonstrates the practical application on the principles which are strongly grounded on ecological restoration is *Adyar Poonga* Ecological Restoration Project Phase I (58 acres) (*Poonga* in Tamil means park), which is located in Chennai, India. This project has been initiated by the Government of Tamil Nadu in 1997, and it was envisaged to create an eco-park as *Adyar Poonga* – covering about 358 acres – to restore the ecological balance and raise public awareness on environmental issues. It was a perfect example to prove that people-centric approach is critical to successful environmental projects.

Chennai, formerly known as Madras, is the capital city of Tamil Nadu, located on the Coromandel Coast of the Bay of Bengal. For a burgeoning city which lacks perennial water source especially on the west – it completely depends on the seasonal source of water bodies such as streams, rivers, and lakes, depending on the monsoon. Chennai is marked by two main rivers, the *Cooum* River and Adyar River meander sluggishly through the city, and they play an important role to mitigate flooding during heavy rains. They collect surplus water from tanks: about 75 tanks for the *Cooum* and 450 tanks for Adyar, in their respective catchments. Thus, the flood discharge of the Adyar River is almost three times more than that of the *Cooum* river (Pitchandikulam Forest Consultants Auroville, Tamil Nadu, India 2007).

For the past few decades these waterways are polluted due to the discharge of sewage into water bodies, outfalls from industries, stormwater drains, encroached slums, wastewater from unauthorised drainage systems, etc. This poor quality of water serves as a perfect breeding ground for mosquitoes that transmit viruses that cause malaria, dengue, and life-threatening vector-borne diseases enhancing health risks. The water bodies are used as urban open sewers and act as an eyesore within the city limits (Lakshmi and Ramakrishnan 2011).

A Trust, headed by the Chief Secretary, under the name of 'Adyar Poonga Trust', was formed to create such systems for the restoration of the Adyar Creek and the Estuary area (Lakshmi and Ramakrishnan 2011). The main objective of the project is to demonstrate how 'public-input mechanisms' and 'social or cultural elements of the system' emphasise the need for deep ecological, cultural, and historical links between people and their local ecosystems.

The Chennai River Restoration Trust (CRRT), originally known as Adyar Poonga Trust formed in 2006, was renamed as The Chennai River Restoration Trust in 2010. This CRRT is primarily a state-owned organisation entrusted with the restoration of water bodies in Tamil Nadu. The following are the objectives of CRRT (Cornou 2015):

- To establish best practices to create and develop a replicable model project of international standard to maintain, develop, and conserve Eco Park (Adyar Poonga).
- To attract diverse groups of people to interact with restored Adyar estuary to engage them with nature and include as part of their identity.

- To formulate plans and undertake the implementation of programs, to regenerate the indigenous fauna and flora, maintain the forests, wetlands, and other ecosystems, and towards preservation of ecological and natural resources, such as flora and fauna, waterways, water bodies, and wastewater recycling, preserving rare and all species.
- To formulate, evolve, and identify the suitable mechanism for augmenting necessary revenues through commercial operation or otherwise of some/all facilities of Eco Parks by way of fees, charges, donations, entrance ticket charges, sale of produces, and the like.

The master plan for Adyar *Poonga* was created which covers an area of 58 acres in the first phase along with interim facilities for generating citizen awareness, such that the plan is environmentally sustainable and contributing to the long-term sustainability of the Adyar River system. It is one of the symbolic elements where *Poonga* (park), mudflats, and estuary are considered as one ecological system integrating the biological, physical, and social factors. The main focus is to create awareness on the ecological concerns, enabling the public and possible users of the park to understand the environmental issues by developing live models/areas for observations. These 58 acres were once used to be a waste dumping site, a filthy place with debris strewn around. Since it is a creek, the backwater of the Adyar River is not flowing away, and because of a continuous domestic untreated sewage outlet, there was an excess stagnation of sewage even after light showers. Figure 5.1 shows the conceptual proposal for the area.

The phase-wise goals of the intervention are as follows (Figure 5.2):

- 'To develop the Adyar estuary and its surroundings as an asset to Chennai City.
- To integrate measures to control the proliferation of unplanned development along the riverfront.
- To convert the estuary from a neglected waterfront to a vibrant, recreational, and urban public space'.

The second phase of eco-restoration has started further on the estuary, between the Theosophical Society and Srinivasapuram slum settlement, covering an area of 300 acres. On the same intention as the first phase, there will be water body restoration, removal of invasive exotic species, habitat restoration, monitoring pathways, sanitation, solid waste management, and measures to enhance tidal influx in Adyar estuary and creek (The Hindu 2013). This book chapter focusses on the first phase of the Project (58 acres) (Figure 5.3).

5.3.1.1 Study of Wastewater and Sewage Outfalls

The 58-acre Adyar wetland faces pollution from various points. The study identifies the water quality on various outfall discharges into Adyar *Poonga*, shows the organics and nutrients in the bio-solids of the water bodies that contribute to the internal pollution, and the composition of run-off water – whether it is base flow or stormwater discharge carries significant amounts of pollutants (Samajdar 2013). A study from CMDA on the Cooum and

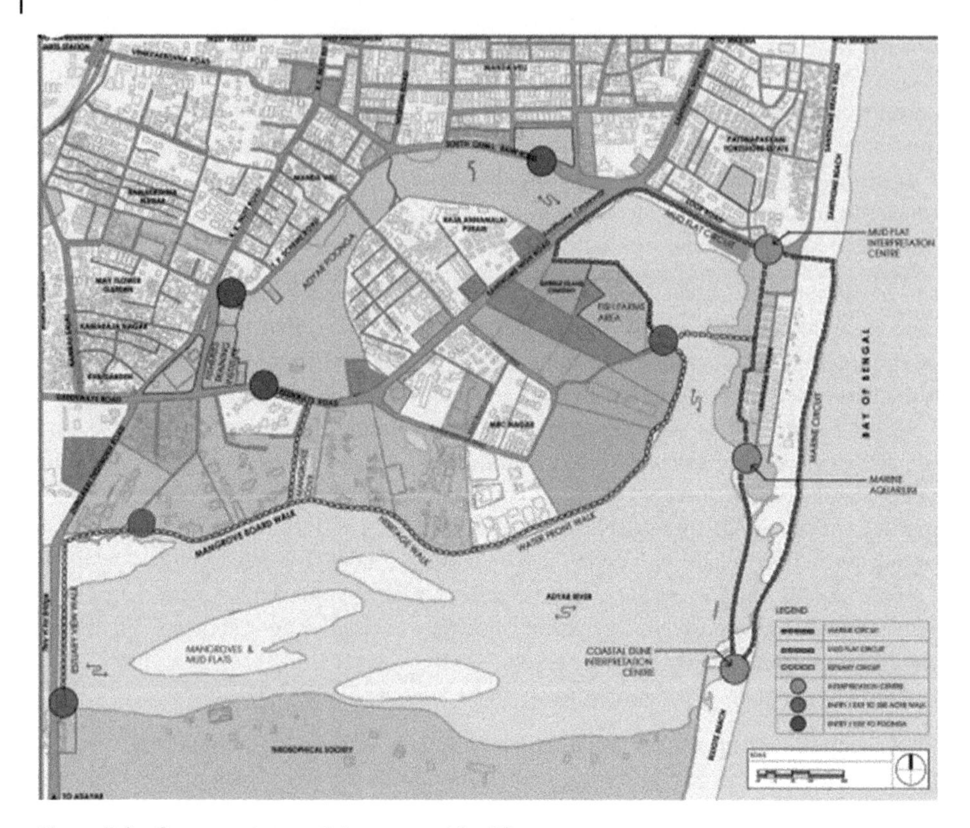

Figure 5.1 Conceptual map of the proposal for 58 acres.

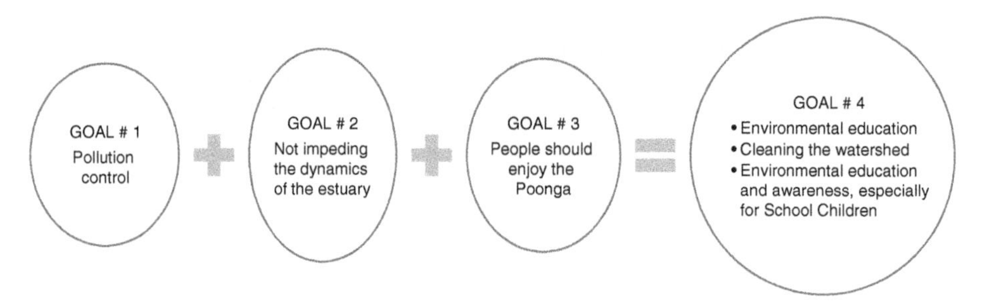

Figure 5.2 Goals of the Adyar Poonga project.

Adyar River looked at the BOD and COD3, which are most widely used as parameters for calculating pollution load applied to both wastewater and surface water (Walther et al. 2008). The biodegradability of the organic compound depends on the BOD:COD ratio in the wastewater. The typical untreated domestic wastewater with high organic content has the BOD5/COD ratio above 0.7. The study concluded that the average BOD:COD ratio obtained in Cooum and Adyar River is in the range of 0.28–0.38, which indicates poor biodegradability and also extensive industrial pollution (Walther et al. 2008). According to the following table (Table 5.1) from the Second Master Plan CMDA, we can notice that the DO

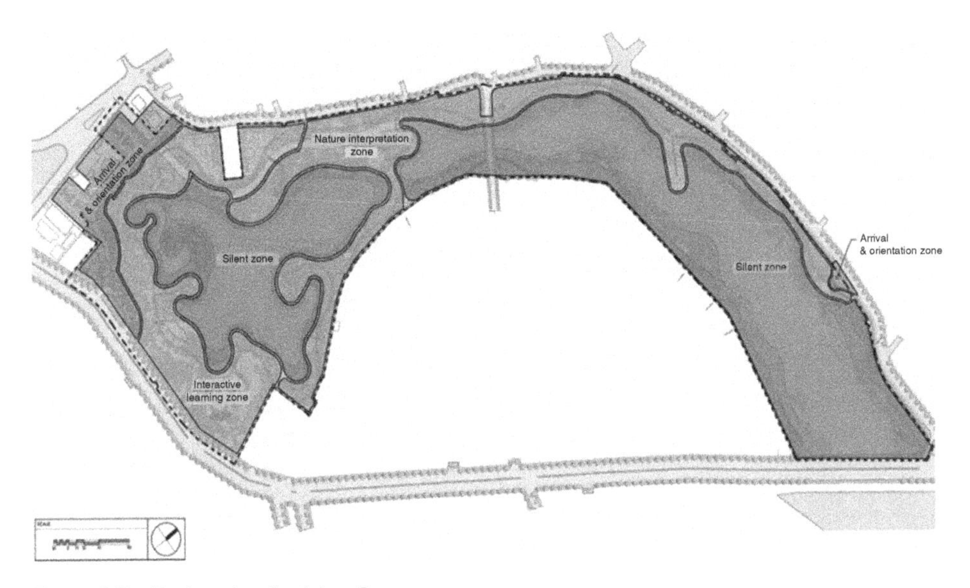

Figure 5.3 Zoning plan for Adyar Poonga.

Table 5.1 Pollution data on the Adyar and Cooum rivers.

	Adyar River			Cooum River		
	Site I – II	Site III– IV	Site V – VI	Site I – II-III-IV	Site V-VI-VII	Site VIII –IX
	Upstream	Middle Stream	Brackish water downstream	Moderate Pollution	High sewage pollution	High Industrial Pollution
DO	2.8–9.8	1.1–9.4	1.3–18.5	0.0–11.7	0.0–0.0	0.0–0.6
pH	8.1	8.0	8.0	7	7.5	8
Cl	47–115	51–130	80–4000	0.8–326	298–336	326–1100
SO_4	30–75	31–75	43–127	70–132	99–132	116–142
NO_2N	0.01–0.32	0.01–4.7	0.01–0.12	0.01–0.12	05–0.07	0.01–0.19
NH_4N	0.16–1.46	0.06–2.86	0.46–18	0.2–1.2	6.9–9.9	5.8–9.5
BOD	1.0–6.6	1.7–32.7	2.3–53.0	8.2–3.1	23–47	62–74
COD	9.5–195	10–72	10–830	29–316	159–5081	289–8 10

Source: CMDA Second Master Plan, Public domain.

of the Adyar River level is not as less as in the Cooum river. It indicates that the river contains still few organisms which can mix with the air for decontaminating naturally.

Moreover, according to Figure 5.4, it is evident that the BOD and DO levels in the Adyar River do not conform to the legal norms instituted by the Government of India.

The concentrations are much higher than the permissible limits for the discharge of sewage effluents into inland surface water (IS 4764). In the case of heavy rainfall during the SW

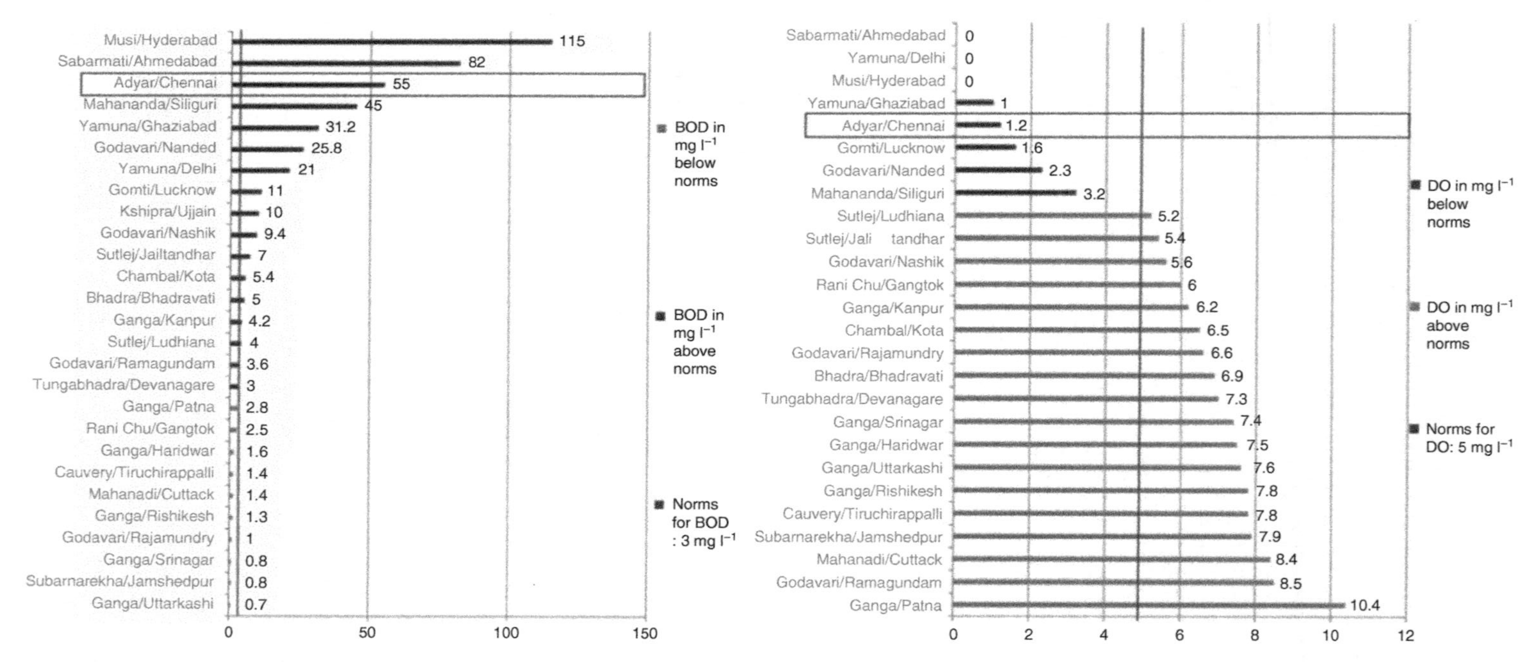

Figure 5.4 The actual BOD and DO (mg l^{-1}) in Indian rivers against the norms of the Government of India. *Source:* Supreme Audit Institute of India 2011.

monsoon, it is estimated that there is a 60% reduction of the major pollutants (Pitchandikulam Forest Consultants Auroville, Tamil Nadu, India 2007). Further reductions in the pollution levels can be expected with stormwater events during the usually stronger NE monsoon.

5.3.1.2 Encroachments Inside the Poonga

Until late May 2007, much of the project site could not be accessed for study and analysis due to encroachment (Kavitha 2013). Over 100 encroached pucca dwellings near the bridge at Karpagam Avenue posed one of the biggest threats to the integrity of the *Poonga*. Other related issues such as continuous garbage dumping and sewage and wastewater inflows into the *Poonga* wetlands from these encroachments exacerbated the problems here. These encroachments had prevented the completion of the compound wall for the 58 acres of the Poonga. Although these encroachers have since been relocated, an estimated 61 families continue (Pitchandikulam Forest Consultants Auroville, Tamil Nadu, India 2007). As a result of the eco-restoration, within the slum area concerned by the Adyar *Poonga* project, hundreds of huts were evicted earlier, and thousands of more homes of the poor threatened with demolition (Coelho and Raman 2010). The government has carefully decided 'making and unmaking communities' (Arabindoo 2012) for justifying a necessary need for eco-restoration that starts with the eviction of the slums, which will be explained later in the chapter.

5.3.1.3 The Ecological Philosophy for the Poonga – Restoration Ecology: An Evolutionary Paradigm

The Indian Scenario can be characterised by various problems such as land degradation due to overexploitation, groundwater depletion, environmental pollution, and the decimation of biological diversity. As these problems influence the survival and livelihood options of the world's poorest especially marginalised and peripheral population, environmentalism in India has altered from voluntary, decentralised, and people-centred resistance to a more centralised and organised activity. Since the members of this new brand of environmentalists have easy access to the media and are financially strong, they have succeeded in disseminating information to influential segments of the public, thus creating a larger support base for their causes (PUCL 2010). Constant input is sort from the community through frequent workshops, awareness campaigns, and public participation and volunteer programmes.

As explained by P. Arabindoo, the government used the 'Association of Friends' of Adyar Poonga for creating a constructed sense of ecology and to promote an 'aestheticised commodification of nature': 'the surrounding middle- and upper-class residents suggested a cosy alliance between the state and a specific form of civil society promoting a post-materialist understanding of nature in a city'. This organisation is made of about 40 members from the neighbourhood of Adyar Poonga and has 'the initiative for promoting environmental conservation. It aims to help in creating environmental awareness and primarily serve to build and support the relationship between the parks, local communities, and interested constituents like volunteers' (Madras Musings 2010).

The proposed interventions are multidimensional and work in conjunction with each other. They will intervene in the damaged ecosystem of the wetland and the waterway and restore it to behave as a natural system.

The plan proposes (Pitchandikulam Forest Consultants Auroville, Tamil Nadu, India 2007)

1) Ecosystem conservation
   a) Water quality restoration
   b) Land and soil quality restoration
   c) Reintroduce local flora
   d) Facilitate return of native fauna
   e) Water management
2) Create a central task that engages the people within the ecosystem to have a meaningful and satisfying relationship with the Poonga and the waterways in a larger context. The human interface strategy plans to
   a) create an education program about the wetlands and sustainability and
   b) position the area as environmentally sustainable – ECOPARK.

5.3.1.4 The physical and social aspects

- Increasing the number of habitat/habitats present in a given area is the primary motivation, and this is the primary consideration for undertaking restoration, particularly where extensive ecosystem modification and destruction has taken place. The types and juxtaposition of habitats, land use, and, in particular, sensitivities of the species in question have been kept in mind during the whole design process.
- Eco Park and the four zones
  - Arrival and Orientation Zone, Interactive Learning Zone, Nature Interpretation Zone, Silent Zone, Green centre
  - The Adyar Watershed Restoration Institute (AWRI): The research building and administration building will be housed in this building. The administrative block and the park warden's office will also be located here.
- Sewage treatment plant: To meet the needs of water for the park, a facility to process the sewage from the New Adyar sewage pumping station will be created.
- Other amenities
  - Interactive education zone
  - Interactive children's learning space
  - Nursery and gardens
  - Walking through the ecological zones
  - View points
  - Temporary bamboo pavilions with stone seating
  - Continuing education program
  - Birdwatchers Study Group (ornithology studies).

5.3.1.5 Significance of the project

The world's biodiversity hotspots are the regions that possess high species richness and a high degree of endemism. They not only serve as building blocks of the earth's life support system but are also critical for human survival as these are the homes for more than 20% of the world's population. We rely on these healthy ecosystems to provide food security, freshwater, clean air, climate stability, and even our happiness and well-being. The ticking bomb of ecological hotspots is India's biggest threat and facing a crisis of historic proportions.

The speed and scale in which the saga of depletion has been inflicted on the environment since the past few decades is the most profound in the history of mankind. In recent times, people are beginning to realise that landscape conservation and restoration can boost the economy. The efforts of landscape restoration are vital to reversing the depletion of natural capital, an approach which has been underutilised. Linking to the concept of multifunctionality, there has been increasing attention and research where the potential of the landscape maximises the benefits to the community beyond commodities production.

For architects and planners, the ultimate goal is to increase the rate of efforts to reverse centuries of damage to wetlands and other ecosystems. However, despite the high level of political and civic participation from public to private sectors, research, and academia, at all levels, from local to global, restoration is not happening at scale across the globe. A recent study (Strassburg and Hawthorn 2019) states that ecological restoration can provide vast benefits for global goals of biodiversity conservation and climate change mitigation. For instance, restoring 15% of converted lands globally could reduce the current global species extinction debt by approximately 65% if concentrated in priority areas for biodiversity. Ecological restoration refers to the process whereby an entire ecosystem is brought back to health by altering a degraded area in such a way as to re-establish an ecosystem's structure and function, usually bringing it back to its original or pre-disturbance state or to a healthy state close to the original (Pitchandikulam Forest Consultants Auroville, Tamil Nadu, India 2007). The benefits of ecological restoration are numerous, including, but certainly not limited to, increasing biodiversity, re-establishing ecosystem services, sustaining livelihoods, and improving the well-being both materially and culturally (Aronson et al. 2016).

5.3.2 Case Study 2: Ganga *Ghats* and *Kunds* of Varanasi

India is a vast geographical entity with many types of ecosystems coexisting within its boundaries. In India, nature has been worshipped since time immemorial (Sinha 2000), trees, water bodies, mountains, and forests, to name a few. The importance of nature and the ecosystem services it provides are of great significance in the Indian cultural context. It is an interesting phenomenon as how nature gets embedded into the sociocultural practices of a community and forms an intrinsic part of it. Here human being is not just a participant but also the catalyst of the various functions which are continuously happening.

Varanasi, the cultural capital of the country, which lies in the northern part of India, is one such city. What makes Varanasi unique is the relationship of the city with its water bodies. In ancient times, the city evolved as a small settlement along the banks of Ganga (Das 2019) (Figure 5.5). It slowly grew into layers of newer settlements as migrants arrived from different parts of the country. The city as we see today is a continuous aggregation of multifarious decisions which has been taking place for the past 3000 years and more (Singh 2007). The composite of a city visible to all has its charm. The city is dotted with 84 ghats (Sinha 2020; Sharma 2018; Singh 1994) which are unique to Varanasi.

In India, the land–water interface has historically acted as dynamic catalysts, as both interconnected climate-adaptive physical and multifunctional sociocultural constructs. It can be stated that the pond (*kund/talaab*) and the stepped banks (*ghats*) in India are examples of two vernacular and culturally integrated water infrastructure landscapes. They are the land–water interfaces that combine the social with the functional aspects (Nawre 2014;

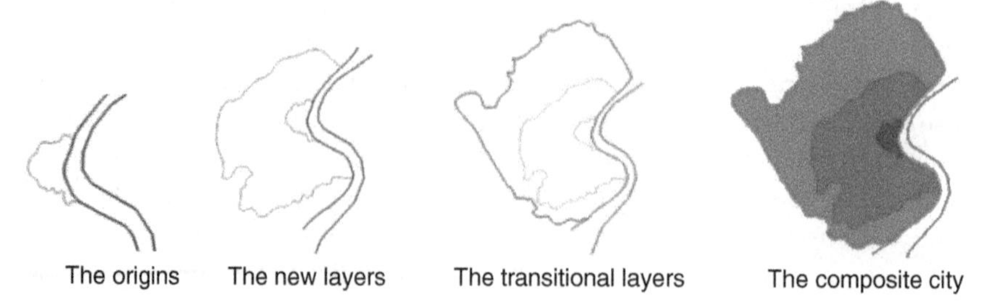

Figure 5.5 Formation of Varanasi.

Praharaj 2014). The banks of the Ganga dotting the palatial mansions form a continuous promenade of ghats (barring a few) (Singh 2010; World Bank Group; Cities Alliance 2014). These stepped embankments are also called ghats. These ghats are part of the water–land interface which forms the urban ecosystem of the city. Here, the realms of social and cultural constructs interact with the natural resources and evolve into a unique form of dynamic interactions (See Table 5.2 and Figure 5.6).

5.3.2.1 River vs. Kunds of Varanasi

In the Hindu religion, nature is considered sacred, be it trees, water bodies, animals, or birds. Sacredness may have ensured that they might be taken care of and would be sustained for future generations. For example, kunds or ponds were the focal points during worship, as the water source for daily household activities, and as a water source for

Table 5.2 Multiple dimensions of Varanasi.

	Historical	Spiritual	Ritualistic	Modern
Historical	Monuments Historical structures of importance *Micro-spaces + meso-spaces* *Organically designed places*	Temples/other places of worship	Ghats /temples/ temples existing in lanes and nooks and crannies (mystic niches)	There is simultaneous existence of structure from various eras/ *coexistence*
Spiritual	Temples/other places of worship	Ashrams/ghats/ mediation centres/ some spaces in the lanes/mystic niches	Ghats	Temples (recent structures)
Ritualistic	Ghats/temples	Ghats	Temples/ghats	Temples (recent structures)
Modern	There is simultaneous existence of structure from various eras *coexistence*	Temples (recent structures)	Temples (recent structures)	The urban ecosystem of modern life which exists beyond the historical interface of ghats *malls/housing/ markets*

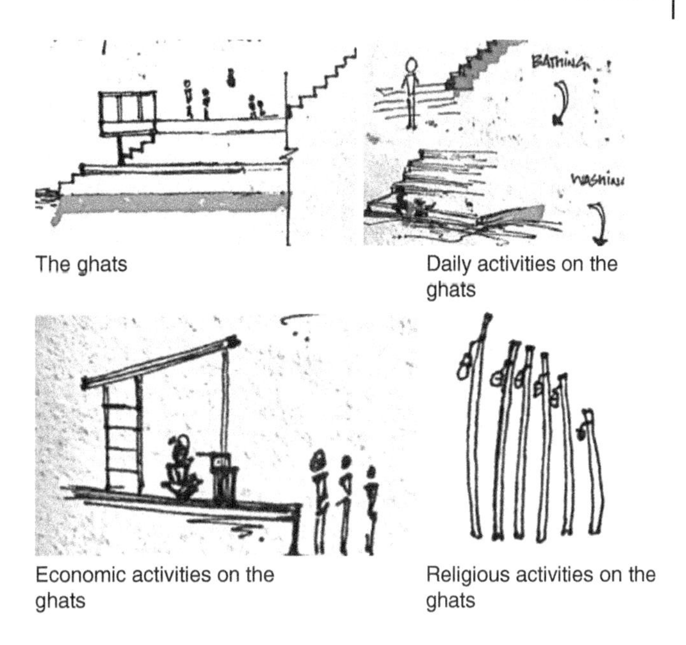

Figure 5.6 Activity on the Ghats (land–water interface).

domestic animals. They were an ecosystem in itself (Azam et al. 2015). It has been documented in ancient texts that kunds in Varanasi were interconnected at subterranean level with each other and with the River Ganga. During the times of flood, these kunds acted as sinks for the overflowing water of Ganga, hence maintaining the balance (Cunha 2018). Slowly with colonisation and the advent of British planning agendas, the native knowledge got lost. The kunds were drained off, and this system of maintaining the water balance between the city edge (river) and the city interiors (kunds) got disconnected. It has been stated in the old documentation of Varanasi that there used to be 118 kunds in the city, of which only a few are still surviving, that too in a very dilapidated condition (Raju and Bhatt 2015; Mishra et al. 2014).

Earlier, both the river and the *kunds* were revered among the citizens, but gradually, this reverence is vanishing as are the cultural practices which once glorified these spaces (Dubey and Dubey 2015). This can be attributed to rapid urbanisation and its consequent changing lifestyles. Historically speaking, the introduction of property rights by the British colonial regime did the most irreversible damage to the ethos and the consciousness of the Indian people (Giri 2005). The regime made all the water bodies the state's property. That is, the rivers, lakes, and ponds which were common property before now no more belonged to the village or its people who inhabited their banks. When rivers, lakes, and ponds were common property, the villages used to maintain these assets by coming together as a community and taking turns to maintain them. This community effort established an intrinsic connection between these water bodies and its people. Though as of now, gradual deviation from the ancient ritualistic practices has led to a slow decline of these places (GSAPP Columbia University 2018; SPA Delhi 2014).

Some of the *kunds* which are still on the mental maps of the citizens are Pushkar Kund near Assi Ghat, Lolark Kund near Tulsi Ghat, Durga Kund, Kurukshetra Kund (see Figure 5.7), etc. In the past few years, restoration work has been done by ONGC to rejuvenate

Lolark Kund, Varanasi

Pushkar Kund, Varanasi

Figure 5.7 Kunds of Varanasi.

these urban ecosystems. The project was taken under the corporate social responsibility (CSR) initiative of ONGC. A fund of 11.46 crore was sanctioned for the renovation works. The renovation of four kunds has been taken up under this project, which are Durga Kund, Lakshmi Kund, Sarangnath Kund, and Lat Bhairav Kund (TNN 2016; GN Bureau 2017).

Some of the works included physical fittings such as railings, electrical fixtures, and bunds to check the ingress of pollutants. Further purification processes such as anaerobic bioremediation have also been used. Some of the *Kunds* have been drained off water for the cleaning of debris, while in some others where there is the presence of large aqua life water was not removed, instead other scientific methods are being adopted for water purification without disturbing the aquatic life (ONGC 2017). The culmination of this project will lead to a maintenance phase which will be done by ONGC for five years.

5.4 Inferences from the Two Case Studies

Table 5.3 looks into the various aspects of the two case studies and tries to take a stock of the understanding of the multifunctional landscape in the context of urban areas and how it differs in definition and function across different geographic settings.

5.5 Conclusion

Water is life. It is the marker of civilisation and remains one of the most precious resources on earth. Water is the basis of physical life, a proponent of cultural practices, and a catalyst of economic activities (Heynen 2013). The two case studies in this study look into the effect of the water body on the physical aspect as well as the sociocultural aspects of the life of citizens in two geographical locations.

The strategies adopted in *Adyar Poonga* embed an outward-looking and inclusive approach that appreciates the local, regional, and national significance. The decision-making process

Table 5.3 Inferences from the case studies on the basis of four parameters.

Parameters	Case 1: Adyar Eco Restoration Project	Case 2: Ganga and *Kunds* of Varanasi
Eco-restoration and sustainability in traditional cities	The study focuses on the Adyar estuary and its phase 1 process. It shows how a multifunctional landscape project if well executed can create a plethora of exquisite experiences for the residents of the city as well as for the tourist.	The river Ganga is the dominating water body, whereas the *kunds* have taken a backseat in the citizens' cognitive planes.
	This can not only lead to economic rejuvenation of the area but also affect the lives of all the people who are stakeholders (in some way) of this project.	Restoration of few places is going on, but the condition is still very bleak and would require a huge effort from the public and private players.
	Such type of project can act as a prototype that can help water bodies like the above case study and use the methodology to develop into a self-sustaining urban ecosystem	Revival of places should be accompanied by their re-introduction to the citizens. They should be made more aware of the existing scenarios, why work is being done? what is its significance, heritage value? Why a particular decision is being made? What effect it will cause to their day-to-day life? Citizens should feel included in the environment they live. Encouragement to community participation.
Multifunctional landscapes	The physical realm of the Adyar estuary is studied to understand its relationship with the social and cultural realms and how citizens interact with such kind of eco reserve. The focus has been on how the conversion of the estuary from a neglected waterfront to a vibrant, recreational, and urban public space happened.	The multifunctional aspect of the above case study lies in the fact that there are a lot of dimensions that play on the edges of such water systems. The interaction of the users and the physical elements such as water, land, natural resources, and man-made resources, and how these factors affect the citizens living in the vicinity.
Water urbanism	Engaging the people within the ecosystem to have a meaningful and satisfying relationship with the Poonga and the waterways in a larger context.	A huge emphasis needs to be given to reviving the lost heritage of *Kunds* such that they can be brought to mainstream urbanism and be useful for the citizens.
Water as an element of multifunctional landscapes	Water is central to such landscapes as they make use of wetlands which are essentially water based and rich in natural resources; land parcels.	Water is essential to such landscapes as it is the central feature, be it the river or the pond (*kund*). It is the basis of tradition (worship, celebration, and death), work (boating and transport), and leisure (water sports, dip in water, and water as a backdrop for conversation).

includes protection of public interest, opportunities for outdoor access for health and well-being, greater civic awareness and engagement in land-related matters, identifying precious landscape resources – recognised, enhanced, and protected for sustainable benefit (Bhan 2016). This case study shows how developing waterfront in India generally and Chennai specifically should focus on purposes other than just environmental preservation. It shows how a multifunctional landscape project if well executed would not only be beneficial for the microclimate but can also create a plethora of exquisite experiences for the residents of the city as well as for the tourist. This in turn will boost the economy of the region and hence be beneficial to all the stakeholders.

Varanasi on the other hand portrays a very close relationship of its rivers, Ganga and its tributaries Assi and Varuna, with its *kunds* which are slowly vanishing due to various factors of urbanisation (Dharmadhikari and Sandbhor 2017). The study aims to make people aware of the huge benefits of developing these urban spaces as they are a complete ecosystem in themselves. These will not only add up to the social interaction of the city but will also create a better environment for the city which will help in mitigating climatic phenomena such as the urban heat island effect. The study bridges the social–cultural perspective differences which are associated with such places and try to understand the social dimension of the multifunctional landscape. The use of two geographical settings to understand how water has played a key role in defining the lives of the people around it is of main concern in this chapter. This also leads to a very important question, that is:

'How to design a community that has a role to play in this multidisciplinary approach?'

This question is trying to unearth the fact that a properly designed multifunctional landscape such as in the Adyar Eco-Restoration Project can lead to a great improvement in the lives of citizens. The viability of such a project would be location dependent apart from other parameters but can be explored more to understand its deep significance. The Varanasi case study highlights how different water bodies react to the changing time in the social aspects. If the urban places of such cities need to be protected, then robust schemes which are continuously evolving must be considered while taking any design decision. These not only help in efficient utilisation of space, which are multifunctional, but are also climatically respondent. These spaces will be part of the urban pockets of the city and would help in the overall augmentation of the environment.

Acknowledgement

We are very thankful to Adyar Poonga Trust and Pitchavaram Auroville who helped in providing us with relevant information to complete this study.

References

Ahern, J. (2013). Urban landscape sustainability and resilience: the promise and challenges of integrating ecology with urban planning and design. *Landscape Ecology* 28: 1203–1212.

Arabindoo P. (2012). Constructed Ecologies, Imagined Communities - The Politics of Adyar Poonga in Chennai. Retrieved from https://www.cprindia.org/events/3976.

Aronson, J.C., Blatt, C.M., and Aronson, T.B. (2016). Restoring ecosystem health to improve human health and well-being: physicians and restoration ecologists unite in a common cause. *Ecology and Society* 21 (4): 1–8.

Azam, M.M., Kumari, M., Singh, A.K., and Tripathi, J.K. (2015). *A Preliminary Study on Water Quality of Ponds of Varanasi City*. Uttar Pradesh: BiogeochemEnvis.

Benedict, M. and McMahon, E. (2006). *Green Infrastructure Linking Landscape and Communities*. Washington, D.C.: Island Press.

Benne, B.C. and Mang, P. (2015). Working regeneratively across scales— Insights from nature applied to the built environment. *Journal of Cleaner Production* 109: 42–52.

Bhan, G. (2016). *In the Public's Interest: Evictions, Citizenship and Inequality in Contemporary Delhi*. Orient Blackswan.

Bollinger, J., Baettig, M.B., Kläy, A., and Stauffacher, M. (2011). Landscape multifunctionality: a powerful concept to identify effects of environmental change. *Regional Environmental Change* 11: 203–206.

Borja, A., Dauer, D., Elliott, M., and Simenstad, C. (2010). Medium- and long-term recovery of estuarine and coastal ecosystems: patterns, rates and restoration effectiveness. *Estuaries and Coasts* 33 (6): 1249–1260.

Coelho, K. and Raman, V. (2010). Salvaging and scapegoating: slum evictions on Chennai's waterways. *Economic & Political Weekly* 45: 22–28.

Cornou, A. (2015). *The Environmental Awareness and Riverfront Development Projects - A Case Study of the Adyar River in Chennai, India*. Tours, France: Polytechnic University of Tours.

Costanza, R., Alperovitz, G., Daly, H. et al. (2012). *Building a Sustainable and Desirable Economy in- Society - in - Nature*. New York: United Nations Division for Sustainable Development.

Cunha, D. D. (2018, January). Water Urbanism Studio-Varanasi (Joint studio by IIT Kharagpur and Columbia University). (I. K. University, Interviewer).

Das, S. (2019, February 28). Brief History of Varanasi-Why Varanasi Could be the World's Oldest City. Retrieved from Learn Religions: https://www.learnreligions.com/brief-history-of-varanasi-banaras-1770408.

Dharmadhikari, S. and Sandbhor, J. (2017). Reclaim Asi and Varuna for Culture of Varanasi. Retrieved March 2021, from https://gangatoday.com/articles/58-reclaim_asi_and_varuna_for_culture_of_varanasi.html.

Dubey, R.S. and Dubey, A.R. (2015). Effect of idols immersion on anthropogenic influenced ritual ponds water quality at holy city Varanasi. *International Journal of Engineering Sciences & Research Technology* 4 (12): 656–665.

Fagerholm, N., Martín-López, B., Torralba, M. et al. (2020). Perceived contributions of multifunctional landscapes to human well-being: Evidence from 13 European sites. *People and Nature* 2 (1): 217–234.

Giri, A.K. (2005). Rule of Law and Indian Society Colonial Encounters, Post-Colonial Experiments and Beyond. Retrieved from www.juragentium.org: https://www.juragentium.org/topics/rol/en/giri.htm.

GN Bureau (2017). ONGC Begins to Restore Varanasi Kunds. Retrieved February 2021, from https://www.governancenow.com/news/psu/ongc-begins-restore-varanasi-kunds.

GSAPP Columbia University (2018). *Water Urbanism in Varanasi - Global Cities and Climatic Change Studio*. GSAPP Columbia University.

Heynen, N. (2013). Urban political ecology I. *Progress in Human Geography* 38 (4): 598–604.
IUCN (1980). *The World Conservation Strategy by IUCN in collaboration with UNEP and WWF*. Gland, Switzerland: International Union for Conservation of Nature and Natural Resources.
Kavitha, A. (2013). Impact of urbanization on rivers of Chennai. *International Journal of Scientific & Engineering Research* 4 (5): 396. ISSN: 2229-5518.
Lakshmi, K., and Ramakrishnan, D. H. (2011, September 29). Untreated sewage pollutes waterways. Retrieved from https://www.thehindu.com/: https://www.thehindu.com/news/cities/chennai/untreated-sewage-pollutes-waterways/article2496148.ece
Lovell, S.T. and Johnston, D.M. (2009). Creating multifunctional landscapes: how can the field of ecology inform the design of the landscape? *Frontiers in Ecology and Environment* 7 (4): 212–220.
Madras Musings (2010). Adyar Poonga Gets Ready for the Public. Retrieved from www.madrasmusings.com: http://madrasmusings.com/Vol%2020%20No%204/adyar-poonga-gets-ready-for-the-public.html.
Mastrangelo, M.E., Villarino, S.H., Barral, M.P. et al. (2013, 2014). Concepts and methods for landscape multifunctionality and a unifying framework based on ecosystem services. *Landscape Ecology* 29: 345–358.
Mishra, S., Singh, A.L., and Tiwary, D. (2014). Studies of physico-chemical status of the ponds at Varanasi Holy City under Anthropogenic influences. *International Journal of Environmental Research and Development* 4 (3): 261–268.
Nawre, A. (2014). INFRASTRUCTURE AS PUBLIC SPACE- Ghats and Talaabs in India. *My Liveable City-The Art and Science of It*. Retrieved from vidnext.com: http://www.vidnext.com/.
ONGC (2017, January 12). *Oil and Natural Gas Corporation Limited*. Retrieved from https://www.ongcindia.com/: https://www.ongcindia.com/wps/wcm/connect/en/media/press-release/ongc+restoring+the+ancient+kunds+of+varanasi.
Pitchandikulam Forest Consultants Auroville, Tamil Nadu, India (2007). *Adyar Poonga Ecological Restoration Plan*, vol. *1*. Executive Summary.
Pitchandikulam Forest Consultants Auroville, Tamil Nadu, India (2008). *Adyar Poonga Ecological Restoration Plan*, vol. *2*. Master Plan.
Praharaj, S. (2014). Rejuvenation of water bodies (Kunds) and restoring active community spaces in the cultural capital of India: Varanasi. *Neo-International Conference on Habitable Environments*.
PUCL (2010). *Fact Finding Team on Forced Eviction and Rehabilitation of Slum Dwellers in Chennai*. Chennai: People's Union for Civil Liberties-Tamil Nadu and Puducherry. https://fdocuments.in/reader/full/final-report-of-pucl-ff-team-on-slum-eviction-n-rr-13th-jan-2011.
Raju, K.P. and Bhatt, D. (2015). Water in ancient Indian perspective and ponds of Varanasi as water harvesting structures. In: *Management of Water, Energy and Bio-resources in the Era of Climate Change: Emerging Issues and Challenges*. Springer.
Samajdar, P. (2013, August 6). *Chennai faces a unique pollution challenge – pollution levels that appear to be low or moderate but are not so*. CSE India.
Sharma, A. (2018, September 28). *Modern plaques to adorn all 84 ghats along the Ganga in Varanasi*. Retrieved from https://economictimes.indiatimes.com/: https://economictimes.indiatimes.com/news/politics-and-nation/modern-plaques-to-adorn-all-84-ghats-along-the-ganga-in-varanasi/articleshow/65996844.cms.

Singh, R.P. (1994). Water symbolism and sacred landscape in Hinduism: A study of Benares (Varanasi). *Erdkunde* 48 (3): 210–227.
Singh, R.P. (2007). Urban Planning of the Heritage City of Varanasi (India) and its role in Regional Development. *REAL CORP 007 Proceedings*, (pp. 259-268). Vienna.
Singh, R.P. (2010). Varanasi, India's cultural heritage city: contestation, conservation & planning. In: *Heritagescapes and Cultural Landscapes* (ed. R.P.B. Singh). New Delhi: Shubhi Publications.
Sinha, A. (2000). Ghats on the Ganga in Varanasi, India: a sustainable approach to heritage conservation. *Third International Workshop on Sustainable Land Use Planning 2000*.
Sinha, A. (2020). Ghats on the Ganga in Varanasi, India. In: *The Routledge Handbook on Historic Urban Landscapes in the Asia-Pacific* (ed. K.D. Silva). Taylor & Francis Group.
SPA Delhi (2014). *Sustainable Development of Heritage city Varanasi*, Executive Summary. New Delhi: School of Planning and Architecture. http://spa.ac.in/writereaddata/executive-summary-varanasi.pdf.
Strassburg, B. and Hawthorn, B.L. (2019). Strategic approaches to restoring ecosystems can triple conservation gains and halve costs. *Nature Ecology & Evolution* 3: 62–70.
Supreme Audit Institute of India (2011). © *Content Owned by Comptroller and Auditor General of India* 68–69.
The Hindu (2013). One Lakh Mangroves in Adyar Creek Soon. Retrieved from The Hindu: https://www.thehindu.com/news/cities/chennai/one-lakh-mangroves-in-adyar-creek-soon/article4696453.ece.
TNN (2016). ONGC to Clean and Beautify Four Kunds in Varanasi. Retrieved March 2021, from https://timesofindia.indiatimes.com/ongc-to-clean-and-beautify-four-kunds-in-varanasi/articleshow/53316757.cms.
Walther, D., Renuka, R., Vivek, B., and Thanasekaran, K. (2008). Pollution Management of an Urban Wetland in Chennai City. *11th International Conference on Wetland Systems for Water Pollution Control*. Indore, India: International Water Association.
WCED (1987). *Report of the World Commission on Environment and Development: Our Common Future*. Oxford: Oxford University Press.
World Bank Group; Cities Alliance (2014). openknowledge.worldbank.org. Retrieved from Inclusive Heritage-Based City Development Program in India: https://openknowledge.worldbank.org/handle/10986/20800.
Yang, B., Li, M.-H., and Li, S. (2013). Design-with-nature for multifunctional landscapes: environmental benefits and social barriers in community development. *International Journal of Environmental Research and Public Health* 10 (11): 5433–5458.
Young, R.F. (2016). Modernity, postmodernity, and ecological wisdom: Toward a new framework for landscape and urban planning. *Landscape and Urban Planning* 155: 91–99.

6

Urban Landscape Change Detection Using GIS and RS: Chattogram City Corporation, Bangladesh

Mohammad Ali and Gul-e-Noor T. Hasnat

Institute of Forestry and Environmental Sciences, University of Chittagong, Chittagong, Bangladesh

6.1 Introduction

Land is a finite asset for most anthropogenic activities such as agriculture, industry, forestry, energy production, settlement, recreation, and water catchments and storage (Sarwar et al. 2016; Andrews-Speed et al. 2014; Slee et al. 2014). Land use and land cover (LULC) change is the prime modifier of the landscape, ultimately affecting the socioeconomic, biological, climatic, and hydrologic systems in nature (Hussain et al. 2016; Zhang et al. 2016; Sohl and Sohl 2012). Humans have been altering land to obtain food and other purposes for thousands of years (Chamberlain et al. 2020; Genet 2020; Kareiva et al. 2007). But current rates and intensities of LULC are far higher than ever in history, resulting in unwanted changes in the ecosystem and environment on local, national, and global scales (Schirpke et al. 2020; Figueroa et al. 2009).

LULC changes have been the most explicit indicator of the human footprint and the most important reason for biodiversity loss and other forms of land degradation (Clerici et al. 2019; Sharma et al. 2018; Alemu 2015). During the past millennium, humans have taken a large role in the modification of the global environment (Ojima et al. 1994). With the increase in population and developing technologies, human has become the most powerful environmental change maker (Dalby 2009; Drayton 2006). LULC change is the key driver of environmental change (Morales-González et al. 2020; Riebsame et al. 1994). Anthropogenic influences on land-use patterns are the primary component of many recent environmental concerns. Urban sprawl is one of the major anthropogenic causes for LULC changes (Wu et al. 2016; Balogun et al. 2011).

Urban areas are the most dynamic places and have a considerable impact on the surrounding ecosystems (Mansour et al. 2020; Yuan et al. 2005). Urban growth is a common phenomenon mostly in developing countries over the world (Cohen 2004). Currently, these are the major environmental concerns that have to be analysed and monitored carefully for effective land-use management. Land-use change has transformed a vast area of the

Urban Ecology and Global Climate Change, First Edition. Edited by Rahul Bhadouria, Shweta Upadhyay, Sachchidanand Tripathi, and Pardeep Singh.

natural landscapes of the developing world for the past 50 years (Rounsevell et al. 2006; Foley et al. 2005). In the developing countries, a large human population directly depend on natural resources for their livelihoods, which results in rapid LULC changes (Sunderlin et al. 2005).

Bangladesh is experiencing an increased rate of land-use change in urban areas as a result of population dynamics, economic development, climate change, improved accessibility, and technological and industrial developments (Hassan and Nazem 2016; Dewan and Yamaguchi 2009). Bangladesh faces a remarkable change in land-use patterns, mostly in the case of forest lands (Hasan et al. 2020; Islam et al. 2020; Kurowska et al. 2020; Khan et al. 2015; Mamun et al. 2013). People have to depend on forest land and natural resources for their food, oxygen, and livelihood. Urbanisation has a great impact on changes in these land-use patterns (Dadashpoor et al. 2019; Deng et al. 2009). It should be critically mentioned that human beings do not consider the consequences of land-use changes, which may be detrimental in longer run (Dale et al. 2000).

LULC changes is a dynamic and continuous process (Azizi et al. 2016; Melendez-Pastor et al. 2014; Käyhkö et al. 2011), and therefore, extensive research on LULC pattern is important along with their social and environmental implications at different spatial and temporal scales (López et al. 2001). Information relating to LULC can play a vital role in natural resources management (Shaharum et al. 2018; Kantakumar and Neelamsetti 2015; Sohel et al. 2015). Land-use/land-cover inventories are assuming increasing importance in various sectors such as agricultural planning, urban planning, and infrastructural development (Szarek-Iwaniuk 2021; Sallustio et al. 2016; Imura et al. 1999). The technique of remote sensing has been widely applied in various land-use researches. Remote sensing technology and GIS can give reliable information on land cover (Wang et al. 2020; Shalaby and Tateishi 2007). The spatial extent and temporal change of land cover are analysed by remotely sensed data (Li et al. 2017; Gómez et al. 2016). Land-use data are needed for analysing environmental processes and problems (Nagendra et al. 2004). Applying remote sensing and GIS for understanding LULC change may help to create appropriate planning for improved living conditions in the future (Anand and Oinam 2020; Wang et al. 2020; Nath et al. 2018).

Chattogram is the major coastal city and trade zone of southeastern Bangladesh (Ahamed et al. 2020; Mia et al. 2015). Due to population growth, industrialisation, settlement land use, and land cover of Chattogram City Corporation are changing for the past couple of decades (Rai et al. 2017; Hassan and Nazem 2016). This study considered 1990 as the base year and examined the LULC from 1990 to 2020, so that recent changes can be identified. In these 30 years, all the changes that occurred in agriculture, barren land, settlement, water body, and vegetation land-use categories of Chattogram City Corporation will be classified and analysed using Remote Sensing and GIS.

6.2 Materials and Methods

Chattogram City Corporation is located on the banks of the Karnaphuli River and the Bay of Bengal. This is a major coastal city and financial centre of Bangladesh. Chattogram City Corporation comprises an area of 155.40 sq. km. and 2068082 population (BBS 2011).

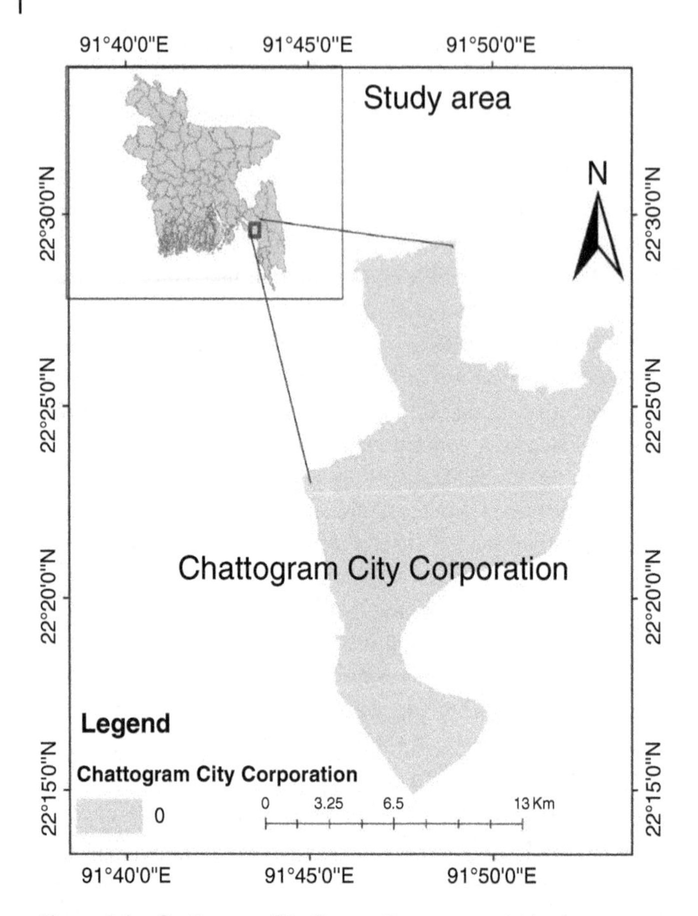

Figure 6.1 Chattogram City Corporation area as study site.

The study area is located in the centre of Chattogram district between 22°13' and 22°27' North latitude and 91°40' and 91°53' East longitude (Figure 6.1). Chattogram City area was formed on 31 July 1990 (BBS 2011) and has a unique combination of hillocks in the north, flat land limited by river Karnaphuli at south and east, coastline on the west, and low-lying Halda Valley on the northeast. It has 11 thanas, 41 wards, and 211 mahallas and is responsible for governing the municipal areas in the Chattogram Metropolitan Area. Chattogram Development Authority (CDA) operates development programmes for the improvement and expansion of this city.

6.2.1 Data Collection

The latest high-resolution satellite imageries of 2020, 2010, 2000, and 1990 were downloaded from the United States Geological Survey (USGS) website. In the study, Landsat 8 Operational Land Imager/Thermal Infrared Sensor (OLI-TIRS), Landsat 7 Thematic Mapper (TM), and Landsat 5 Multispectral Scanner-Thematic Mapper (MSS-TM) imageries were used for visual image analysis, land-use identification, and classification. Landsat 8 imagery was downloaded for the year 2020, Landsat 7 for 2000, and Landsat 5 for 2010 and 1990 respectively. The spatial resolution of Landsat images was 30 m. In image selection, land cloud cover and scene cloud coverwere set below 10% in the additional criteria to get cloud and unwanted shadefree imageries, as imageries having cloud or shade could substantially reducethe accuracy of the classification. A total of 38 400.8 acres (155.403 sq. km.) of land area was estimated for the whole Chattogram City Corporation (CCC) after supervised image classification using ERDAS Imagine 2015.

The secondary information was collected from different books, journals, published papers, and browsing websites such as Google Scholar, Wikipedia, and Bing. Three GIS software, i.e. ArcGIS, ERDAS Imagine, Google Earth Pro, were used for processing and mapping. ArcGIS version 10.4 was used for GIS analysis and mapping, ERDAS Imagine 15 for image processing classification and calculation, and Google Earth Pro for matching, digitising, and Kappa accuracy assessment.

6.2.2 Shape File Preparation and Image Processing

The shape file of Chattogram City Corporation was prepared for image processing and land classification. Image processing and performing supervised classification after processes help to extract information from imageries. At first, layer stack for band combination, the option of ERDAS Imagine 15, was used to convert blue, green, red, and NIR bands into a single layer for an image. After layer stacking, radiometric and atmospheric corrections were done (Zerga et al. 2021). Radiometric and atmospheric corrections are important processing steps and need to do as a signal when measured from the satellite could be affected by the presence of gases, solid, and liquid particles from the atmosphere. Due to less atmospheric effects, the correction processes were not done for Landsat 8 imageries in the study. From the corrected stacked layer, the study area was clipped using a subset tool and shape file of Chattogram City Corporation. The vector layer of Chattogram City Corporation was made from the collected shape file. Thus, four images for the year 1990, 2000, 2010, and 2020 were made ready for supervised classification.

6.2.3 Supervised Classification and Map Preparation

Supervised classification is the technique for the quantitative analysis of remotely sensed image data. Supervised classification was done using the signature editor. After adding and editing the signature for the desired number of land-use classes, the signature file was saved. During classification, the saved signature file was used (Figure 6.2). For classification, Chattogram City Corporation land uses were categorised into five groups – vegetation, agricultural land, settlement, water bodies, and barren land. After classification, the classified images from ERDAS Imagine were saved in .img format and opened in ArcMap. Then, they are exported as .png format with inserted necessary items such as legends, north arrow, and scale bar. The whole process was done by following Sarif and Gupta (2021).

6.2.4 Land Use and Land Cover (LULC) Change Detection

LULC change can be detected by the maximum-likelihood classification (MLC) approach (Hishe et al. 2021; Ogato et al. 2021). Maximum-likelihood classification is a mostly used and convenient process to apply with satisfactory accuracy. In the present study, the

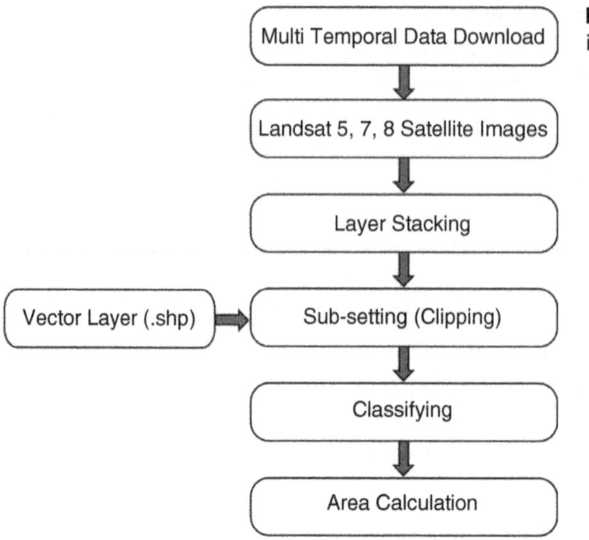

Figure 6.2 Procedure of satellite image classification.

magnitude change for each land-use class was calculated by following Eq. (6.1). In the equation, the area covered in the initial year was subtracted from the area of the second year (Islam et al. 2018),

$$\text{Magnitude} = \text{Magnitude of the new year} - \text{Magnitude of the previous year} \tag{6.1}$$

The magnitude change was divided by the number of years and then multiplied by 100 to calculate the percentage change trend for each land-use type as shown in Eq. (6.2) (Islam et al. 2018),

$$\text{Percentage change} = \frac{\text{Magnitude of change} \times 100}{\text{No. of year}} \tag{6.2}$$

6.2.5 Accuracy Assessment

Accuracy assessment quantifies the reliability of a classified image and constructs an error matrix. In the present study, the pixels derived from the classified map image were compared with the same location in the geospatial map, and an overall accuracy of each map for each decade was calculated. The overall accuracy percentage was calculated using the following formula (6.3):

$$\text{Overall accuracy} = \frac{\text{Total number of correctly classified pixels (diagonal)}}{\text{Total number of reference pixels}} \times 100 \tag{6.3}$$

The accuracy of individual classes in each map was measured for finding the overall accuracy. For individual accuracy assessment, the user's accuracy and producer's accuracy were two approaches applied (Pal and Ziaul 2017). The user accuracy measures the proportion of each land-use/land-cover class which is correct. On the other hand, the producer accuracy measures the proportion of each land base that is correctly classified. Producer's accuracy (6.4) and user's accuracy (6.5) are calculated using the following formulae:

Table 6.1 Strength of agreement for Kappa statistics.

Kappa statistics	Strength of agreement
0	Poor
0.00–0.20	Slight
0.21–0.40	Fair
0.41–0.55	Moderate
0.55–0.70	Good
0.70–0.85	Substantial
0.85–1	Almost perfect/perfect

$$\text{Producer accuracy} = \frac{\text{Number of correctly classified pixels in each category}}{\text{Total number of reference pixels in that category}} \times 100 \quad (6.4)$$

$$\text{User accuracy} = \frac{\text{Number of correctly classified pixels in each category}}{\text{Total number of classified pixels in that category}} \times 100 \quad (6.5)$$

$$\text{Kappa coefficient} = \frac{(\text{TP} \times \text{TCP}) - \sum(\text{Column total} \times \text{Row total})}{\text{TP}^2 - \sum(\text{Column total} \times \text{Row total})} \times 100 \quad (6.6)$$

Kappa coefficient is a measure of the proportion or improvement by the classifier's overall purely random assignment to the class. It is used to understand the accuracy of the classification. The value of Kappa varied between 0 and 1 (Hua and Ping 2018; Pal and Ziaul 2017). In the present study, the following strength of agreements was used for accuracy assessment (Table 6.1), and equation (6.6) was used to calculate Kappa coefficient, where TP is the total pixels, and TCP is the total corrected pixels.

6.3 Results and Discussion

The land-use maps are derived from different Landsat imageries in the present study. The aerial distribution of various land-use classes at 10-years intervals for the years 1990, 2000, 2010, and 2020 and their change scenarios in between different time frames are represented. The results of land-use mapping of Chattogram City Corporation provide information on the aerial distribution of land-use categories and identify and estimate land-use changes over the past 30 years (Figure 6.3, Table 6.2).

In order to know the past land-use patterns of the study area, Landsat imageries were used. The land-use pattern was identified in five categories: vegetation, settlement, agricultural land, water body, and barren land. Since the Chattogram City Corporation established in 1990, the present study aimed to detect the land-use and land-cover changes from 1990 to present at 10 years of interval (Figures 6.3 and 6.4).

Results from the study revealed that in 1990, vegetation coverage was dominating land-use type with almost half (47.68%) of the total land. At that time, vegetation coverage was sporadically distributed throughout the city corporation area but mostly in the northern

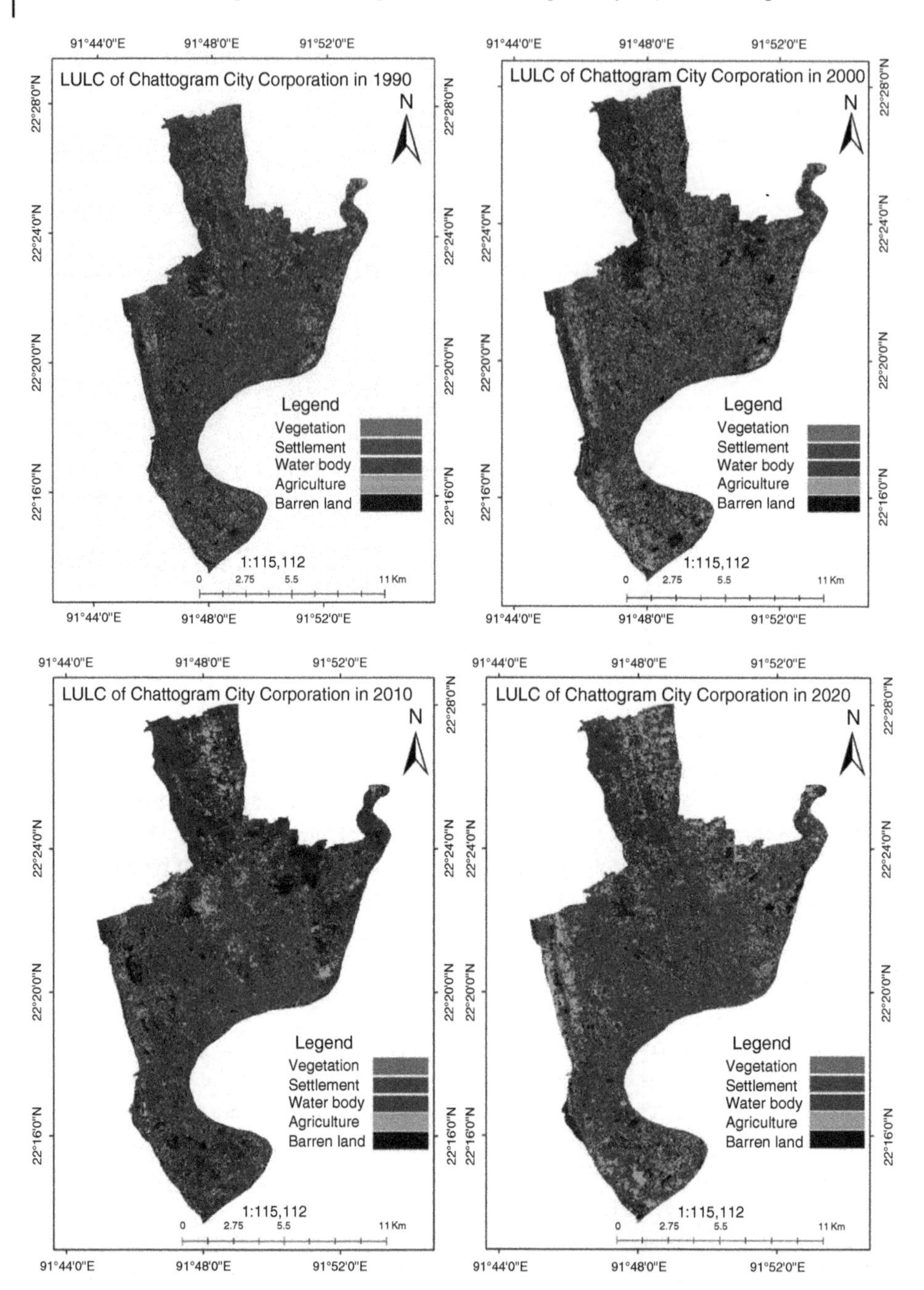

Figure 6.3 Land-use pattern of Chattogram City Corporation at different time periods.

Table 6.2 Overall land-use area from 1990 to 2020.

Land-use category	Area (Acre)			
	1990	2000	2010	2020
Vegetation	18 308.92	13 017.69	12 324.32	9 928.05
Settlement	7 302.53	8 875.54	13 631.04	17 155.14
Water body	4 585.62	4 366.41	3 048.70	2 085.89
Agriculture	3 939.95	6 566.22	3 862.04	5 679.86
Barren land	4 263.75	5 574.94	5 534.7	3 551.87

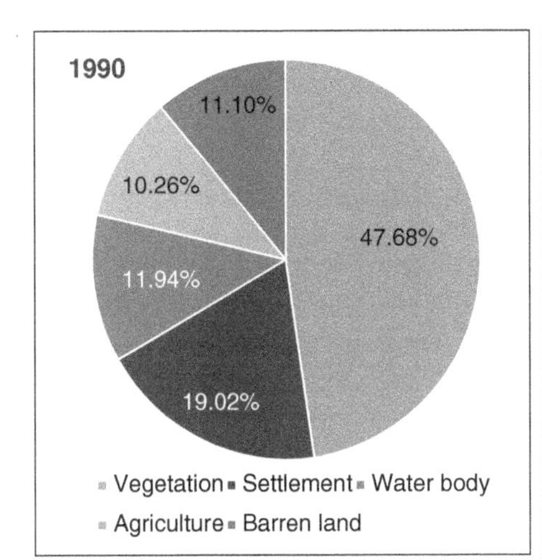

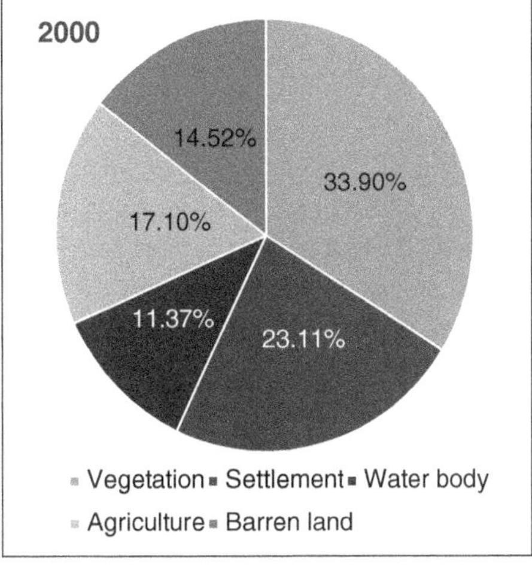

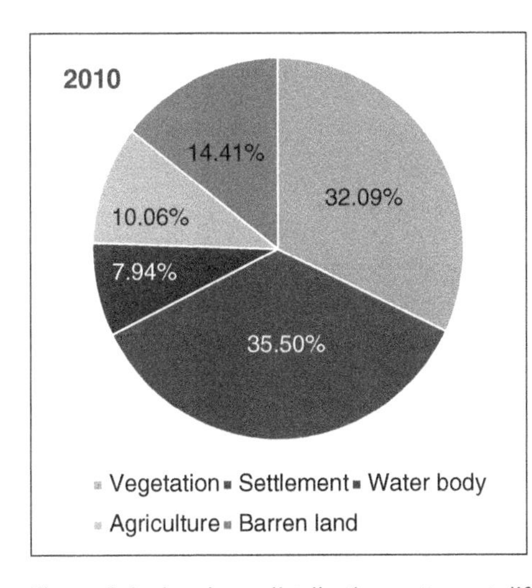

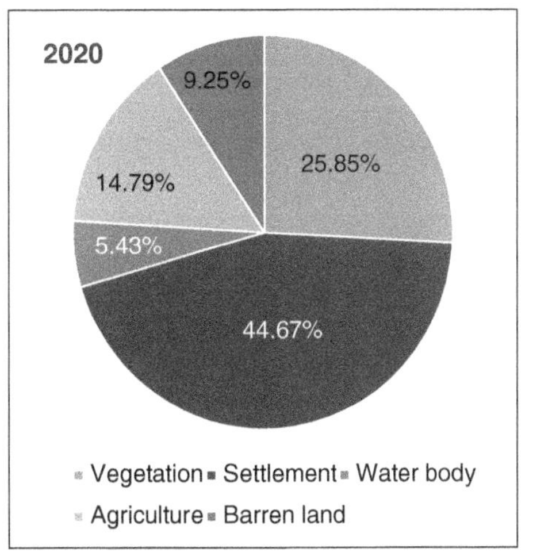

Figure 6.4 Land-use distribution pattern at different time periods at 10-years interval.

part. Settlement was the second largest land-use type, and it covered nearly 19.02%. Water body held the third largest land-use type (11.94%), and surprisingly barren land (11.10%) was higher than the agriculture (10.26%) in 1990.

A rapid loss of vegetation and a sudden increase in agricultural and barren lands were observed in 2000. In the year 2000, vegetation area rapidly reduced to 33.90% (13017.69 acres) compared to 47.68% (18308.92 acres) in 1990 (Figure 6.4, Table 6.3). The settlement area was increased slightly to 23.11% (8875.54 acres) in 2000 compared to 19.02% (7302.53 acres) in 1990. The agricultural land increased from 10.26% (3939.95 acres) in 1990 to 17.10% (6566.22 acres) in 2000. Throughout the 30 years of study time, it was the highest area for using as agricultural land. Comparing barren land uses since 1990–2020, the maximum barren land (14.52%) was also observed in 2000. A very slight decrease in waterbody area was found from 1990 (11.94%) to 2000 (11.37%).

A sudden increase in the settlement area was observed in 2010 from 2000, and an upward trend detected later up to 2020. In 2010, the settlement area was 35.50% (13631.04 acres) compared to 23.11% (8875.54 acres) in the year 2000 (Figure 6.3, Table 6.3). In 2020, the settlement area has increased to 44.67% (17155.14 acres) from the year 2010, which was 35.50% (13631.04 acres). Most settlement increase was observed in the middle to the southern portion of the city corporation.

The vegetation area is seen reduced to 25.85% (9928.05 acres) in 2020 compared to 2010, which was 32.10% (12324.32 acres). Almost half of the total vegetation reduced since 1990–2020, while the settlement area increased more than double (Figure 6.4). The overall vegetation coverage has seen decreased at the middle to the southern portion of the city corporation area. Along with vegetation, a moderate decrease in barren land and water body was also observed since 1990–2020. Almost 21.82% vegetation coverage, 6.52% water body area, and 1.85% barren land have converted to settlement growth during 1990–2020. Besides settlement, some vegetation coverage has also converted to agricultural land as slight increase in agricultural land was also found in 2020 compared with 1990 (Figure 6.4, Table 6.3). The rest of the vegetation basically confined to the northern area (Figure 6.3).

From the identified land-use categories, the vegetation coverage found after 30 years was 25.85%, drastically reduced from 47.67%; settlement area 44.67%, dramatically expand from 19.02%; water body 5.43%, moderately decreased from 11.95%; agricultural land area 14.79%,

Table 6.3 Land-use percentage change from 1990 to 2020.

Land-use category	Change in area (%)			
	1990	2000	2010	2020
Vegetation	47.67	33.90	32.10	25.85
Settlement	19.02	23.11	35.50	44.67
Water body	11.95	11.37	7.94	5.43
Agriculture	10.26	17.10	10.06	14.79
Barren land	11.10	14.52	14.41	9.25
Total	100	100	100	100

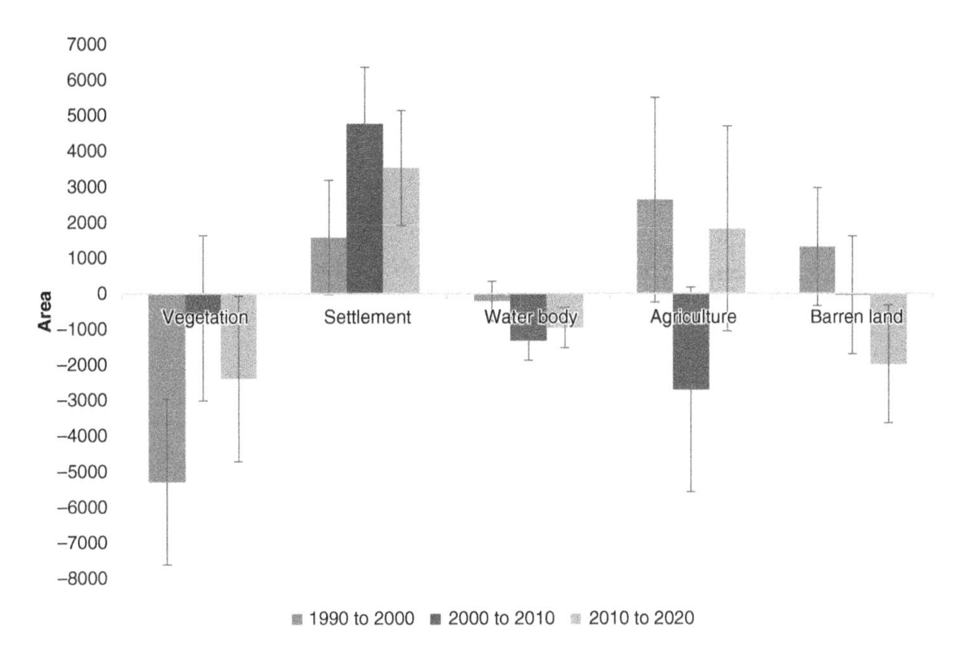

Figure 6.5 Change in land-use pattern from 1990 to 2020.

slightly increased from 10.26%; and barren land 9.25%, slightly declined from 11.10% (Figure 6.4, Table 6.3). With the time frame, the urban settlement expanded mostly and increased by 25.65%, while the agricultural land increased a little by 4.53%. At the same time, the decline in vegetation was significant (21.82%) along with a slight decrease in water bodies (6.52%) and barren lands (1.85%) (Gazi et al. 2020; Hassan and Nazem 2016; Hussain et al. 2016). Thus, in 2020, the highest land use observed settlement area compared to vegetation coverage in 1990.

The land-use patterns from 1990 to 2020 showed drastic changes, especially in the vegetation cover area and the settlement area (Tables 6.2, 6.3 and Figures 6.3–6.5). A significant level of changes were also observed in agricultural land from 1990 to 2020 (Figure 6.5). The changes in agricultural land fluctuated throughout the decades. An increase in the agricultural land was observed during 1990–2000. After that, a sudden decrease was found from 2000 to 2010. The agricultural land dramatically rises in the next decade (Figure 6.5). From Tables 6.2 and 6.3 and Figures 6.3–6.5, it was observed that a considerable amount of vegetation area and agricultural land were converted into the settlement area (Zhang and Xu 2014), and it is highly noticeable during 2000–2010. Changes in the other two land-use patterns remain relatively constant. From the results, it is easily estimated that vegetation land area decreased to a greater extent due to population pressure, industrialisation, and unplanned destruction of forests for people's settlement and accommodation (Gazi et al. 2020; Hassan and Nazem 2016; Hussain et al. 2016).

From the present study, the average rate of losing vegetation area per year was found to be about 2.79%, whereas the average rate of increasing settlement area per year was observed to be about 3.28%. Again, an average reduction in waterbodies and barren lands was calculated at 0.83 and 0.24% per year, respectively. The average growth rate of agricultural land was 0.58% each year.

In the studied time period of 30 years, the vegetation area of Chattogram City Corporation reduced heavily due to increased settlement area. A few LULC changes studies in Bangladesh (Gazi et al. 2020; Mamnun and Hossen 2020; Hussain et al. 2016; Dewan et al. 2012) and Kenya (Kogo et al. 2019) revealed the similar type of results on increased settlement that supports the present study. On the other hand, waterbody also reduced with the passage of time, which is similar with the study by Mamnun and Hossen (2020) and Kogo et al. (2019). The total agricultural land increased slightly in 2020 as compared to 1990, which means people are still involved in agricultural activities and agricultural production. Similar types of reports were found from the researches by Mamnun and Hossen (2020) in Bangladesh, Boakye et al. (2020) in Ghana, Kogo et al. (2019) in Kenya, Parsa et al. (2016) in Iran, and Zhang and Xu (2014) in China. It is evident from Figure 6.6 that vegetation land area decreased in a large number due to population pressure, industrialisation, and rampant destruction of forest and trees for the settlement and accommodation of people. The findings of the study are in conformity with the findings of Boakye et al. (2020) in Ghana, Prasad and Ramesh (2019) in India, Parsa et al. (2016) in Iran, and Ejaro and Abdullahi (2013) in Nigeria.

Accuracy assessment is prerequisite in land-use classification. To remove the ambiguousness regarding accuracy of the land-use type, each land class of the classified image was matched with the data source. In the present study, a randomly selected sample point of the classified map was compared with the Google Earth map, and thus, the accuracy of classification was proven. From Table 6.4, Kappa statistics for the year 1990 is 0.73, which would imply that the classification process avoided 73% of errors. The results indicate a significant agreement between the classified and reference maps (Mamnun and Hossen 2020). Kappa statistics for the year 2000, 2010, and 2020 were 0.80, 0.78, and 0.79, which means that the classification avoided 80, 78, and 79% error, respectively (Mamnun and Hossen 2020;

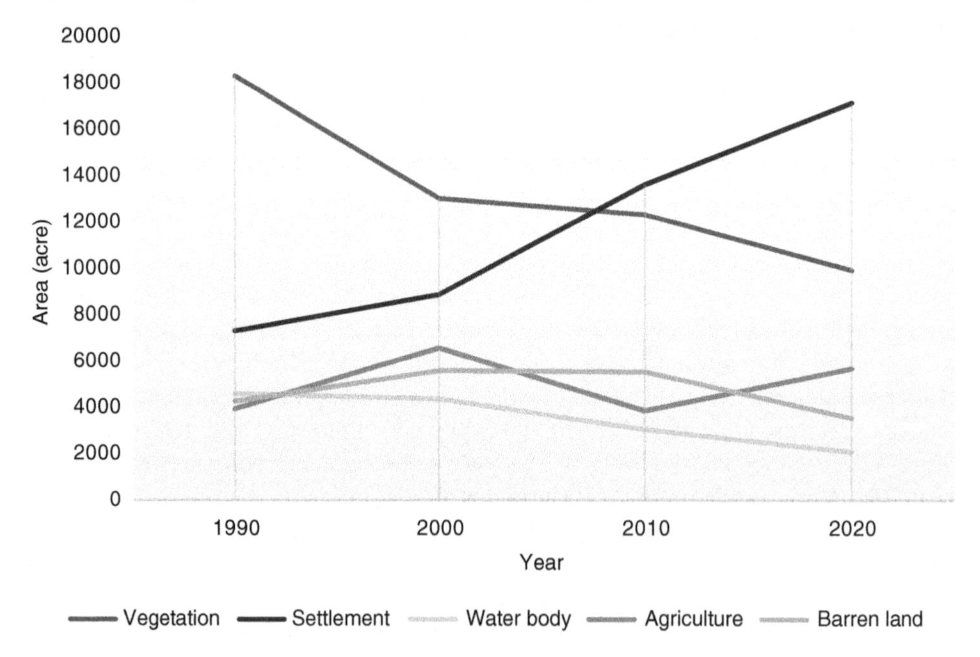

Figure 6.6 Temporal distribution of land-use pattern.

Table 6.4 Classification accuracy and overall Kappa statistics.

	Overall classification accuracy				Overall Kappa statistics			
	Year				Year			
Accuracy	1990	2000	2010	2020	1990	2000	2010	2020
Assessment	82%	86%	84%	84%	0.73	0.80	0.78	0.79

Hassan and Nazem 2016). This study reveals that supervised land-use classification was a better option for land-use study. The accuracy of the land-use classification found from the study was satisfactory. The highest accuracy was obtained for the Landsat OLI-TIRS data, while the lowest accuracy was attained for the MSS-TM image.

The present study took the advantages of GIS and remote sensing techniques to quantify the land-cover change in the study area over the past 30 years. The results of the study revealed that as built-up areas increase, the area with different natural/environmental benefits is decreasing. The rapid decline in the forest and vegetative cover areas is pointing to environmental deterioration and ecological imbalances.

6.4 Conclusion

Land-cover change has a great impact on the environment. For this reason, GIS, remote sensing, and satellite-derived images are some effective tools not only to detect LULC change of an area from time to time but also to make planning for using land properly. Remote sensing-based land-cover mapping is a very effective mapping system. This study quantitatively examined the changes in land use/land cover of the Chattogram City Corporation area, using Landsat data and supervised land-use classification of 1990, 2000, 2010, and 2020 time period. The Chattogram City Corporation area highly indicates rapid urbanisation and industrialisation as the quick transformation of vegetation-covered lands into the built-up area. The built-up area has been growing up, resulting in a loss of vegetation cover mostly and loss of water bodies to some extent. For every country, urban sprawl is an obvious process with population growth and civilisation. In this situation, more multistoried buildings could be developed in the future by maintaining building codes rather than lower buildings to give proper spaces for the vegetation and other natural resources. Further researches could be done for developing a model on balanced urban sprawl with highly maintained vegetation coverage to save the nature.

References

Ahamed, A., Harun-Or-Rashid, M., and Rahman, M.S. (2020). A critical study on the eastern coastal zone of Bangladesh: policy implication for development. *Research, Society and Development* 9 (10): 1-23. e8659109037-e8659109037. DOI: https://doi.org/10.33448/rsd-v9i10.9037

Alemu, B. (2015). The effect of land use land cover change on land degradation in the highlands of Ethiopia. *Journal of Environment and Earth Science* 5 (1): 1–13.

Anand, V., and Oinam, B. (2020). Future land use land cover prediction with special emphasis on urbanization and wetlands. *Remote Sensing Letters* 11 (3): 225-234. Doi: https://doi.org/10.1080/2150704X.2019.1704304

Andrews-Speed, P., Bleischwitz, R., Boersma, T. et al. (2014). *Want, Waste or War?: The Global Resource Nexus and the Struggle for Land, Energy, Food, Water and Minerals.* Routledge.

Azizi, A., Malakmohamadi, B., and Jafari, H.R. (2016). Land use and land cover spatiotemporal dynamic pattern and predicting changes using integrated CA-Markov model. *Global Journal of Environmental Science and Management* 2: 223–234.

Balogun, I.A., Adeyewa, D.Z., Balogun, A.A., and Morakinyo, T.E. (2011). Analysis of urban expansion and land use changes in Akure, Nigeria, using remote sensing and geographic information system (GIS) techniques. *Journal of Geography and Regional Planning* 4 (9): 533-541. Doi: https://doi.org/10.5897/JGRP.9000020

BBS (2011). *Bangladesh Bureau of Statistics (BBS).* Statistics and Informatics Division (SID) Ministry of Planning Government of the People's Republic of Bangladesh.

Boakye, E., Anyemedu, F.O.K., Quaye-Ballard, J.A., and Donkor, E.A. (2020). Spatio-temporal analysis of land use/cover changes in the Pra River Basin, Ghana. *Applied Geomatics* 12 (1): 83-93. Doi: https://doi.org/10.1007/s12518-019-00278-3

Chamberlain, J.L., Darr, D., and Meinhold, K. (2020). Rediscovering the contributions of forests and trees to transition global food systems. *Forests* 11 (10): 1098. Doi: https://doi.org/10.3390/f11101098

Clerici, N., Cote-Navarro, F., Escobedo, F.J. et al. (2019). Spatio-temporal and cumulative effects of land use-land cover and climate change on two ecosystem services in the Colombian Andes. *Science of the Total Environment* 685: 1181-1192. Doi: https://doi.org/10.1016/j.scitotenv.2019.06.275

Cohen, B. (2004). Urban growth in developing countries: a review of current trends and a caution regarding existing forecasts. *World Development* 32 (1): 23-51. Doi: https://doi.org/10.1016/j.worlddev.2003.04.008

Dadashpoor, H., Azizi, P., and Moghadasi, M. (2019). Land use change, urbanization, and change in landscape pattern in a metropolitan area. *Science of the Total Environment* 655: 707-719. Doi: https://doi.org/10.1016/j.scitotenv.2018.11.267

Dalby, S. (2009). *Security and Environmental Change.* Cambridge, UK: Polity Press.

Dale, V.H., Brown, S., Haeuber, R.A. et al. (2000). Ecological principles and guidelines for managing the use of land sup> 1. *Ecological Applications* 10 (3): 639-670. Doi: https://doi.org/10.1890/1051-0761(2000)010[0639:EPAGFM]2.0.CO;2

Deng, J.S., Wang, K., Hong, Y., and Qi, J.G. (2009). Spatio-temporal dynamics and evolution of land use change and landscape pattern in response to rapid urbanization. *Landscape and Urban Planning* 92 (3-4): 187-198. Doi: https://doi.org/10.1016/j.landurbplan.2009.05.001

Dewan, A.M., and Yamaguchi, Y. (2009). Land use and land cover change in Greater Dhaka, Bangladesh: using remote sensing to promote sustainable urbanization. *Applied Geography* 29 (3): 390-401. Doi: https://doi.org/10.1016/j.apgeog.2008.12.005

Dewan, A.M., Yamaguchi, Y., and Rahman, M.Z. (2012). Dynamics of land use/cover changes and the analysis of landscape fragmentation in Dhaka Metropolitan, Bangladesh. *GeoJournal* 77 (3): 315-330. Doi: https://doi.org/10.1007/s10708-010-9399-x

Drayton, B. (2006). Everyone a changemaker. In: *Innovations*. 111 River Street, Hoboken, NJ: John Wiley and Sons, Inc.

Ejaro, S.P. and Abdullahi, U. (2013). Spatiotemporal analyses of land use and land cover changes in Suleja local government area, Niger State, Nigeria. *Journal of Environment and Earth Science* 3 (9): 72–83.

Figueroa, F., Sanchez-Cordero, V., Meave, J.A., and Trejo, I. (2009). Socioeconomic context of land use and land cover change in Mexican biosphere reserves. *Environmental Conservation* 36 (3): 180-191. Doi: https://doi.org/10.1017/S0376892909990221

Foley, J.A., DeFries, R., Asner, G.P. et al. (2005). Global consequences of land use. *Science* 309 (5734): 570-574. DOI: https://doi.org/10.1126/science.1111772

Gazi, M.Y., Rahman, M.Z., Uddin, M.M., and Rahman, F.A. (2020). Spatio-temporal dynamic land cover changes and their impacts on the urban thermal environment in the Chittagong metropolitan area, Bangladesh. *GeoJournal* 1-16. DOI: https://doi.org/10.1007/s10708-020-10178-4

Genet, A. (2020). Population growth and land use land cover change scenario in Ethiopia. *International Journal of Environmental Protection and Policy* 8 (4): 77-85. DOI: https://doi.org/10.11648/j.ijepp.20200804.12

Gómez, C., White, J.C., and Wulder, M.A. (2016). Optical remotely sensed time series data for land cover classification: a review. *ISPRS Journal of Photogrammetry and Remote Sensing* 116: 55-72. Doi: https://doi.org/10.1016/j.isprsjprs.2016.03.008

Hasan, S.S., Sarmin, N.S., and Miah, M.G. (2020). Assessment of scenario-based land use changes in the Chittagong Hill Tracts of Bangladesh. *Environment and Development* 34: 100463. Doi: https://doi.org/10.1016/j.envdev.2019.100463

Hassan, M.M., and Nazem, M.N.I. (2016). Examination of land use/land cover changes, urban growth dynamics, and environmental sustainability in Chittagong city, Bangladesh. *Environment, Development and Sustainability* 18 (3): 697-716. Doi: https://doi.org/10.1007/s10668-015-9672-8

Hishe, H., Giday, K., Van Orshoven, J. et al. (2021). Analysis of land use land cover dynamics and driving factors in Desa'a forest in Northern Ethiopia. *Land Use Policy* 101: 105039. Doi: https://doi.org/10.1016/j.landusepol.2020.105039

Hua, A.K., and Ping, O.W. (2018). The influence of land-use/land-cover changes on land surface temperature: a case study of Kuala Lumpur metropolitan city. *European Journal of Remote Sensing* 51 (1): 1049-1069. Doi: https://doi.org/10.1080/22797254.2018.1542976

Hussain, M., Alak, P., and Azmz, I. (2016). Spatio-temporal analysis of land use and land cover changes in Chittagong city corporation, Bangladesh. *International Journal of Advancement in Remote Sensing, GIS and Geography* 4: 56–72.

Imura, H., Chen, J., Kaneko, S., et al. (1999). Analysis of industrialization, urbanization and land-use change in East Asia according to the DPSER framework. *Proceedings of 1999 NIES Workshop on Information Bases and Modeling for Land-use and Land-cover Changes in East Asia* (p. 227).

Islam, K., Jashimuddin, M., Nath, B., and Nath, T.K. (2018). Land use classification and change detection by using multi-temporal remotely sensed imagery: the case of Chunati wildlife sanctuary, Bangladesh. *The Egyptian Journal of Remote Sensing and Space Science* 21 (1): 37–47. Doi: https://doi.org/10.1016/j.ejrs.2016.12.005

Islam, M.M., Jannat, A., Dhar, A.R., and Ahamed, T. (2020). Factors determining conversion of agricultural land use in Bangladesh: farmers' perceptions and perspectives of climate change. *GeoJournal* 85 (2): 343-362. Doi: https://doi.org/10.1007/s10708-018-09966-w

Kantakumar, L.N., and Neelamsetti, P. (2015). Multi-temporal land use classification using hybrid approach. *The Egyptian Journal of Remote Sensing and Space Science* 18 (2): 289–295. Doi: https://doi.org/10.1016/j.ejrs.2015.09.003

Kareiva, P., Watts, S., McDonald, R., and Boucher, T. (2007). Domesticated nature: shaping landscapes and ecosystems for human welfare. *Science* 316 (5833): 1866-1869. DOI: https://doi.org/10.1126/science.1140170

Käyhkö, N., Fagerholm, N., Asseid, B.S., and Mzee, A.J. (2011). Dynamic land use and land cover changes and their effect on forest resources in a coastal village of Matemwe, Zanzibar, Tanzania. *Land Use Policy* 28 (1): 26-37. Doi: https://doi.org/10.1016/j.landusepol.2010.04.006

Khan, M.M.H., Bryceson, I., Kolivras, K.N. et al. (2015). Natural disasters and land-use/land-cover change in the southwest coastal areas of Bangladesh. *Regional Environmental Change* 15 (2): 241-250. Doi: https://doi.org/10.1007/s10113-014-0642-8

Kogo, B.K., Kumar, L., and Koech, R. (2019). Analysis of spatio-temporal dynamics of land use and cover changes in Western Kenya. *Geocarto International* 36 (4): 376-391. Doi: https://doi.org/10.1080/10106049.2019.1608594

Kurowska, K., Kryszk, H., Marks-Bielska, R. et al. (2020). Conversion of agricultural and forest land to other purposes in the context of land protection: evidence from Polish experience. *Land Use Policy* 95: 104614. Doi: https://doi.org/10.1016/j.landusepol.2020.104614

Li, X., Ling, F., Foody, G.M. et al. (2017). Generating a series of fine spatial and temporal resolution land cover maps by fusing coarse spatial resolution remotely sensed images and fine spatial resolution land cover maps. *Remote Sensing of Environment* 196: 293-311. Doi: https://doi.org/10.1016/j.rse.2017.05.011

López, E., Bocco, G., Mendoza, M., and Duhau, E. (2001). Predicting land-cover and land-use change in the urban fringe: a case in Morelia city, Mexico. *Landscape and Urban Planning* 55 (4): 271–285. Doi: https://doi.org/10.1016/S0169-2046(01)00160-8

Mamnun, M., and Hossen, S. (2020). Spatio-temporal analysis of land cover changes in the evergreen and semi-evergreen rainforests: a case study in Chittagong Hill Tracts, Bangladesh. *International Journal of Forestry, Ecology and Environment* 2 (2): 87-99. DOI: https://doi.org/10.18801/ijfee.020220.10

Mamun, A.A., Mahmood, A., and Rahman, M. (2013). Identification and monitoring the change of land use pattern using remote sensing and GIS: a case study of Dhaka City. *IOSR Journal of Mechanical and Civil Engineering* 6 (2): 20–28.

Mansour, S., Al-Belushi, M., and Al-Awadhi, T. (2020). Monitoring land use and land cover changes in the mountainous cities of Oman using GIS and CA-Markov modelling techniques. *Land Use Policy* 91: 104414. Doi: https://doi.org/10.1016/j.landusepol.2019.104414

Melendez-Pastor, I., Hernández, E.I., Navarro-Pedreño, J., and Gómez, I. (2014). Socioeconomic factors influencing land cover changes in rural areas: the case of the Sierra de Albarracín (Spain). *Applied Geography* 52: 34-45. Doi: https://doi.org/10.1016/j.apgeog.2014.04.013

Mia, M.A., Nasrin, S., Zhang, M., and Rasiah, R. (2015). Chittagong, Bangladesh. *Cities 48*: 31–41.

Morales-González, A., Ruiz-Villar, H., Ordiz, A., and Penteriani, V. (2020). Large carnivores living alongside humans: brown bears in human-modified landscapes. *Global Ecology and Conservation* 22, e00937. Doi: https://doi.org/10.1016/j.gecco.2020.e00937

Nagendra, H., Munroe, D.K., and Southworth, J. (2004). From pattern to process: landscape fragmentation and the analysis of land use/land cover change. *Agriculture, Ecosystems and Environment 101* (2004): 111–115.

Nath, B., Niu, Z., and Singh, R.P. (2018). Land use and land cover changes, and environment and risk evaluation of Dujiangyan city (SW China) using remote sensing and GIS techniques. *Sustainability* 10 (12): 4631. Doi: https://doi.org/10.3390/su10124631

Ogato, G.S., Bantider, A., and Geneletti, D. (2021). Dynamics of land use and land cover changes in Huluka watershed of Oromia Regional State, Ethiopia. *Environmental Systems Research* 10 (1): 1-20. Doi: https://doi.org/10.1186/s40068-021-00218-4

Ojima, D.S., Galvin, K.A., and Turner, B.L. (1994). The global impact of land-use change. *Bioscience* 44 (5): 300-304. Doi: https://doi.org/10.2307/1312379

Pal, S., and Ziaul, S. (2017). Detection of land use and land cover change and land surface temperature in English Bazar urban centre. *The Egyptian Journal of Remote Sensing and Space Sciences* 20 (1): 125–145. Doi: https://doi.org/10.1016/j.ejrs.2016.11.003

Parsa, V.A., Yavari, A., and Nejadi, A. (2016). Spatio-temporal analysis of land use/land cover pattern changes in Arasbaran Biosphere Reserve: Iran. *Modeling Earth Systems and Environment* 2 (4): 1-13. Doi: https://doi.org/10.1007/s40808-016-0227-2

Prasad, G., and Ramesh, M.V. (2019). Spatio-temporal analysis of land use/land cover changes in an ecologically fragile area—Alappuzha District, Southern Kerala, India. *Natural Resources Research* 28 (1): 31-42. Doi: https://doi.org/10.1007/s11053-018-9419-y

Rai, R., Zhang, Y., Paudel, B. et al. (2017). A synthesis of studies on land use and land cover dynamics during 1930–2015 in Bangladesh. *Sustainability 9*: 1866.

Riebsame, W.E., Meyer, W.B., and Turner, B. (1994). Modeling land use and cover as part of global environmental change. *Climatic Change* 28 (1–2): 45–64. Doi: https://doi.org/10.1007/BF01094100

Rounsevell, M.D.A., Reginster, I., Araújo, M.B. et al. (2006). A coherent set of future land use change scenarios for Europe. *Agriculture, Ecosystems and Environment* 114 (1): 57-68. Doi: https://doi.org/10.1016/j.agee.2005.11.027

Sallustio, L., Munafò, M., Riitano, N., et al. (2016). Integration of land use and land cover inventories for landscape management and planning in Italy. *Environmental Monitoring and Assessment* 188 (1): 1-20. Doi: https://doi.org/10.1007/s10661-015-5056-7

Sarif, M.O., and Gupta, R.D. (2021). Spatiotemporal mapping of land use/land cover dynamics using Remote Sensing and GIS approach: a case study of Prayagraj City, India (1988–2018). *Environment, Development and Sustainability* 1-33. Doi: https://doi.org/10.1007/s10668-021-01475-0

Sarwar, M.I., Billa, M., and Paul, A. (2016). Urban land use change analysis using RS and GIS in Sulakbahar ward in Chittagong city, Bangladesh. *International Journal of Geomatics and Geosciences* 7 (1): 1–10.

Schirpke, U., Tscholl, S., and Tasser, E. (2020). Spatio-temporal changes in ecosystem service values: effects of land-use changes from past to future (1860–2100). *Journal of Environmental Management* 272: 111068. doi: https://doi.org/10.1016/j.landusepol.2020.104868

Shaharum, N.S.N., Shafri, H.Z.M., Gambo, J., and Abidin, F.A.Z. (2018). Mapping of Krau Wildlife Reserve (KWR) protected area using Landsat 8 and supervised classification algorithms. *Remote Sensing Applications: Society and Environment* 10: 24-35. Doi: https://doi.org/10.1016/j.rsase.2018.01.002

Shalaby, A. and Tateishi, R. (2007). Remote sensing and GIS for mapping and monitoring land cover and land-use changes in the Northwestern coastal zone of Egypt. *Applied geography 27* (1): 28–41.

Sharma, R., Nehren, U., Rahman, S.A. et al. (2018). Modeling land use and land cover changes and their effects on biodiversity in Central Kalimantan, Indonesia. *Land* 7 (2): 57. Doi: https://doi.org/10.3390/land7020057

Slee, B., Brown, I., Donnelly, D. et al. (2014). The 'squeezed middle': identifying and addressing conflicting demands on intermediate quality farmland in Scotland. *Land Use Policy* 41: 206-216. Doi: https://doi.org/10.1016/j.landusepol.2014.06.002

Sohel, M.S.I., Mukul, S.A., and Burkhard, B. (2015). Landscape' s capacities to supply ecosystem services in Bangladesh: a mapping assessment for Lawachara National Park. *Ecosystem Services* 12: 128-135. Doi: https://doi.org/10.1016/j.ecoser.2014.11.015

Sohl, T.L., and Sohl, L.B. (2012). Land-use change in the Atlantic Coastal Pine Barrens Ecoregion. *Geographical Review* 102 (2): 180–201. Doi: https://doi.org/10.1111/j.1931-0846.2012.00142.x

Sunderlin, W.D., Angelsen, A., Belcher, B. et al. (2005). Livelihoods, forests, and conservation in developing countries: an overview. *World Development* 33 (9): 1383-1402. Doi: https://doi.org/10.1016/j.worlddev.2004.10.004

Szarek-Iwaniuk, P. (2021). A comparative analysis of spatial data and land use/land cover classification in urbanized areas and areas subjected to anthropogenic pressure for the example of poland. *Sustainability* 13 (6): 3070. Doi: https://doi.org/10.3390/su13063070

Wang, S.W., Gebru, B.M., Lamchin, M. et al. (2020). Land use and land cover change detection and prediction in the Kathmandu district of Nepal using remote sensing and GIS. *Sustainability* 12 (9): 3925. Doi: https://doi.org/10.3390/su12093925

Wu, Y., Li, S., and Yu, S. (2016). Monitoring urban expansion and its effects on land use and land cover changes in Guangzhou city, China. *Environmental Monitoring and Assessment* 188 (1): 54. Doi: https://doi.org/10.1007/s10661-015-5069-2

Yuan, F., Sawaya, K.E., Loeffelholz, B.C., and Bauer, M.E. (2005). Land cover classification and change analysis of the Twin Cities (Minnesota) Metropolitan Area by multitemporal Landsat remote sensing. *Remote Sensing of Environment* 98 (2–3): 317–328. Doi: https://doi.org/10.1016/j.rse.2005.08.006

Zerga, B., Warkineh, B., Teketay, D. et al. (2021). Land use and land cover changes driven by expansion of eucalypt plantations in the Western Gurage Watersheds, Central-south Ethiopia. *Trees, Forests and People* 5: 100087. Doi: https://doi.org/10.1016/j.tfp.2021.100087

Zhang, F., Tiyip, T., Kung, H. et al. (2016). Dynamics of land surface temperature (LST) in response to land use and land cover (LULC) changes in the Weigan and Kuqa river oasis, Xinjiang, China. *Arabian Journal of Geosciences* 9 (7): 1-14. Doi: https://doi.org/10.1007/s12517-016-2521-8

Zhang, Y., and Xu, B. (2014). Spatiotemporal analysis of land use/cover changes in Nanchang area, China. *International Journal of Digital Earth* 8 (4): 312-333. Doi: https://doi.org/10.1080/17538947.2014.894145

7

Emerging Techniques for Urban Resource Restoration of Various Ecosystem: Bioremediation, Phytoremediation, Habitat Enhancement

Riddhi Shrivastava[1,2], Jabbar Khan[1], Govind Gupta[1], and Naveen K. Singh[1]

[1] *Environmental Science Discipline, Department of Chemistry, Manipal University Jaipur, Jaipur, Rajasthan, India*
[2] *Poornima College of Engineering, Jaipur, Rajasthan, India*

7.1 Introduction

Since 1950, urban population is growing day by day all over the world. For a better life and improved living conditions, people usually are attracted to the urban areas. Migration from rural to urban areas is the common and main factor for a rapidly growing urban population (Jorgenson and Burns 2007). Presently urban population shares almost more than 50% of the world population, this process is responsible for the urban environmental degradation and depletion of natural resources, which is becoming a threat to sustainable urban development. As the population grows due to industrialisation a high amount of waste is also generated which creates a high risk of environmental components and ecosystem. Today, due to overpopulation and high consumption rate of natural resources, human generates a high amount of waste (Lebreton and Andrady 2019), which lead to environmental degradation and related social, economic, and human health issues. Due to the urban lifestyles, people consume more natural resources and generate more wastes (Hao et al. 2018; Yang and Liu. 2018). In the current era, to fulfil all the basic needs, industries are using all the natural resources in different ways to generate new materials which are also responsible for generating waste and contamination of water, soil, and air. In the last 10 years in India, gross domestic products have increased by 7% and solid waste generated by 45% from a total of 48 to 70 million tons. Population growth, as well as industrialisation, comes with an increased rate of consumption of natural resources and waste generation (Swim et al. 2011; Gutberlet and Uddin 2017). Increased income also enables consumers to fulfil all their needs. This is one of the major factors to generate waste (Gutberlet and Uddin 2017). Waste generation in urban areas is a byproduct of different socio-economic activities. More common areas like industrial, commercial, domestic, and institutional sectors generate a huge amount of waste and affect the urban environment (Singh and Sharma 2002; Minghua et al. 2009). As compared to other waste, emissions, and release of more greenhouse gases

Urban Ecology and Global Climate Change, First Edition. Edited by Rahul Bhadouria, Shweta Upadhyay, Sachchidanand Tripathi, and Pardeep Singh.

(GHGs) especially carbon dioxide continuously adding into the urban atmosphere by burning fossil fuel which leads to climate change (Hoornweg et al. 2013; Widmer et al. 2005). Due to the overpopulation in India after 2001, urbanisation rate has been increased by about 31.16% (Bhagat 2011; Chandrasekhar and Sharma 2015). Urban wastes must be treated for safe disposal and discharge into the environment which needs the development of sustainable treatment and disposal technologies (Gomez et al. 2009; Misra and Pandey 2005). Characteristics of urban waste depend on the state as well as the reusability and recyclability of waste according to the health standards. Three key system elements have been identified to treat the waste, i.e. public health, environmental protection, and resource management. Unregulated and direct disposal of waste affects the environment detrimentally, designing and developing more viable, economic, and sustainable technologies is needed for waste treatment and their safe disposal to ensure social, environmental, and economic growth with environmental conservation. Further, better urban planning, sustainable utilisation of resources with more technological development would help achieve sustainable development goals for humans with environmental conservation.

7.2 Urban Resources and Waste Generation

As the overconsumption of natural resources is growing rapidly, the amount of waste generation has also been increased significantly in recent decades. With advancement and economic development, more transformation has occurred in humans in terms of environmental degradation (Wilson et al. 2015; Kumar et al. 2016). To minimise human impact on the environment and achieve sustainable development, we need to develop and use viable technologies for resource utilisation and to reduce the amount of waste generation. During the last twenty to the twenty-first century, it has been observed by researchers that technological advances have facilitated human beings by providing basic needs from food to shelter but due to the extensive developmental activities of humans and overexploitation of natural resources, more amount of wastes are generating which lead to more environmental degradation (Kastenhofer 2007; Yakovleva et al. 2017). Further, with a rapidly growing population, urbanisation, and industrialisation especially from metallurgical fields, nuclear power plants release more wastewater, and sewage directly into the aquatic environment which causes deterioration of the environment and affects flora and fauna (Zhang et al. 2009; Patel and Kasture 2014). Discharge of wastewater containing metals and other toxic chemicals now becomes a serious concern of humans and other living beings for health risks. Metal contaminations in water exhibit toxic effects on living organisms even at low concentrations (Deng et al. 2007). The toxicity of a contaminant depends on its long-term persistence like lead is one of the most persistent metals in soil in which completed its half life cycle (at least 10–18 years) in the human body (Khalid et al. 2017). Other elements like arsenic, cadmium, and mercury may present in trace amounts; however, these elements can be uptaken by plants from water and soil and accumulate in living organisms at a higher trophic level including human being through food chain contamination and bio-magnification. Toxicity of these metals increases in an acidic medium, nutrient-deficient ecosystem, and poor physical conditions. A degraded ecosystem directly affects the life support system like land, water, forest resources, biota, and ecosystem processes and services like mineralisation, nutrients cycling as well as energy budget directly affects human health and development (Batayneh 2012). Increasing human population and extensive urbanisation lead to more environmental degradation especially deforestation, depletion of water resources, changes in land use pattern, encroachment, emission of more GHG and other gaseous pollutants, discharge of more sewage, generation of solid waste, etc. Muñoz and Cohen (2016) predicted the more than two-thirds population of the world will live in the cities at the end of 2050. Due to the overpopulation in cities as the population densities increases, the level of pollution also increases (Goel 2006). Where their densities offer a more sustainable lifestyle (Muñoz and Cohen 2016), and they can take advantage of the economy that is directly linked to environmental services (UNEP 2017). Starting with its primary components, people need to ensure that every single aspect of society addresses urban pollution. Minimising the waste resources by using fewer inputs (Newman 1999) in any operation by all members of the population of a community would help in reducing waste generation which involves how individuals should behave in their personal and professional lives. To minimise adverse impacts of extensive development, we must develop sustainably without compromising our environment with more technological advancement for resource utilisation, conservation of water and land resources, use of renewable and alternate energy resources, better urban planning, treatment, and recycling of waste.

7.3 Composition of Urban Solid Waste

Any discarded material, which is not liquid or gas, is known as solid waste. On the basis of the characteristics, solid waste mainly can be categorised into two types, i.e. organic waste and inorganic waste. Organic waste decays with time but evolve highly offensive odour and gases which are highly detrimental to health whereas inorganic wastes mainly consist of non-combustible materials such as grit, dust, mud, metal pieces, metal containers, broken glass, waste building material, etc. which are not subjected to decay and decomposition. Basically, urban solid waste includes waste from different sources particularly municipal solid waste from domestic use, biomedical waste from hospitals and dispensaries, industrial waste or sludge, fisheries waste, and E-waste from electronic waste. The common characteristics of solid waste depend upon the living standards, food habits, topographical, and climatic conditions (Jin et al. 2006). Apart from this, the waste that is generated from low- and middle-income countries has high moisture content and density. At present in India, due to the high population and urbanisation, amount of waste is almost eight times more than in 1947 (Sharholy et al. 2008). In India, there are great variations in waste characteristics due to the composition and hazardous nature of the waste (Gupta et al. 1998; Sharholy et al. 2008). In which organic wastes contribute a major part of all. Mumbai has the highest organic wastes that are almost about 62%. In India, Municipal Solid Waste Management (MSW) composition around 40–60% biodegradable material, 30–50% inert material, and 10–30% recyclable material. Indian trash contains $0.64 \pm 0.8\%$ Nitrogen, $0.67 \pm 0.15\%$ Phosphorus, and $0.68 \pm 0.15\%$ Potassium, according to the National Environmental Engineering Research Institute (NEERI), and has a 26 ± 5 C : N ratio (Gupta et al. 2015; Joshi and Ahmed 2016).

7.4 Threats from Urban Wastes

Disposal of urban solid wastes without any treatment gives so many negative impacts on all the components of the environment, i.e. soil, air, water, human health, etc. Major impacts are as follows:

7.4.1 Health Impacts

Due to the mismanagement and poor public attitude for its reduction public is directly exposed to health risks. Poor management, treatment, and disposal of the waste directly or indirectly affect public health (Giusti 2009). Wastes that are created by people directly throw on the road which creates water stagnant conditions and clogs drains too that gives favourable conditions to mosquitoes and insects for breeding hence possess high health risks (Castro et al., 2010). Apart from this dumping of waste without prior treatment creates groundwater pollution and health issues (Kathiravale and Yunus 2008). According to World Health Organisation (WHO), the common risks of wastes landfills and incineration are cancer, infertility, and a high rate of mortality. Due to the airway inflammation, landfill workers are more vulnerable to tissue damage.

7.4.2 Environmental Impacts

Due to the mismanagement and unscientific techniques of waste disposal it directly deteriorates the surroundings like air, water, soil, etc. common impacts are as follows.

7.4.2.1 Soil

According to UNEP (2017), natural sources or anthropogenic sources can be responsible for soil contamination. Soil is the base of every structure, whether of natural or social origin. In urban areas, land plays an important role to promote urban communities in the environment (Cachada et al. 2018). Soil which is the topmost layer on earth is mainly composed of organic minerals and materials on which plant grows (Brady and Weil 1974). It also acts as a protective layer for groundwater which mitigates the impacts of pollutants (Venkatesan and Swaminathan 2009). Urban soil depletion is growing because of sealing, erosion, desertification, acidification, loss of soil microbes, and contamination (European Environment Agency 2015). Water quality and human health are directly affected by the characteristics of the soil. Since they reflect the interface between the biosphere, the atmosphere, and the hydrosphere (Cachada et al. 2018). In developing countries, due to the urbanisation and increased rate of industrialisation, the burden of MSW is continuously increasing which affects the soil properties.

7.4.2.2 Water

Water is a basic element for living organisms and sustainable development. Oceans that hold most of the water on earth so their exploitation is also a big problem at present as the percent of potable water is very less on earth and the basic need for water is mainly fulfilled either by groundwater or by rainwater (Goel 2006). On one side, the amount of pure water is already very less, but on the other side, the inadequate distribution of industries and urbanisation also polluting the water resources. As per WHO and Bureau of Indian Standards (BIS), groundwater quality like total dissolved solids (TDS), total hardness (TH), total alkalinity (TA), presence of Sodium (Na^+), Chloride (Cl^-), Magnesium (Mg^+), and Flouride (F^-) has been affected due to high permissible limit of TDS, TH, TA, and presence of Na^+, Mg^+, Cl^-, etc.

Due to leaching, the by-product of organic wastes percolates through the soil and directly pollutes the groundwater. This leaching directly affects the groundwater quantity in terms of electrical conductivity (EC), TDS, chloride (Cl^-), sulphate (SO_4^{2-}), etc. (Nagarajan et al. 2012). Water contamination is caused by anthropogenic sources of organic compounds, marine debris and environmental plastics, metals, and metalloids, bacterial runoff, and algal toxins so there is an urgent need to design proper landfill sites to prevent groundwater contamination.

7.4.2.3 Air

Natural sources like volcanoes, forest fires, dust storms, etc., are continuously polluting urban air or from anthropogenic activities like nuclear power plants, manufacturing industries, transport, use of pesticides in agriculture fields, and waste disposal mainly responsible to decrease the air quality (UNEP 2017). Whereas in urban areas high rate of population growth, industrialisation, and traffics are the major sources to pollute the air and environment. As for air pollutants, depending on their provenance, there are two broad categories. The contaminants that are directly released into the environment are known as primary pollutants whereas when two or more primary pollutants reacted with each other then it forms secondary pollutants by their chemical reactions (Zhang and Batterman 2013). In developing countries, due to the high population rate and urbanisation, high amount of solid waste is generated which directly emits a high amount of landfill gas due to the anaerobic decomposition of solid waste. These landfill gases are CH_4 and CO_2 with some volatile organic compounds and other gases (Hegde et al. 2003) that directly pollute the air. However, due to the lack of available data, the quantity of landfill gas is not decided yet but to control its environmental impacts proper dumping site with adequate treatment of waste must be predecided.

7.5 Emerging Techniques for Waste Treatment and Ecological Restoration

Increasing pollution with overpopulation and urbanisation needs more appropriate preventive and control measures for ecological restoration of an urban environment. More emerging technologies are being used for waste treatment and safe disposal to minimise the environmental impact of urban pollution. Emerging technological are viable, economic, and more sustainable uses the most effective and innovative methods to eliminate contaminants from water, air, and soil especially by using natural chemical and biological processes. Efficient microorganisms and plants are being used to treat contaminated sites for their reclamation (Chakraborty et al. 2012; Awasthi et al. 2019).

7.5.1 Bioremediation

Bioremediation is a technique that use living organisms especially efficient plants and microorganisms to degrade or detoxify environmental contaminants into less toxic forms. In bioremediation process, various pollutants from the soil and water get degrade, remove, immobilise or detoxify by the activities of living organisms like bacteria, fungi, algae, and plants. Bioremediation process utilises the biological mechanisms and metabolite of plants and microbes to eliminate toxicants and restore the environment to its original state. As bioremediation technology is solar driven, require no energy, operate with minimum maintenance cost, it is sustainable, cost-effective, and eco-friendly technology for removal of contaminants including metals (Ojuederie and Babalola 2017; Yeung 2009; Tandon and Singh 2016). As compared to conventional treatment methods, bioremediation is effective technique and can save 50–60% of cost (Blaylock et al. 1997). Metals are a common non-biodegradable contaminants in water and soil which may be removed by using bioremediation techniques via the use of plants and microorganisms. Coelho et al. (2015) reported that bacteria and fungi are the most common microorganisms used in heavy metals removal in contaminated soil, although other microorganisms are also frequently used. Vangronsveld et al. (2009) proposed that the combination of microorganisms and plants results in the fast and more efficient removal of heavy metals in contaminated soil. Verma and Jaiswal (2016) proposed that the microorganism ability depends on the environmental condition suitability such as pH, moisture, and temperature. Ok et al. (2015) reported that biochar is the product of biomass pyrolysis collected from crop residues, and solid waste that can be used in bioremediation to activate microorganisms by making the environment more beneficial. Biochar has the capacity either abiotically or via biological pathways within its environment to send, accept or transfer electrons (Klüpfel et al. 2014; Saquing et al. 2016). Use of biochar reduced the bioavailability of metals by increasing the pH of contaminated soil (Bolan and Duraisamy 2003). Many elements, such as Cr, Se, Pb, As, Cu, and Ni, depend on the mobility and toxicity of their oxidation states, which in turn are controlled by redox mechanisms (Tandon and Singh 2016; Violante et al. 2010). The amount of polluted soil, the metal contaminant's bioavailability, and the plant's retention of metals as biomass are crucial to the success of phytoremediation as a means of plant-based eradication of heavy metals from polluted sites (Tak et al. 2013). In situ bioremediation is a method of on-site cleaning of contaminated ecosystems that include supplementing contaminated soil with nutrients to promote the degradation of contamination by microorganisms, introducing new microorganisms to the ecosystem, or improving indigenous microorganisms to use genetic engineering to degrade similar contaminants (Mani and Kumar 2014; Rayu et al. 2012). The absence of sufficient environmental conditions at contaminated sites affects the use of natural microorganisms in the ecosystem for in situ bioremediation (Lu et al. 2014; Smith et al. 2015). According to Obiri-Nyarko et al. (2014), the biological reaction is one of various pollutant removal mechanisms (degradation, sorption, and precipitation) in the permeable reactive barrier process. Frutos et al. (2010) reported that the bioventing treatment technique to be beneficial in the remediation of phenanthrene-contaminated soil and recorded 93% removal of contaminant.

7.5.2 Phytoremediation

Phytoremediation is one of the most effective metal removal technologies. The term phytoremediation comes from the Greek word Phyto (meaning a plant) and the Latin word remedy (meaning curing an evil) (Prasad 2003). Phytoremediation is an evolving technique for eliminating contaminants from the atmosphere using selected plants. Phytostabilisation,

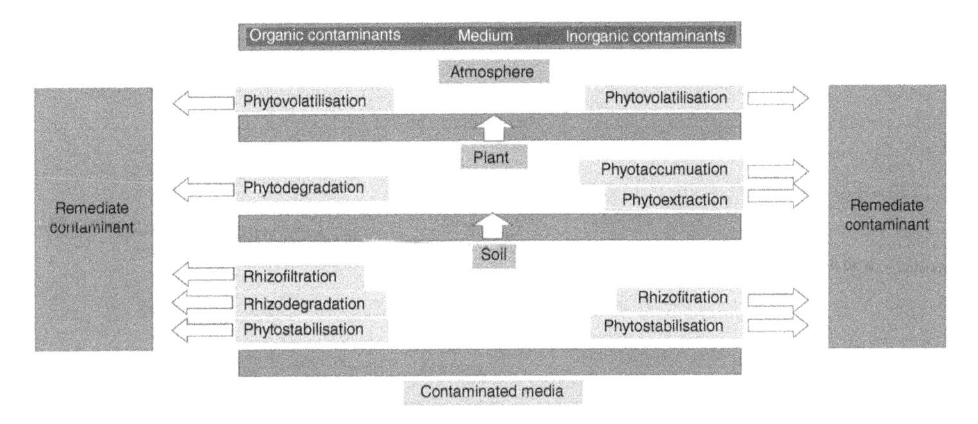

Figure 7.1 Uptake mechanisms of phytoremediation technology.

phytoextraction, phytodegradation, rhizofilteration, and rhizodegradation are some essential processes in phytoremediation technology (Prasad 2003; USEPA 2000). Pivetz (2001) reported that phytoremediation is a green remediation technology which includes the use of plant to remove the contaminants from soil, sludge, groundwater, and wastewater. The phytoremediation process is based on the different adsorption pathways shown in Figure 7.1 (Ghosh and Singh 2005; Halim et al. 2003; Yang et al. 2005).

Phytoextraction is the adsorption of metals by the plant (root) in soil and water into the aboveground (Ghosh and Singh 2005; Kotrba et al. 2009). The capacity of phytoextraction is limited required for root growth of the plant (from 30 to 90 cm). Sheoran et al. (2009) reported that the low productivity, slow growth, and root system of surface are also limited for the use for hyperaccumulator plants. Phytoextraction output is also correlated to a great extent with the seasonal environment and climatic conditions (Bhargava et al. 2012). Phytoextraction allows large-scale cleaning of soils with an irregular pattern of emissions. The use of plant biomass appears to be a major challenge. Phytoextraction must be enhanced for the realistic application and production of the heaviest metals of phytoextraction, including the investment from *ex situ* to *in situ*. Phytofilteration consists of three types, i.e. rhizofiltration (using plant roots), blasto filtration (using seedlings), and caulofiltration (using excised plant shoots) as reported by Dixit et al. (2015). Further, experiments must be performed to identify the areas of the plant that would be most effective in collecting metal contaminants so that phytofiltration may be used successfully as a phytoremediation technique. Phytostabilisation is used primarily for water, mud, and sludge remediation, also known as in-place inactivation (EPA 2000). The phytostabilisation cycle depends on a plant's capacity for tolerance to a contaminant. Phytodeposition or Phyto sequestration process is based on the ability of plants to absorb heavy metals. Islam et al. (2013) reported that the plants have low mobility of metals from roots to shoot if having wide roots. Phytovolatisation is the absorption and transpiration by a plant of a contaminant, with a discharge of the contaminant from the plant into the atmosphere. Tak et al. (2013) reported that the contaminants may be absorbed by the root, and then retained or metabolised by the plant. During phytovolatisation, the contaminants convert into volatile compounds due to plants metabolic potential in union with a microorganism that reside in the rhizosphere.

Table 7.1 Efficient plants and their metal accumulation potential.

Plant species	Metal	Metal accumulation (mg kg^{-1})	References
Aeolanthus biformifolius	Cu	13 700	Chaney et al. (2010)
Achillea millefolium	Hg	18 275	Jianxu et al. (2012)
Euphorbia cheiradenia	Pb	1138	Chehregani and Malayeri (2007)
Azolla pinnata	Cd	740	Rai (2008)
Brassica nigra	Pb	9400	Koptsik (2014)
Salix viminalis	Hg	0.66	Wang et al. (2005)

Phytodegradation is the breaking up of organic pollutants by plant enzymes into non-hazardous forms (Ali et al. 2013). Plants use different enzymes such as nitro and dehalogenases to degrade organic pollutants (Favas et al. 2014). Such enzymes must be used for the effective degradation of pollutants at optimum temperature and pH conditions. Rhizodegradation is known as the purification of polluted water because of the absorption, accumulation, and precipitation of metals by plant roots (Vishnoi et al. 2008). Rhizofiltration requires the removal by filtration of plant roots of harmful compounds or contaminants from groundwater. The rhizofiltration process is based on the mechanism of the plant's rhizospheric accumulation. Phytofiltration is also the same as this process. The only difference is that the remediation plants are grown in greenhouses, with their roots in the soil. It is the degradation of organic contaminants in water by increased microbial activity in the rhizosphere of the plants. Phytoremediation methods can also be more socially appropriate, aesthetically pleasing, and less perturbing than conventional physicochemical methods (Salido et al. 2003). The advantages of this technology are its efficiency in wastewater treatment, its low cost, its applicability to a wide variety of contaminants, and, overall, it is eco-friendly. Phytoremediation can be used to restore established hazardous sites as the cleanest and cheapest possible process (USEPA 2000). Phytoremediation is considered a new method for cleaning up contaminated water, soil, and surface air (Ginneken et al. 2007). Phytoextraction is used as an environmentally safe method of extracting metals from in situ polluted soils. Wang and Greger (2006) had used this procedure for other heavy metals in larger cleanup operations by using *Salix viminalis* and *Salix schwerinii* clone. Some plant species and their metal hyperaccumulation potential are depicted in the Table 7.1.

7.6 Mitigation and Remedial Measures for Urban Environmental Problems

Over the years, due to the overpopulation in urban areas as well as industrialisation, global environment is directly affected (Belussi and Barozzi 2015). To control the waste generation that affects the natural, economic, and social spheres differently and to provide the natural resources into their original forms some steps have been reported. To achieve the basic needs some industries like the construction industry consumes high energy and emit more pollutants, therefore, there is a need to developing and implementing more strategies to continue developmental processes with minimum cost on the environment through sustainable development. More precisely, planned urbanisation with sustainable way of development like green cities with green building, well-equipped drainage system for water use and recycling, energy balance, and management, industrial set up in separate industrial area in the cities with effective norms of waste discharge, emission and disposal, etc. Application of such strategies for urban planning would make it possible to achieve a double result: on the one hand, both from an environmental and economic point of view, a particular benefit relating to the life cycle of the product concerned and, on the other, a large-scale benefit relating to the positive impact on the environment (Akbari et al. 2001). While no widely accepted definition exists, the environmental pollution concept refers to the existence or introduction of toxic or hazardous substances in cities and urban areas. Public contamination can be from natural sources, but the most dangerous are the contaminants connected to human activities. Due to the local concentration of humans and human activities, anthropogenic sources like manufacturing processes, industries, transportation, and so on, are usually intensified in rural areas which are mainly responsible for urban pollution like global warming, waste generation through industrial activities (Martínez-Bravo and Martínez-del-Río 2019). Cities are significant contributors to pollution as there is a clear correlation between the levels of pollution and the density of population (Goel 2006; Whiteman et al. 2011). Mainly cities are responsible for pollution as well as waste generation so it is a challenge to eliminate the pollutants. Cities that consume more than 70% of the world's electricity are also responsible for 80% of GHG throughout the world (Whiteman et al. 2011; Martos et al. 2016).

7.6.1 Waste Management Practices

Due to the overpopulation and uses of new techniques, the design of cities is continuously changing, and their design expertise is rapidly expanding (Bibri and Krogstie 2017) which provides resilience that may be used to resolve issues related to pollution. 'Resilience' this term was used from an ecological point of view that can withstand disturbances and survive (Holling 1973). To solve emerging pollution problems, cities may adapt appropriate techniques to maintain urban environment. In this context, there have already been major advances in tackling urban emissions from different sources of vehicular and industrial units, which could be further promoted by intervention (UNEP 2017). Similarly, more advanced practices are available and are being used for managing solid waste.

7.6.1.1 Waste to Energy

7.6.1.1.1 Incineration

It is a thermal process to minimise the waste in which combustion of waste is done in the presence of air under controlled conditions at 850 °C (Rani et al. 2008). This is done in an enclosed chamber. The by-products like CO_2, sulphur dioxide, CO, heat, and ash, where ash contains the least amount of carbon. This process is suitable to minimise the large volume of waste. As this process is an exothermic reaction so heat can be utilised for the production of heat and electricity (Vergara and Tchobanoglous 2012). Production waste associated with the production of organochlorine has been gradually disposed of by hazardous waste incinerators in developed countries since the 1970/1980s. Due to this process, there is a major reduction in the amount of waste generation and effective waste disposal which also regulate the air pollution. However, it has also been found that after a long-time solid residue releases into the environment that leads to pollution. Similarly, a case study documented polychlorinated dibenzofurans (PCDD/F) emission from an incinerator used at Coalite Chemicals, Bolsover, UK to dispose of waste products of chlorophenol. These operations directly affect the agricultural fields (Weber et al. 2008). However, incineration process is highly vulnerable to human health due to emission and subsequent air pollution, therefore, the chamber must be far from the residential area.

7.6.1.1.2 Pyrolysis and Gasification

This process is also done for the minimisation of waste volume, in which combustion of waste is done in the deficiency of oxygen. Both processes are endothermic. Pyrolysis is the degradation of organic matter under the temperature of about 400–1000 °C in the absence of oxygen whereas gasification is done at 1000–1400 °C in the presence of oxygen for the partial combustion of organic matter (Mohan et al. 2006). Through both the processes, by-products can be reused as fuel in the internal combustion engine. For example, nickel-based catalysts were reported to be particularly successful for tar conversion in the secondary reactor at around 700–800 °C, removing approximately 98% of the tar from the produced gas (Caballero et al. 2000). Carlson et al. (2008) are reported zeolite catalyst in pyrolysis at intermediate temperature 400–600 °C.

7.6.1.2 Composting

Composting is the process of aerobic biological decomposition of organic matter to convert it into compost under controlled conditions like temperature, pressure, and humidity (Hashemimajd et al. 2004). This compost as the soil can be reused in the agricultural or horticultural field (Srivastava et al. 2015; Neher et al. 2013). The amount and quality of compost depend upon the category of waste.

7.6.1.3 Vermicomposting

It is an ecofriendly and biotechnological process where earthworms and other microorganisms are used to convert the waste product into useful by-products. In this process, earthworms play an important role as mediators for microorganisms by increasing the accessibility of surface area thus it increases the enzymatic reaction (Malley et al. 2006; Fornes et al. 2012). This compost is commonly rich in N, S, C, P thus it improves the quality of soil.

7.6.1.4 Landfilling/Dumping

Usually, the solid residues coming directly from the chemical factories are released into the landfills for disposal. Many of the dumped chemicals are permanent. Sometimes landfills and dumping have not been properly secured. That represents a high risk of release of pollutants as well as cleaning and remediating the landfills. However, some chlorinated organics can be easily released into the environment while manufacturing different pesticides. The German Environmental Agency has previously released a similar list, and the International persistent organic pollutants (POPs) Elimination Network has compiled a further list (IPEN 2004). In India, Liu (2008) reported the 91.1% of collected waste was landfilled and the remaining 8.6% incinerated and composted. Nagarajan et al. (2012) conducted studies in Erode, Tamil Nadu, India, on the impacts of landfill leachate on groundwater. They observed that groundwater near waste sites had a high EC and a high concentration of dissolved solids (Swati et al. 2018).

7.6.1.5 Disposal into Aquatic Systems

Recently, detailed studies have been published (Heinisch et al. 2004, 2006, 2007) In European rivers, monitoring techniques are used to control the POPs pollution. High concentration of chlorine POPs level found in various environmental components and reported in the Baltic Sea, the Laggi Maggiore/Italy, and the Venice lagoon (Isosaari et al. 2000, 2002a,b; Verta et al. 2007; Binelli et al. 2004; Hilscherova et al. 2003; Wilken et al. 2006). Examples of such cases include releases from the chlorophenol/chloralkali processing facility in Finland, resulting in river contamination receiving approximately 28 kg of toxic equivalent (TEQ), part of which eventually reached the Baltic Sea. In an instance, a former pesticide processing plant near Sydney, Australia, caused a wide area of the port of Sydney to become polluted, resulting in recent restrictions on commercial and recreational fishing (Birch et al. 2007; Rudge et al. 2008). In India, over 1.3 billion litres per day of domestic sewage goes directly into the river Ganga (Singh et al. 2004). Direct discharge of sewage into the river without treatment or partial treatment affect the quality of Ganga water, which contribute to 75% of total river pollution. Sewage contains metals such as Cd, Zn, Ni, Pb, Cr, Co, and Cu are also released into the river water and sediments and their persistence in the aquatic environment leads to their bioaccumulation and biomagnification into the food chain affecting the aquatic flora and fauna (Gochfeld 2003). Recently, constructed wetland (CW) have reported to effectively reduce 90% in biological oxygen demand (BOD) and improved dissolved oxygen (DO) in treated sewage by employing selected plants and suggested that CW may serve as an eco-friendly, simple, requires no chemicals and energy and an effective ecological tool in sewage treatment over other conventional methods for conservation of the rivers (Rai et al. 2013, 2015).

7.7 Conclusion

With the increased rate of the population, extensive development and urbanisation lead to a generation of more waste, emission of GHGs, changing landscape and climate, depleting resources, etc. Direct discharge of wastewater on land and water leads to contamination of aquatic and terrestrial ecosystems. Further, rapid urbanisation resulting in shrinking of forest ecosystem, habitat loss of wildlife, encroachment on lake and pond ecosystem and degradation of air and water quality, etc. The present study enlightens major changes and impacts due to human development on the urban ecology and their mitigation measures especially emerging waste management practices and sustainable treatment methods for urban ecosystem restoration. Increasing demand for energy and other requirements for growing population creating more burden on natural resources, therefore, we are using resources at a higher rate of their regeneration leads to depletion and more environmental degradation. Now depletion of groundwater, scarcity of safe drinking water, poor quality of ambient air, non-availability of landfills site, pollution of the aquatic ecosystem (Lake, Pond,

River) are being reported from cities especially from metropolitan cities across the globe due to extensive urbanisation. Therefore, to ensure the sustainable use of natural resources for fulfilling the requirements present generation and conservation of resources for future generation to meet theirs, there is an urgent need to promote more use of renewable energy resources, conservation of water and soil, using more efficient engines and appliances, using recyclable and repairable materials, conservation of habitat and biodiversity, etc. Further, better urban planning with green buildings and green cities could be helpful in the restoration of the urban ecosystem by reducing energy demand, recycling, and reuse of waste to maintain the air, water, and soil quality for better human health. Besides, emerging more efficient, sustainable and eco-friendly treatment technologies could be employed to treat and recycle the urban waste for the protection and conservation of the urban environment.

Acknowledgements

The authors are thankful to Manipal University Jaipur, India for providing facilities and continuous encouragement.

References

Akbari, H., Pomerantz, M., and Taha, H. (2001). Cool surfaces and shade trees to reduce energy use and improve air quality in urban areas. *Solar Energy* 70: 295–310.

Ali, H., Khan, E., and Sajad, M.A. (2013). Phytoremediation of heavy metals—concepts and applications. *Chemosphere* 91: 869–881.

Awasthi, A.K., Li, J., Pandey, A.K., and Khan, J. (2019). An overview of the potential of bioremediation for contaminated soil from the municipal solid waste site. In: *Emerging and Eco-friendly Approaches for Waste Management* (eds. N. Bharagava and P. Chowdhary), 59–68. Singapore: Springer.

Batayneh, A.T. (2012). Toxic (aluminum, beryllium, boron, chromium and zinc) in groundwater: health risk assessment. *International Journal of Environmental Science and Technology* 9 (1): 153–162.

Belussi, L. and Barozzi, B. (2015). Mitigation measures to contain the environmental impact of urban areas: a bibliographic review moving from the life cycle approach. *Environment Monitoring and Assessment* 187: 1–13.

Bhagat, R.B. (2011). Emerging pattern of urbanization in India. *Economic and Political Weekly* 46: 10–12.

Bhargava, A., Carmona, F.F., Bhargava, M., and Srivastava, S. (2012). Approaches for enhanced phytoextraction of heavy metals. *Journal of Environmental Management* 105: 103–120.

Bibri, S.E. and Krogstie, J. (2017). Smart sustainable cities of the future: an extensive interdisciplinary literature review. *Sustainable Cities and Society* 31: 183–212.

Binelli, A., Ricciardi, F., and Provini, F. (2004). Present status of POP contamination in Lake Maggiore (Italy). *Chemosphere* 57: 27–34.

Birch, G.F., Harrington, C., Symons, R.K., and Hunt, J.W. (2007). The source and distribution of polychlorinated dibenzo-*p*-dioxin and polychlorinated dibenzofurans in sediments of Port Jackson, Australia. *Marine Pollution Bulletien* 54: 295–308.

Blaylock, M.J., Salt, D.E., Dushenkov, S. et al. (1997). Enhanced accumulation of Pb in Indian mustard by soil-applied chelating agents. *Environmental Science and Technology* 31: 860–865.

Bolan, N.S. and Duraisamy, V.P. (2003). Role of inorganic and organic soil amendments on immobilization and phytoavailability of heavy metals: a review involving specific case studies. *Australian Journal of Soil Research* 41: 533.

Brady, N.C. and Weil, R.R. (1974). Glossary of soil science terms. In: *The Nature and Properties of Soils*, 8e (ed. N.C. Brady), 593–621. New York, NY: Macmillan.

Caballero, M.A., Corella, J., Aznar, M.P., and Gil, J. (2000). Biomass gasification with air in fluidized bed. Hot gas cleanup with selected commercial and full-size nickel-based catalysts. *Industrial & Engineering Chemistry Research* 39: 1143–1154.

Cachada, A., Rocha-Santos, T., and Duarte, A.C. (2018). Soil and pollution: an introduction to the main issues. In: *Soil Pollution* (eds. A. Duarte, A. Cachada and T. Rocha-Santos). Cambridge, MA: American Press.

Carlson, T.R., Vispute, T.P., and Huber, G.W. (2008). Green gasoline by catalytic fast pyrolysis of solid biomass derived compounds. *ChemSusChem* 1: 397–400.

Castro, M.C., Kanamori, S., Kannady, K. et al. (2010). The importance of drains for the larval development of lymphatic filariasis and malaria vectors in Dar es Salaam, United Republic of Tanzania. *PLoS Neglected Tropical Diseases* 4: 693.

Chakraborty, R., Wu, C.H., and Hazen, T.C. (2012). Systems biology approach to bioremediation. *Current Opinion in Biotechnology* 23: 483–490.

Chandrasekhar, S. and Sharma, A. (2015). Urbanization and spatial patterns of internal migration in India. *Spatial Demography* 3: 63–89.

Chaney, R.L., Broadhurst, C.L., and Centofanti, T. (2010). Phytoremediation of soil trace elements. In: *Trace Elements in Soils* (ed. P.S. Hooda), 311–352. Chichester: Wiley.

Chehregani, A. and Malayeri, B.E. (2007). Removal of heavy metals by native accumulator plants. *International Journal of Agriculture and Biology* 9: 462–465.

Coelho, L.M., Rezende, H.C., Coelho, L.M. et al. (2015). Bioremediation of polluted waters using microorganisms. In: *Advances in Bioremediation of Wastewater and Polluted Soil*, vol. 10 (ed. N. Shiomi), 60–70. Shanghai: InTech.

Deng, L., Su, Y., Su, H. et al. (2007). Sorption and desorption of lead (II) from wastewater by green algae *Cladophora fascicularis*. *Journal of Hazardous Materials* 143: 220–225.

Dixit, R., Malaviya, D., Pandiyan, K. et al. (2015). Bioremediation of heavy metals from soil and aquatic environment: an overview of principles and criteria of fundamental processes. *Sustainability* 7: 2189–2212.

EPA (2000). Wastewater technology sheet: chemical precipitation. United State Environmental Protection, EPA 832-F-00-018. http://www.epa.Gov/own/mtb/chemical_precipitation.pdf (accessed 7 July 2010).

Favas, P.J., Pratas, J., Varun, M. et al. (2014). Phytoremediation of soils contaminated with metals and metalloids at mining areas: potential of native flora. In: *Environmental Risk Assessment of Soil Contamination* (eds. C. Maria and S. Hernandez), 485–517. Shanghai, China: InTech.

Fornes, F., Mendoza-Hernández, D., García-de-la-Fuente, R. et al. (2012). Composting versus vermicomposting: a comparative study of organic matter evolution through straight and combined processes. *Bioresource Technology* 118: 296–305.

Frutos, F.J.G., Escolano, O., García, S. et al. (2010). Bioventing remediation and ecotoxicity evaluation of phenanthrene-contaminated soil. *Journal of Hazardous Materials* 183: 806–813.

Ghosh, M. and Singh, S.P. (2005). A review on phytoremediation on heavy metals and utilization of its byproducts. *Applied Ecology and Environmental Research* 3: 1–18.

Ginneken, V.L., Meers, E., Guisson, R. et al. (2007). Phytoremediation for heavy metal-contaminated soils combined with bioenergy production. *Journal of Environmental Engineering and Landscape Management* 15: 227–236.

Giusti, L. (2009). A review of waste management practices and their impact on human health. *Waste Management* 29: 2227–2239.

Gochfeld, M. (2003). Cases of mercury exposure, bioavailability, and adsorption. *Ecotoxicology and Environmental Safety* 56: 174–179.

Goel, P.K. (2006). *Water Pollution: Causes, Effects, and Control*. New Delhi: New Age International.

Gomez, E., Rani, D.A., Cheeseman, C.R. et al. (2009). Thermal plasma technology for the treatment of wastes: a critical review. *Journal of Hazardous Materials* 161: 614–626.

Gupta, S., Mohan, K., Prasad, R. et al. (1998). Solid waste management in India: options and opportunities. *Resources, Conservation and Recycling* 24: 137–154.

Gupta, N., Yadav, K.K., and Kumar, V. (2015). A review on current status of municipal solid waste management in India. *Journal of Environmental Sciences (China)* 37: 206–217.

Gutberlet, J. and Uddin, S.M.N. (2017). Household waste and health risks affecting waste pickers and the environment in low-and middle-income countries. *International Journal of Occupational and Environmental Health* 23: 299–310.

Halim, M., Conte, P., and Piccolo, A. (2003). The potential availability of heavy metals to phytoextraction from contaminated soils induced by exogenous humic sub-stances. *Chemosphere* 52: 265–275.

Hao, Y., Liu, S., Lu, Z.N. et al. (2018). The impact of environmental pollution on public health expenditure: dynamic panel analysis based on Chinese provincial data. *Environmental Science and Pollution Research* 25: 18853–18865.

Hashemimajd, K., Kalbasi, M., Golchin, A., and Shariatmadari, H. (2004). Comparison of vermicompost and composts as potting media for growth of tomatoes. *Journal of Plant Nutrition* 27: 1107–1123.

Hegde, U., Chang, T.C., and Yang, S.S. (2003). Methane and carbon dioxide emissions from Shan-Chu-Ku landfill site in northern Taiwan. *Chemosphere* 52: 1275–1285.

Heinisch, E., Kettrup, A., Bergheim, W., and Wenzel, S. (2004). Persistent chlorinated hydrocarbons, source-oriented monitoring in aquatic media. 1. Methods of data processing and evaluation. *Fresenius Environmental Bulletin* 15: 148–169.

Heinisch, E., Kettrup, A., Bergheim, W. et al. (2006). Persistent chlorinated hydrocarbons, source-oriented monitoring in aquatic media. 4. The chlorobenzenes. *Fresenius Environmental Bulletin* 15: 148–169.

Heinisch, E., Kettrup, A., Bergheim, W., and Wenzel, S. (2007). Persistent chlorinated hydrocarbons, source-oriented monitoring in aquatic media. 6. Strikingly high contaminated sites. *Fresenius Environmental Bulletin* 16: 1248–1273.

Hilscherova, K., Kannan, K., Nakata, H. et al. (2003). Polychlorinated dibenzo-*p*-dioxin and dibenzofuran concentration profiles in sediments and flood-plain soils of the Tittabawassee River, Michigan. *Environmental Science and Technology* 37: 468–474.

Holling, C. (1973). Resilience and stability of ecological systems. *Annual Review of Ecology and Systematics* 4: 1–23.
Hoornweg, D., Bhada-Tata, P., and Kennedy, C. (2013). Environment: waste production must peak this century. *Nature News* 502: 615.
IPEN (2004). Comments on the standardized toolkit for identification and quantification of dioxin and furan releases. http://www.ipen.org/ipenweb/library/4_2_dpcbw_doc_9.html (accessed 31 March 2004).
Islam, M.S., Ueno, Y., Sikder, M.T., and Kurasaki, M. (2013). Phytofiltration of arsenic and cadmium from the water environment using Micranthemumumbrosum (jf GMEL)sfblake as a hyperaccumulator. *International Journal of Phytoremediation* 15: 1010–1021.
Isosaari, P., Kohonen, T., Kiviranta, H. et al. (2000). Assessment of levels, distribution, and risks of polychlorinated dibenzo-*p*-dioxins and dibenzofurans in the vicinity of a vinyl chloride monomer production plant. *Environmental Science and Technology* 34: 2684–2689.
Isosaari, P., Kankaanpää, H., Mattila, J. et al. (2002a). Spatial distribution and temporal accumulation of polychlorinated dibenzo-*p*-dioxins, dibenzofurans, and biphenyls in the Gulf of Finland. *Environmental Science and Technology* 36: 2560–2565.
Isosaari, P., Kankaanpää, H., Mattila, J. et al. (2002b). Amounts and sources of PCDD/Fs in the Gulf of Finland. *Organohalogen Compound* 59: 195–198.
Jianxu, W., Xinbin, F., Christopher, W.N.A. et al. (2012). Remediation of mercury contaminated sites – a review. *Journal of Hazardous Materials* 221–222: 1–18.
Jin, J., Wang, Z., and Ran, S. (2006). Solid waste management in Macao: practices and challenges. *Waste Management* 26: 1045–1051.
Jorgenson, A.K. and Burns, T.J. (2007). Effects of rural and urban population dynamics and national development on deforestation in less-developed countries, 1990–2000. *Sociological Inquiry* 77: 460–482.
Joshi, R. and Ahmed, S. (2016). Status and challenges of municipal solid waste management in India: a review. *Cogent Environmental Science* 2: 1–18.
Kastenhofer, K. (2007). Converging epistemic cultures? A discussion drawing on empirical findings. *Innovations* 20: 359–373.
Kathiravale, S. and Yunus, M.M. (2008). Waste to wealth. *Asia Europe Journal* 6: 359–371.
Khalid, S., Shahid, M., Niazi, N.K. et al. (2017). A comparison of technologies for remediation of heavy metal contaminated soils. *Journal of Geochemical Exploration* 182: 247–268.
Klüpfel, L., Keiluweit, M., Kleber, M., and Sander, M. (2014). Redox properties of plant biomass-derived black carbon (biochar). *Environmental Science and Technology* 48: 5601–5611.
Koptsik, G.N. (2014). Problems and prospects concerning the phytoremediation of heavy metal polluted soils: are view. *Eurasian Soil Science* 47: 923–939.
Kotrba, P., Najmanova, J., Macek, T. et al. (2009). Genetically modified plants in phytoremediation of heavy metal and metalloid soil and sediment pollution. *Biotechnology Advances* 27: 799–810.
Kumar, S., Kumar, N., and Vivekadhish, S. (2016). Millennium development goals (MDGS) to sustainable development goals (SDGS): addressing unfinished agenda and strengthening sustainable development and partnership. *Indian Journal of Community Medicine: Official Publication of Indian Association of Preventive & Social Medicine* 41: 1.
Lebreton, L. and Andrady, A. (2019). Future scenarios of global plastic waste generation and disposal. *Palgrave Communications* 5: 1–11.

Liu, C.F., Yuan, X.Z., Zeng, G.M. et al. (2008). Prediction of methane yield at optimum pH for anaerobic digestion of organic fraction of municipal solid waste. *Bioresource Technology* 99: 882–888.

Lu, L., Huggins, T., Jin, S. et al. (2014). Microbial metabolism and community structure in response to bioelectrochemical enhanced remediation of petroleum hydrocarbon-contaminated soil. *Environmental Science and Technology* 48: 4021–4029.

Malley, C., Nair, J., and Ho, G. (2006). Impact of heavy metals on the enzymatic activity of substrate and on composting worms *Eisenia fetida*. *Bioresource Technology* 97: 1498–1502.

Mani, D. and Kumar, C. (2014). Biotechnological advances in bioremediation of heavy metals contaminated ecosystems: an overview with special reference to phytoremediation. *International Journal of Environmental Science and Technology* 11: 843–872.

Martínez-Bravo, M. and Martínez-del-Río, J. (2019). Urban pollution and emission reduction. In: *Sustainable Cities and Communities. Encyclopedia of the UN Sustainable Development Goals* (eds. W.L. Filho, A. Azul, L. Brandli, et al.). Cham: Springer.

Martos, A., Pacheco-Torres, R., Ordóñez, J., and Jadraque-Gago, E. (2016). Towards successful environmental performance of sustainable cities: intervening sectors. A review. *Renewable and Sustainable Energy Reviews* 57: 479–495.

Minghua, Z., Xiumin, F., Rovetta, A. et al. (2009). Municipal solid waste management in Pudong new area, China. *Waste Management* 29: 1227–1233.

Misra, V. and Pandey, S.D. (2005). Hazardous waste, impact on health and environment for the development of better waste management strategies in the future in India. *Environment International* 31: 417–431.

Mohan, D., Pittman, C.U. Jr., and Steele, P.H. (2006). Pyrolysis of wood/biomass for bio-oil: a critical review. *Energy & Fuels* 20: 848–889.

Muñoz, P. and Cohen, B. (2016). The making of the urban entrepreneur. *California Management Review* 59: 71–91.

Nagarajan, R., Thirumalaisamy, S., and Lakshumanan, E. (2012). Impact of leachate on groundwater pollution due to non-engineered municipal solid waste landfill sites of Erode city, Tamil Nadu, India. *Iranian Journal of Environmental Health Science and Engineering* 9: 1–12.

Neher, D.A., Weicht, T.R., Bates, S.T. et al. (2013). Changes in bacterial and fungal communities across compost recipes, preparation methods, and composting times. *PLoS One* 8: 79512.

Newman, P.W. (1999). Sustainability and cities: extending the metabolism model. *Landscape and Urban Planning* 44: 219–226.

Obiri-Nyarko, F., Grajales-Mesa, S.J., and Malina, G. (2014). An overview of permeable reactive barriers for *in situ* sustainable groundwater remediation. *Chemosphere* 111: 243–259.

Ojuederie, O.B. and Babalola, O.O. (2017). Microbial and plant-assisted bioremediation of heavy metal polluted environments: a review. *International Journal of Environmental Research and Public Health* 14: 1504.

Ok, Y.S., Uchimiya, S.M., Chang, S.X., and Bolan, N. (2015). *Biochar: Production, Characterization, and Applications*. New York, NY: CRC Press.

Patel, S. and Kasture, A. (2014). E (electronic) waste management using biological systems-overview. *International Journal of Current Microbiology and Applied Sciences* 3: 495–504.

Pivetz, B.E. (2001). Phytoremediation of contaminated soil and groundwater at hazardous waste sites. EPA Ground Water Issue, EPA/540/S-01/500.

Prasad, M.N.V. (2003). Phytoremediation of metal-polluted ecosystems: hype for commercialization. *Russian Journal of Plant Physiology* 50: 686–700.

Rai, P.K. (2008). Phytoremediation of Hg and Cd from industrial effluents using an aquatic free floating macrophyte *Azolla pinnata*. *International Journal of Phytoremediation* 10: 430–439.

Rai, U.N., Tripathi, R.D., Singh, N.K. et al. (2013). Constructed wetland as an ecotechnological tool for pollution treatment for conservation of Ganga River. *Bioresource Technology* 148: 535–541.

Rai, U.N., Upadhyay, A.K., Singh, N.K. et al. (2015). Seasonal applicability of horizontal sub-surface flow constructed wetland for trace elements and nutrient removal from urban wastes to conserve Ganga River water quality at Haridwar, India. *Ecological Engineering* 81: 115–122.

Rani, D.A., Boccaccini, A.R., Deegan, D., and Cheeseman, C.R. (2008). Air pollution control residues from waste incineration: current UK situation and assessment of alternative technologies. *Waste Management* 28: 2279–2292.

Rayu, S., Karpouzas, D.G., and Singh, B.K. (2012). Emerging technologies in bioremediation: constraints and opportunities. *Biodegradation* 23: 917–926.

Rudge, S., Staff, M., Capon, A., and Paepke, O. (2008). Serum dioxin levels in Sydney Harbour commercial fishers and family members. *Chemosphere* 73: 1692–1698.

Salido, A.L., Hasty, K.L., Lim, J.M., and Butcher, D.J. (2003). Phytoremediation of arsenic and lead in contaminated soil using Chinese Brake ferns (*Pteris vittata*) and Indian mustard (*Brassica juncea*). *International Journal of Phytoremediation* 5: 89–103.

Saquing, J.M., Yu, Y.-H., and Chiu, P.C. (2016). Wood-derived black carbon (biochar) as a microbial electron donor and acceptor. *Environmental Science and Technology Letters* 3: 62–66.

Sharholy, M., Ahmad, K., Mahmood, G., and Trivedi, R.C. (2008). Municipal solid waste management in Indian cities–a review. *Waste Management* 28: 459–467.

Sheoran, V., Sheoran, A.S., and Poonia, P. (2009). Phytomining: a review. *Minerals Engineering* 22: 1007–1019.

Singh, A. and Sharma, S. (2002). Composting of a crop residue through treatment with microorganisms and subsequent vermicomposting. *Bioresource Technology* 85: 107–111.

Singh, K.P., Mohan, D., Sinha, S., and Dalwani, R. (2004). Impact assessment of treated/untreated wastewater toxicants discharged by sewage treatment plants on health, agricultural, and environmental quality in the wastewater disposal area. *Chemosphere* 55: 227–255.

Smith, E., Thavamani, P., Ramadass, K. et al. (2015). Remediation trials for hydrocarbon-contaminated soils in arid environments: evaluation of bioslurry and biopiling techniques. *International Biodeterioration and Biodegradation* 101: 56–65.

Srivastava, V., Ismail, S.A., Singh, P., and Singh, R.P. (2015). Urban solid waste management in the developing world with emphasis on India: challenges and opportunities. *Reviews in Environmental Science and Biotechnology* 14: 317–337.

Swati, T.I.S., Vijay, V.K., and Ghosh, P. (2018). Scenario of landfilling in India: problems, challenges, and recommendations. In: *Handbook of Environmental Materials Management* (ed. C. Hussain), 1–16. Cham: Springer https://doi.org/10.1007/978-3-319-58538-3_167-1.

Swim, J.K., Clayton, S., and Howard, G.S. (2011). Human behavioral contributions to climate change: psychological and contextual drivers. *American Psychologists* 66: 251.

Tak, H.I., Ahmad, F., and Babalola, O.O. (2013). Advances in the application of plant growth-promoting rhizobacteria in phytoremediation of heavy metals. In: *Reviews of Environmental Contamination and Toxicology* (ed. D.M. Whitacre), 33–52. New York, NY: Springer.

Tandon, P.K. and Singh, S.B. (2016). Redox processes in water remediation. *Environmental Chemistry Letters* 14: 15–25.

UNEP (2017). *Towards a Pollution-Free Planet Background Report*. Nairobi: United Nations Environment Programme.

USEPA (2000). Methods for measuring the toxicity and bioaccumulation of sediment-associated contaminants with freshwater invertebrates seconded. EPA 600/R-99/064.

Vangronsveld, J., Herzig, R., Weyens, N. et al. (2009). Phytoremediation of contaminated soils and groundwater: lessons from the field. *Environmental Science and Pollution Research* 16: 765–794.

Venkatesan, G. and Swaminathan, G. (2009). Review of chloride and sulfate attenuation in groundwater nearby solid-waste landfill sites. *Journal of Environmental Engineering and Landscape Management* 17: 1–7.

Vergara, S.E. and Tchobanoglous, G. (2012). Municipal solid waste and the environment: a global perspective. *Annual Review of Environment and Resources* 37: 277–309.

Verma, J.P. and Jaiswal, D.K. (2016). Book review: advances in biodegradation and bioremediation of industrial waste. *Frontiers in Microbiology* 6: 1555.

Verta, M., Salo, S., Korhonen, M. et al. (2007). Dioxin concentrations in sediments of the Baltic Sea—a survey of existing data. *Chemosphere* 67: 1762–1775.

Violante, A., Cozzolino, V., Perelomov, L. et al. (2010). Mobility and bioavailability of heavy metals and metalloids in soil environments. *Journal of Soil Science and Plant Nutrition* 10: 268–292.

Vishnoi, S., Srivastava, P.N., and Shekhawat, N.S. (2008). Removal of color from textile effluent using cyanobacterial biomass. *Journal of Environmental Science and Engineering* 50: 93–96.

Wang, Y., Stauffer, C., and Keller, C. (2005). Changes in Hg fraction at ion in soil induced by willow. *Plant and Soil* 275: 67–75.

Wang, Y. and Greger, M. (2006). Use of iodide to enhance the phytoextraction of mercury-contaminated soil. *Science of the Total Environment* 368: 30–39.

Weber, R., Gaus, C., Tysklind, M. et al. (2008). Dioxin- and POP-contaminated sites—contemporary and future relevance and challenges. *Environmental Science and Pollution Research* 15: 363–393.

Whiteman, G., de Vos, D.R., Chapin, F.S. et al. (2011). Business strategies and the transition to low-carbon cities. *Business Strategy and the Environment* 20: 251–265.

Widmer, R., Oswald-Krapf, H., Sinha-Khetriwal, D. et al. (2005). Global perspectives on e-waste. *Environmental Impact Assessment Review* 25: 436–458.

Wilken, M., Martin, G., Lamparski, L. et al. (2006). Pattern recognition in floodplain samples. *Organohalogen Compound* 68: 22371–22374.

Wilson, D.C., Rodic, L., Modak, P. et al. (2015). *Global Waste Management Outlook*. UNEP.

Yakovleva, N., Kotilainen, J., and Toivakka, M. (2017). Reflections on the opportunities for mining companies to contribute to the United Nations Sustainable Development Goals in sub–Saharan Africa. *Extractive Industries and Society* 4: 426–433.

Yang, X., Feng, Y., He, Z., and Stoffella, P.J. (2005). Molecular mechanisms of heavy metal hyperaccumulation and phytoremediation. *Journal of Trace Elements in Medicine and Biology* 18: 339–353.

Yang, T. and Liu, W. (2018). Does air pollution affect public health and health inequality? Empirical evidence from China. *Journal of Cleaner Production* 203: 43–52.

Yeung, A.T. (2009). Remediation technologies for contaminated sites. In: *Advances in Environmental Geotechnics* (eds. Y. Chen, X. Tang and L. Zhan), 328–369. New York, NY: Springer.

Zhang, X.Y., Lin, F.F., Wong, M.T. et al. (2009). Identification of soil heavy metal sources from anthropogenic activities and pollution assessment of Fuyang County, China. *Environmental Monitoring and Assessment* 154: 439–449.

Zhang, K. and Batterman, S. (2013). Air pollution and health risks due to vehicle traffic. *Science of the Total Environment* 450: 307–316.

8

Phytoremediation of Urban Air Pollutants: Current Status and Challenges

Anina James

Department of Zoology, Deen Dayal Upadhyaya College, University of Delhi, Delhi, India

8.1 Introduction

The industrial revolution in the early nineteenth century and the use of fossil fuels was the advent of air pollution. There are numerous evidences from epidemiologic research on the adverse effects of air pollutants on human health, such as chronic obstructive pulmonary diseases, lung cancer, premature mortality (Newby et al. 2015; Saravia et al. 2013, 2014; Nawrot et al. 2011; Atkinson et al. 2001). It has also been reported that autism spectrum disorder (ASD) may be linked to exposure to particulate matter (PM) during pregnancy or early life (Raz et al. 2015; Becerra et al. 2013). Levels of PM, PM 10 (PM $\leq$ 10 μm), correlate with enhanced death rates due to harmful effects on cardiovascular and respiratory systems (Samet et al. 2000).

Most of the current mitigation strategies are merely stopgaps focusing on specific technical measures and are not sufficient to meet challenges posed by the deteriorating environment (Adedejia et al. 2020). Despite several measures undertaken, populous cities like Delhi are severely polluted throughout the year. Upgradation of combustion technology is reducing the overall emissions, and consequent exposure, but densely populated regions of the world continue to be crippled by the magnitude of emissions. Policy-makers face a tough challenge trying to balance economic development and associated negative impacts on climate. The sustainability of the energy sector and targeted cutback of high carbon footprint fuel use has been a major apprehension. In technologically advanced countries like the United Kingdom, a projection from 2002 to 2020 reported a reduction of 64% of sulphur dioxide emissions, 26% reduction of volatile organic compounds (VOCs), 19% reduction of PM (PM10), and 4% reduction of CO_2 emissions (Air Quality Expert Group, 2007). Hydrocarbons production and usage in countries with reserves are major contributors to high levels of pollution emissions in

Urban Ecology and Global Climate Change, First Edition. Edited by Rahul Bhadouria, Shweta Upadhyay, Sachchidanand Tripathi, and Pardeep Singh.

the atmosphere causing global warming and associated climatic changes such as drought, heatwaves, and floods. Coastal regions and islands are particularly prone (Adedejia et al. 2020).

Ambient air pollution is constituted of several toxicants, such as PM, inorganic pollutants (NO_x, SO_2, CO_2, O_3), and VOCs including benzene, toluene, ethylbenzene, xylene, polyaromatic hydrocarbons (PAHs), formaldehyde; many of these are both outdoor and indoor air pollutants, sometimes indoor concentrations exceeding the outdoors (Myers and Maynard 2005). Cooking (especially on gas appliances), cigarette smoking inside and usage of indoor cleaning products and synthetic materials contribute to indoor levels of and personal exposure to PM 2.5, carbon monoxide and VOCs (Myers and Maynard 2005). In industrialised nations, citizens spend 22 hours per day indoors (Klepeis et al. 2001), leading to chronic exposure to indoor air pollutants (Bernstein et al. 2008).

Phytoremediation is a technique that deploys plants, mostly together with their associated microorganisms, to mitigate sundry environmental contaminants in the air, soil, and water. We have been using plants as environment pollution mitigators since the past few decades, such as, for treating wastewater using constructed wetlands and floating plant systems, removal of vehicular pollutants at roadsides; but recent years have seen more vigour in understating the molecular and biochemical mechanisms within plants during the metabolism of various toxins (Morikawa and Erkin 2003).

Several plant enzymes have been listed by Morikawa and Erkin (2003) as potentially functional during phytoremediation. (i) Nitroreductase, type I and type II: Type I enzyme catalyses two-electron transfers yielding nitroso, amino, or hydroxylamino derivatives, irrespective of oxygen. Type II enzyme catalyses one-electron reductions, resultant products differ based on presence or absence of oxygen, such that they could be superoxide or products same as type I (Schenzle et al. 1999). (ii) Dehalogenase: two types of bacterial dehalogenases function; the aerobic one produces glycolate from perchloroethylene, and the reductive ones are involved in dehalorespiration such that electron acceptors are the halogenated compounds. (iii) Laccase is *p*-diphenol oxidase, a member of the blue copper oxidases. These enzymes have been reported in fungi, insects, higher plants, and bacteria (LaFayette et al. 1995) but fungal laccases and not plant ones, are observed to be functional in the breakdown of lignins, dioxins polychlorinated biphenyls (PCBs) (Li et al. 1999; Larsson et al. 2001). (iv) Peroxidase: two types of fungal peroxidases; lignin peroxidase (LiP) and manganese-dependent peroxidase (MnP). LiP oxidisation of nonphenolic phenylpropanoid units causes polymer fragmentation. MnP oxidisation of Mn^{2+} to Mn^{3+} could target phenolic structures in lignin and could also target nonphenolic structures via lipid peroxidation (Caramelo et al. 1999). These peroxidases partake in the breakdown of dioxins and PAHs (Bumpus et al. 1985; Takada et al. 1996). Transgenic tobacco plants bearing the engineered *Coriolus versicolor* MnP gene have also been reported (Iimura et al. 2002). (v) Nitrilase: three types of bacterial nitrilases have been reported; one that hydrolyses nitriles to the corresponding carboxylic acids and ammonia, nitrile hydratases that add water to give corresponding amides, and amidases that hydrolyse amides to corresponding carboxylic acids and ammonia (Tauber et al. 2002). *Arabidopsis* has been reported to have three nitrilase genes (*NIT1* to *NIT3*), involved in the conversion of indole-3-acetonitrile to indole-3-acetic acid, the plant hormone (Normanly et al. 1997; Eckardt 2001).

8.2 Advantages of Phytoremediation

Some of the apparent merits of phytoremediation are (i) decontamination of a broad range of pollutants; (ii) comparatively cheaper than conventional physico-chemical methods; (iii) negligible environmental disturbance; (iv) particularly useful in regions with low levels of toxins; (v) in situ remediation; (vi) renewable solar energy-driven; (vii) could further improve the conditions, such as by preventing water run-off and soil erosion; (viii) aesthetically pleasing (Gerhardt et al. 2017).

8.3 Disadvantages of Phytoremediation

Gerhardt et al. found that the number of research papers on phytoremediation has increased steadily since the early 1990s, nonetheless, the number of patents on techniques involving phytoremediation remained stagnant each year (Gerhardt et al. 2017). Phytoremediation is not the panacea of environmental pollution as it has limitations. It is a protracted process taking several growing cycles/seasons to decontaminate a site. Plants that accumulate toxic heavy metals or persistent chemicals could potentially contaminate the food web, consequently, these contaminants and their metabolites could harm other plants and animals. Hence, phytoremediation remains largely relegated to the laboratory (Beans 2017). Comprehensive understanding of the metabolic processing of toxin and their mass balance could help overcome the aforementioned impediments.

8.4 Processes Encompassing Phytoremediation

Hitherto research in phytoremediation has revealed the following basic processes involved in it: (i) *Phytostabilisation*: Some pollutants such as certain heavy metals and organic contaminants are held or contained in the root zone. This process may not degrade the pollutant but would restrict their mobility preventing diffuse contamination to other regions. (ii) *Rhizodegredation*: plant secretion of photosynthates, such as sugars, organic acids, and amino acids, to the rhizosphere (Campbell and Greaves 1990) stimulates the growth of microorganisms. These rhizospheric soil microorganisms degrade or mineralise organic contaminants such as PAHs and PCBs (Donnelly and Fletcher 1994). (iii) *Phytoacummulation/ Phytoextraction*: is the accretion of pollutants by plant roots from soil and water and their translocation to aboveground shoots. Metal hyperaccumulator plants amass more than 1.0% (Mn) or 0.1% (Co, Cu, Pb, Ni, Zn), or 0.01% (Cd) of leaf dry matter (Baker et al. 2000). Chernobyl Nuclear Power Plant accident site in Ukraine has been using such plants for remediation (Dobson et al. 1997). (iv) *Phytodegradation*: many contaminants such as total petroleum hydrocarbons (TPHs), PAHs, and PCBs, and inorganics including air pollutants nitrogen oxides and sulphur oxides can be taken up by plants and metabolised (transformed/ degraded). A low octanol–water partition coefficient (log K_{ow}) in the range of 1–3.5 determines the potential of plant cell uptake of organics (Schnoor et al. 1995). (v) *Phytovolatilisation*: volatile contaminants taken up by plants escapes through the stomata. Few pollutants have been reported to undergo phytovolatilisation, such as trichloroethylene (TCE) via poplar

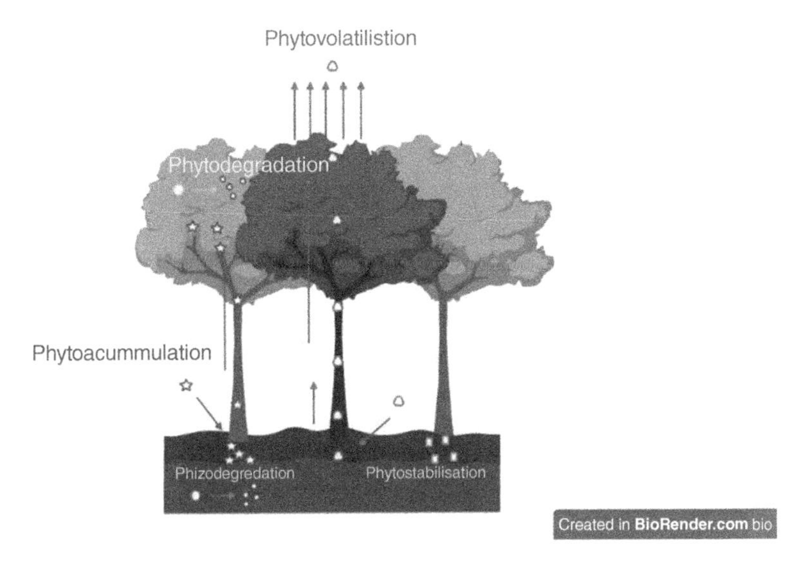

Figure 8.1 Phytoremediation encompasses several processes: pollutants such as certain heavy metals and organic contaminants are held or contained in the root zone (phytostabilisation); rhizospheric soil microorganisms can degrade or mineralise organic contaminants (rhizoderadation); accretion of pollutants by plant roots from soil and water and their translocation to aboveground shoots (phytoaccumulation); many contaminants can be taken up by plants and metabolised (phytodegradation); pollutants inside plant tissues can volatilise into the atmosphere (phytovolatilisation).

(Chappell 1998), methyl mercury via tobacco (Heaton et al. 1998) and yellow poplar (Rugh et al. 1998), methyl tertiary butyl ether (MTBE) via eucalyptus (Newman et al. 1999), and selenium via Indian mustard (de Souza et al. 2000). In the atmosphere, these released compounds may be degraded by hydroxyl radicals or remain as air pollutants. (vi) *Evapotranspiration/hydraulic control*: the release of water from plant leaves, particularly by phreatophytic trees and plants that have the ability to transpire large volumes of water, influences the water balance at the site; hence, evapotranspiration exerts hydraulic control of groundwater (Viessman et al. 1989). Mature phreatophyte trees such as poplar, *Eucalyptus* and river cedar, are deep-rooted and transpire 200–1100 litres of water per day. Hardwood trees can transpire about half the amount of water of that of a phreatophyte. This transpiration may reduce infiltration of precipitation and consequently reduce the leaching of pollutants deeper into the soil or increases transpiration of groundwater, thus eliminating the source of contamination. Hydraulic control can therefore be used to ameliorate contaminants in soil, sediment or groundwater. Figure 8.1 represents the major processes involved in phytoremediation.

8.5 Phytoremediation of Urban Air Pollutants

Plants have been reported to scavenge considerable amounts of air pollutants and to a certain extend, metabolise them (Nowak et al. 2006; Brack 2002). Despite the complexity in composition, phytoremediation is known to be environmentally friendly biotechnology that can reduce and decontaminate/degrade indoor and outdoor air pollutants.

8.5.1 Particulate Matter (PM)

The exposed surface of plants, such as leaves and bark, form a natural sink for PM, as they offer the site for gravity or wind-blown settlement of particulates (Vora and Bhatnagar 1987; Varshney and Mitra 1993). In India, there have been few preliminary studies emphasizing the PM remediating potential of plants. Shetey and Chephekar (1978) made a pollution map of Bombay using plants for biomonitoring of dust load in different localities of the city. Das et al. (1981) reported a comparative study of the dust filtering capabilities of some common Indian trees in Calcutta. Varshney and Mitra (1993) determined the particulate abatement capacity (PAC) of three commonly grown hedge species, *Bougainvillea spectabilis* Willd, *Duranta plumieri* Jacq., and *Nerium indicum* Mill in New Delhi. The PAC of the species was found in the following decreasing order *D. plumieri* > *B. spectabilis* > *N. indicum*. Further, they reported that roadside hedges trapped nearly 40% of PM arising from the vehicular exhaust. In a study carried out in the city, Lucknow by Khan et al. (1989), the dust trapping potential of ten plant species growing along the roadside was examined, and among all these species, the maximum dust load was observed on the leaves of *Nyctanthes arbortristis* L. The dust filtering potential of plant species was attributed to foliar surface characteristics. Some morphological characteristics that determine the accumulation of dust load from the ambience are the orientation of leaf on the main axis, size (leaf area in cm^2) and shape, surface nature (smooth/striate), the presence or absence of trichomes, and wax deposition (Verma 2003). Rai et al. (2013) studied the air pollution tolerance index (APTI) of six common roadside plant species (*Ficus bengalensis, Mangifera indica, B. spectabilis, Psidium guajava, Hibiscus rosa-sinensis*, and *Lantana camara*) growing along industrial (Rourkela) and non-industrial areas (Aizawl), India. The APTI was determined using data of leaf relative water content (RWC), ascorbic acid content (AA), total leaf chlorophyll (TCh), and pH of leaf extract. They reported a reduction in total chlorophyll content and pH in the leaf samples of all selected plants collected from Industrial sites (Rourkela) when compared with samples from non-industrial sites (Aizawl); however, APTI, AA, and RWC were found to be higher in the plant samples of Rourkela as compared to Aizawl. *F. bengalensis* was found to be tolerant (8.64) in industrial sites (Rourkela) and *M. indica* (7.95) in non-industrial sites (Aizawl). They concluded that *M. indica* and *B. spectabilis* may be considered as tolerant for both industrial and non-industrial) sites. Bharti et al. (2018) estimated the APTI of several trees growing at Talkatora Industrial Area, Lucknow Uttar Pradesh, India. They reported biochemical properties of trees; AA: 0.6–19.6 $mg\,g^{-1}$, RWC 41.34–98.62%, pH 4.5–8.2, and chlorophyll content 0.59–1.49 $mg\,g^{-1}$. Further, they reported that among the 25 plant species, *F. bengalensis* > *Ficus religiosa* > *Eucalyptus globus* > *Azadirachta indica* juss > *Heveabra brasiliensis* were tolerant towards air pollution. In addition, the dust capturing potential of the leaves was also determined; *Moringa oleifera* leaves showed the best dust capturing potential (5.7 $mg\,cm^{-2}$), whereas *Acacia nilotica* (0.10 $mg\,cm^{-2}$) showed the lowest potential.

On roadsides, particularly in urban areas, plant foliage can adsorb PM and the waxes on it can stabilise the pollutants (Popek et al. 2013; Saebo et al. 2012; Dzierzanowski et al. 2011). Trees are the most efficient for PM reduction because of their high leaf surface area; moreover, the complex structure of foliage causes movement of air, increasing the scavenging of PM (McDonald et al. 2007; Freer-Smith et al. 2003; Fowler et al. 1989). Herbaceous vegetations also are efficient in the removal of PM from the atmosphere (Weber et al. 2014). The reduction of PM can be improved by features of plants such as trichomes and epicuticular waxes; these waxes effectively immobilise and phytostabilise the adsorbed PM (Kaupp et al. 2000). Surprisingly, the rate of photosynthesis in some species like *Ilex rotunda* and *Sorbaria sorbifolia* was higher in more polluted areas, possibly due to the reduced photoinhibition rendered by PM and in general, better tolerance for the PM-induced oxidative stress (Takagi and Gyokusen 2004; Przybysz et al. 2014). Several laboratory-based experiments and software modellings have been done to determine the potential of urban vegetation to mitigate PM. Nowak et al. developed the i-Tree model, which is the most frequently, used to estimate the urban forest structures and the ecosystem services they provide, such as pollution removal. Studies have estimated the quantity of PM10 removed by trees in urban cities of the world; in Beijing (China) and Chicago (USA) 772 and 234 tons of PM10 were decontaminated by trees annually respectively (Yang et al. 2005; Nowak 1994), and in the United States in total, vegetation in urban locales adsorb around 215 000 tons of PM10 every year, saving the economy 969 million dollars (Nowak et al. 2006).

8.5.2 Volatile Organic Compounds (VOCs)

Among VOCs, benzene, toluene, ethylbenzene, xylene, PAHs and formaldehyde are the most used and consequently most studied; plants also generate VOCs such as chloromethane, isoprene, and monoterpenes (Harrison et al. 2012). VOCs form the major constituents of indoor air pollutants (Mendes et al. 2013; Wolkoff 2003). Reports conclude that VOCs can induce adverse health effects on humans; sensory irritation and nasopharyngeal cancer have been reported to be caused by formaldehyde, blood dyscrasias by benzene (Bluyssen et al. 2011; Jones 1999; WHO 2010). Studies show that plants remove VOCs primarily via uptake through leaf stomata particularly during daylight; some polluting gases are removed through the plant cuticle as well (Kim et al. 2008; Treesubsuntorn and Thiravetyan 2012; Yoo et al. 2006). Weyens et al suggested that crassulacean acid metabolism (CAM) and facultative CAM plants, such as species of *Sedum* genera, which open stomata during the night are beneficial for phytoremediation of air pollutants; these plants are grown extensively of urban green roofs. *Zamioculcas zamiifolia* is also a facultative CAM plant that is efficient at removing xylene from indoor air (Sriprapat and Thiravertyan 2013). Studies done to ascertain the uptake of VOCs through stomata and cuticles concluded stomatal uptake (^{14}C labelling studies) quantitatively significant than cuticular uptake (measured in the wax layer) (Sriprapat and Thiravertyan 2013; Treesubsuntorn and Thiravertyan 2012; Ugrekhelidze et al. 1997). Some reports have revealed the potential of passive botanical biofilters, such as potted plants, to phytoremediate indoor VOCs (Godish and Guindon 1989; Wolverton and Wolverton 1993; Torpy et al. 2015).

Further, the properties of VOCs dictate their access into plants, for example a lipophilic VOC has a better chance of entering the cuticle consisting of lipids (Dela Cruz et al. 2014). Some studies surmised that after cuticular uptake, VOCs are translocated to various plant parts such as seeds, roots via the phloem (Su and Liang 2013; Hanson and Roje 2001). The VOCs potentially diffuse into intercellular spaces being absorbed by water molecules forming different organic acids or react with plant tissues (Smith 1990). Inside the plant, VOCs can get degraded, stored, or excreted; for instance, formaldehyde is metabolised into 2-C skeletons serving as fuel for anabolism (Giese et al. 1994). VOCs like benzene and toluene enter the Calvin cycle and get incorporated into organic and amino acids (Ugrekhelidze et al. 1997). Korte et al. (2000) reviewed the degradation of organic air pollutants by plants. In congruence with the Green Liver Model, in which plants use metabolic processes to decontaminate toxic compounds inside their tissues, Setsungnern et al. (2017) benzene was oxidised to phenol by the cytochrome P450 monooxygenase system in plant cells, before being catalysed to catechol and followed by cleavage to produce *cis*, *cis*-muconic acid.

8.5.3 Inorganic Air Pollutants (IAP)

Majority of inorganic air pollutants (IAP) are constituted by NO_x, SO_2, CO_2, CO, and O_3. The primary oxides of nitrogen (NO_x) include nitric oxide (NO) and nitrogen dioxide (NO_2). The study of NO_2 is of importance with relevance to urban ambient air pollution. The US Environmental Protection Agency (US EPA) monitors NO_2 levels for an overall assessment of the atmospheric NO_x levels (EPA US 2005). Vehicular combustion processes are the most contributing source of NO_2 to the atmosphere. Although atmospheric level NO_2 is not toxic to humans (Samoli et al. 2006), NO_2 plays a key role in the photochemical oxidant cycle with hydroxyl radicles to generate ozone, which is detrimental to human health (Ostro et al. 2006). SO_2 is another major constituent of fossil fuel combustion. The gas can impair breathing and causes acid rain. Awareness of the polluting nature of vehicular exhausts has prompted the use of low sulphur fuels (Gheorghe and Ion 2011). Carbon dioxide (CO_2) is a major greenhouse gas released by natural processes but anthropogenic activities have increased the emissions since the industrial revolution, accelerating global warming and associated climate change, severely impacting human health and survival (National Research Council 2010; Frumkin et al. 2008). Ozone (O_3) is generated in the troposphere when UV radiation results in complex photochemical cyclical reactions with NO_x, VOCs, and CO. Many studies have shown significant correlations between concentrations of tropospheric ozone and several adverse health issues such as premature mortality, respiratory impairments, etc. (Ostro et al. 2006; Vagaggini et al. 2002; Borrego-Hernandez et al. 2014)

In a study in the Pudong district of Shanghai, China, SO_2 and NO_2 concentrations decreased by 5.3 and 2.6%, respectively in summer owing to vegetation in urban parks (Yin et al. 2011). Several studies describe the potential of plants to take up atmospheric NO_2 and utilise it for anabolism, indicative of an added benefit of supplemental plant nutrition besides ameliorating air pollution (Vallano and Sparks 2007; Takahashi et al. 2003; Segschneider et al. 1995; Wellburn 1998). Plants have the potential to assimilate the nitrogen in NO_2 to organic compounds such as amino acids (Kaji et al. 1980; Wellburn 1990). Hence, since the 1990s researchers have been attempting to genetically engineer a 'NO_2-philic plant' that can take up atmospheric NO_2 as its sole nitrogen source (Kamada et al. 1992). Morikawa et al. (1998) studied 217 plant species for their potential in NO_2 assimilation and found more than 600-fold difference between the highest (*Eucalyptus viminalis*) and the lowest-performing species (*Tillandsia ionantha* and *Tillandsia caput-medusae*). In plants, enzymes participating in the primary metabolism of nitrate, nitrate reductase (NR), nitrite reductase (NiR), and glutamine synthetase, potentially play an important role in the metabolism of NO_2-nitrogen (Morikawa et al. 2003; Ostro et al. 2006). Overexpression of NiR gene in transgenic *Arabidopsis* plants showed positive correlations for NiR gene expression and NO_2 assimilation (Takahashi et al. 2001). N_2O from vehicular exhausts in urban areas is known to be an important atmospheric greenhouse gas causing global warming and depletion of stratospheric ozone. Studies on gas–gas-converting plants that convert N_2O to N_2 holds promise. Smart and Bloom estimated that 0.02–0.2% of the $NO^-{}_3$–N assimilated by wheat was released as N_2O–N (Smart and Bloom 2001).

SO_2 enters the plants through stomata; inside the plant, the gas could be detoxified and utilised to make sulphur-containing amino acids needed for growth and development. Reddy and Dubey (2000) reported that in regions with air pollution, the plants had higher sulphate content. This could be attributed to foliar absorption of SO_2 gas and increased uptake of sulphur from the soil (Pandey and Rao 1980; Pawar 1981; Kumar and Dubey 1998). Two other studies from India also reported higher sulphate content in plant tissues growing in SO_2 polluted environments (Singh et al. 1995; Murthy et al. 1989). Agrawal and Singh (2000) in their investigation of plants near two thermal power plants in India found higher accumulation capacity in mango trees of sulphur-rich compounds. Mango trees may be considered bio-indicator species in tropical and subtropical regions because of their wide distribution and accumulation of air pollutants (Chapekar 2000). However, if the concentration of SO_2 is too high, as may be the case in several urban centres, SO_2 could become phytotoxic for plants (Gheorghe and Ion 2011).

Plants use CO_2 to generate biomass and hence act as a sink for the greenhouse gas; the humus generated from dead plants can potentially store CO_2 for thousands of years (Sinha and Singh 2010; Lehmann 2007). The phenomenon of uptake and long-term storage of atmospheric carbon dioxide, carbon sequestration (Sedjo and Sohngen 2012) may reverse the increase of CO_2 in the atmosphere provided earth still has the requisite amount of vegetation (Scheller et al. 2011). Lorenz and Lal (2014) published a review on soil organic carbon sequestration in the environment. They elaborated that soil organic carbon pool was the only terrestrial pool storing carbon which could be increased through agroforestry. Further, they estimated that 2.2 Pg carbon (1 Pg = 1015 g) could be sequestered in 50 years in agroforestry systems.

Several studies indicate that vegetation remediates urban IAP O_3 (Nowak et al. 2000; Taha 1996; Cardelino and Chameides 1990). Nowak et al. (2000) studied the effects of urban tree cover on O_3 concentrations (13–15 July 1995) from Washington, DC, to central Massachusetts through modelling data. They reported that during the daytime, the average decrease in O_3 concentrations in urban areas was greater (1 ppb) than the average increase in O_3 concentrations for the model domain (0.26 ppb). Further, they explain that the trees influence meteorology, dry deposition, VOC emissions, and anthropogenic emissions and cause alterations in dry deposition and meteorology, importantly in air temperatures, wind fields, and boundary layer heights, which affect O_3 concentrations. Bytnerowicz et al. (1999) reported a difference of 40% in O_3 concentrations above and below-forest canopies. O_3 amelioration by plants occurs by cuticle deposition and adsorption through stomatal apertures (Altimir et al. 2006). At high atmospheric moisture, O_3 reacts with waxes, salts, ions (Bytnerowicz et al. 1999; Fares et al. 2010). However, the fate of ozone inside the plant is not fully understood; most likely, biochemical changes generate reactive oxygen species (ROS) (Pell et al. 1997; Oksanen et al. 2004).

Several IAPs are known to commonly occur indoors, such as nitrogen dioxide (Lawson et al. 2011), sulphur dioxide, and ozone (Wisthaler and Weschler 2010). Yet not much research has been done on phytoremediation of indoor IAPs. Few reports indicate that the potential for plants to remove these gases through stomatal uptake has a strong possibility, but these pollutants can be detrimental to plants (Esguerra et al. 1983; Soreanu et al. 2013).

8.6 Plant–microbe Symbiosis in Phytoremediation of Urban Air Pollutants

Plants are known to be associated with thousands of microbes like bacteria and fungi. Studies on the functions of these microorganisms reveal that they aid plants with abiotic and biotic stress tolerance, nutrient and water uptake, and production of plant hormones, siderophores and inhibitory allelochemicals; during phytoremediation, the mutualistic association degrades and sequesters pollutants (Bulgarelli et al. 2013; James et al. 2018; James and Singh 2018, 2021). There is some evidence suggesting that growing plants indoor increases air humidity without the accompanying increase of harmful bacteria as is the case with electric humidifiers; this is probably due to allelochemicals released by the plants' microbiome that inhibit the growth of airborne microorganisms (Berg et al. 2014; Wolverton 2008). Wood et al. (2006) surmised that as indoor air passes over a potted plant and its substrate, air toxicants diffuse into the substrate becoming a carbon nutrient source for the plant-associated microbes. The phyllosphere bacteria residing on the surface of leaves and stems must play an important role in the adsorption of air pollutants via these surfaces. Even leaf endophytes would potentially detoxify pollutants by degradation, transformation, or sequestration. Further, air pollutants dissolved in rain falling in the rhizosphere could be decontaminated by the rhizobacteria and epiphytic root bacteria (Soreanu et al. 2013; Weyens et al. 2015). Root endophytes decrease metal phytotoxicity and enhance their accumulation in above-ground shoot via sequestration (Rajkumar et al. 2012). Phyllosphere bacteria most likely support plants to cope with stresses caused by PM and enhance phytoremediation efficiency. The phyllosphere as a microbial habitat hosts several genera of bacteria, fungi, and to a lesser extent archaea (Rajkumar et al. 2012; Vorholt 2012; Voriskova and Baldrian 2013; Baldotto and Olivares 2008; Lindow and Brandl 2003; Knief et al. 2012; Delmotte et al. 2009). Since plants naturally produce VOCs in their phyllosphere, the occurrence of VOC degrading microorganisms in the phyllosphere can be expected. Weyens et al. (2015) provided a comprehensive tabulated representation of the plant species and phyllosphere microorganisms involved in phytoremediation of both natural and man-made VOCs. Such plants and associated microorganisms hold great potential in the removal of urban indoor and outdoor air pollutants. Rhizospheric and endophytic bacteria have long been established as biodegraders of pollutants (Arslan et al. 2015; McGuinness and Dowling 2009; Barac et al. 2004). Orwell et al. (2004) described in detail the potential mechanism involved in the removal of benzene in dry as well as moist conditions of soil by the indoor plant and substrate microcosm. Besides plant-associated bacteria, mycorrhizal fungi also play a pertinent role in the mineralisation of pollutants (Gao et al. 2010; Mohsenzadeh et al. 2010; Bouwer and Zehnder 1993). Ramos et al. (2009) reported that plant-associated microorganisms produce biosurfactants, extracellular

Table 8.1 Plants and associated microbes decontaminating VOCs based on screening studies and comparison studies since the year 2015.

Plant	Microbe	Pollutant/s	References
Zamioculcas zamiifolia	Endophytic *Bacillus cereus* ERBP	Formaldehyde	Khaksar et al. (2016)
Clitoria ternatea	*Bacillus cereus* ERBP	Ethylebenzene	Daudzai et al. (2018)
Euphorbia mili	*Bacillus thuringiensis*, *Citrobacter amalonaticus* Y19, *Bacillus nealsonii*	Trimethylamine	Siswanto and Thiravetyan (2016)
Sansevieria kirkii	*Bacillus cereus* EN1	Trimethylamine	Treesubsuntorn et al. (2017)
Sorghum x drummondii	*Sphingobium* sp.	Polycyclic aromatic hydrocarbons	Dominguez et al. (2020)
Magnolia grandiflora, *Cedrus deodara*	*Hymenobacter*, *Sphingomonas*, *Methylobacterium*, *Massilia*	Polycyclic aromatic hydrocarbons	Franzettia et al. (2020)
Dracaena sanderiana	*Pantoea* sp. B11, *Staphylococcus* sp. B12	Benzene	Jindachot et al. (2018)
Syngonium podophyllum, *Sansevieria trifasciata*, *Euphorbia milii*, *Chlorophytum comosum*, *Epipremnum aureum*, *Dracaena sanderiana*, *Hedera helix*, *Clitoria ternatea*	*Enterobacter EN2*, *Cronobacter EPL1*, *Pseudomonas EPR2*, *Enterobacter EN2*	Benzene	Sriprapata and Thiravetyan (2016)
Epipremnum pinnatum cv. *Aureum*, *Davallia fejeensis* Hook	Proteobacteria	Benzene, toluene, octane, *p*-xylene, α-pinene, decane, 2-ethylhexano	Mikkonen et al. (2018)

polymeric substances, and biofilm that potentially have a role in enhancing the hydrophobicity of VOCs, making them bioavailable for the plants. Furthermore, the bioremediating activities of these microorganisms are reportedly enhanced by the presence of the plants (Xu et al. 2010, 2013; Wood et al. 2002). Table 8.1 lists plants and associated microbes decontaminating VOCs based on screening studies and comparison studies since the year 2015.

We are aware of the presence of biochemical metabolic cycles for carbon, nitrogen, and sulphur in microbes, by extension, they may be participating in the uptake and utilisation of IAP. Papen et al. (2002) demonstrated that uptake of NH_3 by spruce needles exposed to high levels of atmospheric N is the result of combined activities of the trees as well as the chemolithoautotrophic nitrifiers colonising its needles. In the case of CO_2, autotrophic microorganisms use the gas as a carbon source and, plants' microbiome partakes in humus formation and composition resulting in carbon sequestration (Langley and Hungate 2003; Clemmensen et al. 2013). Clemmensen et al. (2015) reported the significance of mycorrhizae in carbon sequestration in boreal forests. Lesaulnier et al. (2008) showed that increased CO_2 levels in the ambient air significantly affect soil microbial diversity associated with trembling aspen. Ozone is known as an antimicrobial agent. Therefore, Weyens et al. (2015) suggest that the contribution of the microbiome to ozone phytoremediation could be limited; ozone is known to generate ROS, hence bacteria with high antioxidative properties could play a role in ROS detoxification.

8.7 Transgenic Plants for Phytoremediation of Air Pollutants

Recent years have seen a keen interest in the potential use of transgenic plants to remediate air pollution. Several genes and bioactive proteins in plants and microbes facilitating the catabolism of various pollutants have been isolated and identified. Transgenic plants have been used to enhance the metabolism of VOC formaldehyde (Achkor et al. 2003; Chen et al. 2010; Tada and Kidu 2011; Xiao et al. 2011; Nian et al. 2013). One mechanism is the overexpression of dihydroxyacetone synthase and dihydroxyacetone kinase from methylotrophic yeasts in the tobacco plant's chloroplasts (Zhou et al. 2015). Similarly, overexpression of glutathione-dependent formaldehyde dehydrogenase (FALDH) from *Arabidopsis*, rice and *Chlorophytum comosum* (golden pothos) increased the uptake of formaldehyde by about 25–40% in transgenic *Arabidopsis* compared to wild-type plants (Achkor et al. 2003; Tada and Kidu 2011). James et al. (2008) reported the increased decontamination of VOCs benzene and toluene by expression of CYP2E1, a mammalian Cytochrome 450 in transgenic tobacco (*Nicotiana tabacum* cv. Xanthii). Doty et al. (2008) developed transgenic poplars overexpressing mammalian cytochrome P450 2E1 with enhanced uptake and metabolism of toxic volatile pollutants TCE, vinyl chloride, carbon tetrachloride, chloroform, and benzene.

8.8 Conclusion

Air, one of the vital prerequisites for life on Earth, is currently heavily polluted. The deterioration of air quality on a day-to-day basis is the direct result of unsustainable urbanisation and industrialisation. It has become imperative to implement air pollution abetment strategies, particularly in urban habitations. Several technologies involving chemical and physical treatment of polluted air have been applied world over, particularly in the advanced nations. However, there is no significant improvement in air quality, and additionally, these techniques are high maintenance. Reports on phytoremediation of air pollutants over the last two decades and current trends in the field hold promise. Plants and their associated microbes form a natural shield against urban air pollutants. All plants may have an inherent potential to remediate ambient air pollutants, albeit, at varying degrees. Many researchers have identified several species that can be used to decontaminate air, yet, the ever-growing levels of air pollution, particularly in metropolitans like Delhi, make it imperative to identity newer species and bioengineer them to perform better. There is an urgent need to maintain the trees present in and around urban settlements and develop green belts with suitable plant species. There is also a need to initiate the public into this endeavour via relevant awareness campaigns. Hence phytoremediation of urban air pollutants needs joint efforts of people from different walks of life, from scientists and policymakers to administrators and last but not least, the citizens. Such an approach would reduce the severity of catastrophes faced all over the world due to climate change and we may be able to curb the increased rate of global warming, saving millions of lives, and livelihoods.

References

Achkor, H., Díaz, M., Fernández, M.R. et al. (2003). Enhanced formaldehyde detoxification by overexpression of glutathione-dependent formaldehyde dehydrogenase from *Arabidopsis*. *Plant Physiology* 132: 2248–2255.

Adedejia, A.R., Zainia, F., Mathew, S. et al. (2020). Sustainable energy towards air pollution and climate change mitigation. *Journal of Environmental Management* 260: 109978.

Agrawal, M. and Singh, J. (2000). Impact of coal power plant emissions on foliar elemental concentration in plants in a law rainfall tropical region. *Environmental Monitoring and Assessment* 60: 261–282.

Air Quality Expert Group (2007). Air quality and climate change: a UK perspective. Department for Environment, Food and Rural Affairs, London.

Altimir, N., Kolari, P., Tuovinen, J.P. et al. (2006). Foliage surface ozone deposition: a role for surface moisture? *Biogeosciences* 3: 1–20.

Arslan, M., Imran, A., Khan, Q.M., and Afzal, M. (2015). Plant-bacteria partnerships for the remediation of persistent organic pollutants. *Environmental Science and Pollution Research* 24: 4322–4336.

Atkinson, R.W., Anderson, H.R., Sunyer, J. et al. (2001). Acute effects of particulate air pollution on respiratory admissions: air pollution and health: a European approach. *American Journal of Respiratory and Critical Care Medicine* 164: 1860–1866.

Baker, A.J.M., McGrath, S.P., Reeves, R.D., and Smith, J. (2000). Metal hyperaccumulator plants: a review of the ecology and physiology of a biological resource for phytoremediation of metal-polluted soils. In: *Phytoremediation of Contaminated Soils and Water* (eds. N. Terry and G.S. Banuelos), 85–107. Boca Raton, FL: CRC Press.

Baldotto, L.E.B. and Olivares, F.L. (2008). Phylloepiphytic interaction between bacteria and different plant species in a tropical agricultural system. *Canadian Journal of Microbiology* 54: 918–931.

Barac, T., Taghavi, S., Borremans, B. et al. (2004). Engineered endophytic bacteria improve phytoremediation of water-soluble, volatile, organic pollutants. *Nature Biotechnology* 22: 583–588.

Beans, C. (2017). Phytoremediation advances in the lab but lags in the field. *Proceedings of the National Academy of Sciences of the United States of America* 114 (29): 7475–7477.

Becerra, T.A., Wilhelm, M., Olsen, J. et al. (2013). Ambient air pollution and autism in Los Angeles county, California. *Environmental Health and Perspective* 121: 380–386.

Berg, G., Mahmert, A., and Moissl-Eichinger, C. (2014). Beneficial effects of plant-associated microbes on indoor microbiomes and human health? *Frontiers of Microbiology* 5: 1–5.

Bernstein, J.A., Alexis, N., Bacchus, H. et al. (2008). The health effects of nonindustrial indoor air pollution. *Journal of Allergy and Clinical Immunology* 121 (3): 585–591.

Bharti, S.K., Trivedi, A., and Kumar, N. (2018). Air pollution tolerance index of plants growing near an industrial site. *Urban Climate* 24: 820–829.

Bluyssen, P.M., Janssen, S., van den Brink, L.H., and de Kluizenaar, Y. (2011). Assessment of wellbeing in an indoor office environment. *Building and Environment* 46: 2632–2640.

Borrego-Hernandez, O., Garcia-Reynoso, J.A., Ojeda-Ramirez, M.M., and Suarez-Lastra, M. (2014). Retrospective health impact assessment for ozone pollution in Mexico city from 1991 to 2011. *Atmosfera* 27: 261–271.

Bouwer, E.J. and Zehnder, A.J.B. (1993). Bioremediation of organic compounds-putting microbial metabolisms to work. *Trends in Biotechnology* 11: 360–367.

Brack, C.L. (2002). Pollution mitigation and carbon sequestration by an urban forest. *Environmental Pollution* 116: 195–200.

Bulgarelli, D., Schlaeppi, K., Spaepen, S. et al. (2013). Structure and functions of the bacterial microbiota of plants. *Annual Review of Plant Biology* 64: 807–838.

Bumpus, J.A., Tien, M., Wright, D., and Aust, S.D. (1985). Oxidation of persistent environmental pollutants by a white rot fungus. *Science* 228: 1434–1436.

Bytnerowicz, A., Fenn, M.E., Miller, P.R., and Arbaugh, M.J. (1999). Wet and dry pollutant deposition to the mixed conifer forest. In: *Oxidant air pollution impacts in the montane forests of southern California: a case study of the San Bernardino mountains* (eds. P.R. Miller and J.R. McBride), 235–369. New York: Springer.

Campbell, R. and Greaves, M.P. (1990). Anatomy and community structure of the rhizosphere. In: *The Rhizosphere* (ed. J.M. Lynch). West Sussex, England: Wiley.

Caramelo, L., Martinez, M.J., and Martinez, A.T. (1999). A search for ligninolytic peroxidases in the fungus *Pleurotus eryngi* involving a-keto-thiomethylbutyric acid and lignin model dimers. *Applied and Environmental Microbiology* 65: 916–922.

Cardelino, C.A. and Chameides, W.L. (1990). Natural hydrocarbons, urbanization, and urban ozone. *Journal of Geophysical Research* 95: 13971–13979.

Chapekar, S.B. (2000). Phytomonitoring in industrial areas. In: *Environmental Pollution and Plant Responses* (eds. S.B. Agrawal and M. Agarwal), 329–342. Boca Raton, FL: Lewis Publishers.

Chappell, J. (1998). Phytoremediation of TCE in groundwater using *Populus*. Status report prepared for USEPA, Technology Innovation Office.

Chen, L., Yurimoto, H., Li, K. et al. (2010). Assimilation of formaldehyde in transgenic plants due to the introduction of the bacterial ribulose monophosphate pathway genes. *Bioscience, Biotechnology, and Biochemistry* 74: 627–635.

Clemmensen, K.E., Bahr, A., Ovaskainen, O. et al. (2013). Roots and associated fungi drive long-term carbon sequestration in boreal forest. *Science* 339: 1615–1618.

Clemmensen, K.E., Finlay, R.D., Dahlberg, A. et al. (2015). Carbon sequestration is related to mycorrhizal fungal community shifts during long-term succession in boreal forests. *New Phytologist* 205: 1525–1536.

Das, T.M., Bhaumik, A., and Chakravarty, A. (1981). Trees as dust filters. *Science Today* 15 (12): 19–21.

Daudzai, Z., Treesubsuntorn, C., and Thiravetyan, P. (2018). Inoculated *Clitoria ternatea* with *Bacillus cereus* ERBP for enhancing gaseous ethylbenzene phytoremediation: plant metabolites and expression of ethylbenzene degradation genes. *Ecotoxicology and Environmental Safety* 164: 50–60.

de Souza, M.P., Lytle, C.M., Mulholland, M.M. et al. (2000). Selenium assimilation and volatilization from dimethylselenoniopropionate by Indian mustard. *Plant Physiology* 122: 1281–1288.

Dela Cruz, M., Christensen, J.H., Thomsen, J.D., and Müller, R. (2014). Can ornamental potted plants remove volatile organic compounds from indoor air? — a review. *Environmental Science and Pollution Research* 21: 13909–13928.

Delmotte, N., Knief, C., Chaffron, S. et al. (2009). Community proteogenomics reveals insights into the physiology of phyllosphere bacteria. *Proceedings of the National Academy Sciences of the United States of America* 106: 16428–16433.

Dobson, A.P., Bradshaw, A.D., and Baker, A.J.M. (1997). Hopes for the future: restoration ecology and conservation biology. *Science* 277: 515–522.

Dominguez, J.J.J., Inoue, C., and Chien, M.-F. (2020). Hydroponic approach to assess rhizodegradation by sudangrass (*Sorghum x drummondii*) reveals pH- and plant age-dependent variability in bacterial degradation of polycyclic aromatic hydrocarbons (PAHs). *Journal of Hazardous Materials* 387: 121695.

Donnelly, P.K. and Fletcher, J.S. (1994). Potential use of mycorrhizal fungi as bioremediation agents. In: *Bioremediation through Rhizosphere Technology*. ACS Series, vol. 563 (eds. T.A. Anderson and J.R. Coats), 93–99. Washington, DC: American Chemical Society.

Doty, S.L., James, C.A., Moore, A.L., and Vajzovic, A. (2008). Enhanced phytoremediation of volatile environmental pollutants with transgenic trees. *Proceedings of the National Academy of Sciences of the United States of America* 104 (5): 16816–16821.

Dzierzanowski, K., Popek, R., Gawronska, H. et al. (2011). Deposition of particulate matter of different size fractions on leaf surface and waxes of urban forest species. *International Journal of Phytoremediation* 13: 1037–1046.

Eckardt, N.A. (2001). New insights into auxin biosynthesis. *Plant Cell* 13: 1–3.

Environmental Protection Agency, U.S. (2005). National ambient air monitoring strategy, Office of Air Quality Planning and Standards. http://www.epa.gov/ttn/amtic/files/ambient/monitorstrat/ (accessed 2 January 2021).

Esguerra, C., Santiago, E., Aquino, N., and Ramos, M. (1983). The uptake of SO_2 and NO_2 by plants. *Science Diliman* 2: 45–56.

Fares, S., Park, J.-H., Ormeno, E. et al. (2010). Ozone uptake by citrus trees exposed to a range of ozone concentrations. *Atmospheric Environment* 44: 3404–3412.

Fowler, D., Cape, J.N., and Unsworth, M.H. (1989). Deposition of atmospheric pollutants on forests. *Philosophical Transactions of the Royal Society* 324: 247–265.

Franzettia, A., Gandolfi, I., Bestettia, G. et al. (2020). Plant-microorganisms interaction promotes removal of air pollutants in Milan (Italy) urban area. *Journal of Hazardous Materials* 384: 121021.

Freer-Smith, P.H., El-Khatib, A.A., and Taylor, G. (2003). Capture of particulate pollution by trees: a comparison of species typical of semi-arid areas (*Ficus nitida* and *Eucalyptus globulus*) with European and north American species. *Water Air and Soil Pollution* 155: 173–187.

Frumkin, H., Hess, J., Luber, G. et al. (2008). Climate change: the public health response. *American Journal of Public Health* 98: 435–445.

Gao, Y., Cheng, Z., Ling, W., and Huang, J. (2010). Arbuscular mycorrhizal fungal hyphae contribute to the uptake of polycyclic aromatic hydrocarbons by plant roots. *Bioresource Technology* 101: 6895–6901.

Gerhardt, K.E., Gerwing, P.D., and Greenberg, B.M. (2017). Opinion: taking phytoremediation from proven technology to accepted practice. *Plant Science* 256: 170–185.

Gheorghe, I.F. and Ion, B. (2011). The effects of air pollutants on vegetation and the role of vegetation in reducing atmospheric pollution. In: *The Impact of Air Pollution on Health, Economy, Environment and Agricultural Sources* (ed. M. Khallaf), 242–280. London, UK: Intech Open http://cdn.intechopen.com/pdfs-wm/18642.pdf.

Giese, M., Baue-Doranth, U., Langebartels, C., and Sandermann, H. (1994). Detoxification of formaldehyde by the spider plant (*Chlorophytum comosum* L.) and by soybean (*Glycine max* L.) cell-suspension cultures. *Plant Physiology* 104: 1301–1309.

Godish, T. and Guindon, C. (1989). An assessment of botanical air purification as a formaldehyde mitigation measure under dynamic laboratory chamber conditions. *Environmental pollution* 62 (1): 13–20.

Hanson, A.D. and Roje, S. (2001). One-carbon metabolism in higher plants. *Annual Review of Plant Physiology and Plant Molecular Biology* 52: 119–137.

Harrison, S.P., Morfopoulos, C., Dani, K.G.S. et al. (2012). Volatile isoprenoid emissions from plastid to planet. *New Phytologist* 197: 49–57.

Heaton, C.P., Rugh, C.L., Wang, N.-J., and Meagher, R.B. (1998). Phytoremediation of mercury- and methylmercury-polluted soils using genetically engineered plants. *Journal of Soil Contamination* 7: 497–509.

Iimura, Y., Ikeda, S., Sonoki, T. et al. (2002). Expression of a gene for Mn-peroxidase from *Coriolus versicolor* in transgenic tobacco generates potential tools for phytoremediation. *Applied Microbiology and Biotechnology* 59: 246–251.

James, C.A., Xin, G., Doty, S.L., and Strand, S.E. (2008). Degradation of low molecular weight volatile organic compounds by plants genetically modified with mammalian cytochrome P450 2E1. *Environmental Science and Technology* 42 (1): 289–293.

James, A. and Singh, D.K. (2018). Assessment of atrazine decontamination by epiphytic root bacteria isolated from emergent hydrophytes. *Annals of Microbiology* 68: 953–962.

James, A., Singh, D.K., and Khankhane, P.J. (2018). Enhanced atrazine removal by hydrophyte-bacterium associations and *in vitro* screening of the isolates for their plant growth promoting potential. *International Journal of Phytoremediation* 20 (2): 89–97.

James, A. and Singh, D.K. (2021). Atrazine detoxification by intracellular crude enzyme extracts derived from epiphytic root bacteria associated with emergent hydrophytes. *Journal of Environmental Science and Health, Part B* 56: 577–586. https://doi.org/10.1080/03601234.2021.1922043.

Jindachot, W., Treesubsuntorn, C., and Thiravetyan, P. (2018). Effect of individual/co-culture of native phyllosphere organisms to enhance *Dracaena sanderiana* for benzene phytoremediation. *Water Air Soil Pollution* 229: 80.

Jones, A.P. (1999). Indoor air quality and health. *Atmospheric Environment* 33: 4535–4564.

Kaji, M., Yoneyama, T., Totsuka, T., and Iwaki, H. (1980). Absorption of atmospheric NO_2 by plants and soils. *Soil Science and Plant Nutrition* 26 (1): 1–7. https://doi.org/10.1080/00380768.1980.10433207.

Kamada, M., Higaki, A., Jin, Y. et al. (1992). Transgenic "air-pollutant-philic plants" produced by particle bombardment. In: *Research in Photosynthesis*, vol. IV (ed. N. Murata), 83–86. Dordrecht: Kluwer Academic Publishers.
Kaupp, H., Blumenstock, M., and McLachan, M.S. (2000). Retention and mobility of atmospheric particle-associated organic pollutant PCDD/Fs and PAHs in maize leaves. *New Phytologist* 148: 473–480.
Khaksar, G., Treesubsuntorn, C., and Thiravetyan, P. (2016). Endophytic *Bacillus cereus* ERBP—*Clitoria ternatea* interactions: potentials for the enhancement of gaseous formaldehyde removal. *Environmental and Experimental Botany* 126: 10–20.
Khan, A.M., Pandey, V., Yunus, M., and Ahmad, K.J. (1989). Plants as dust scavengers. A case study. *The Indian Foresters* 115 (9): 670–672.
Kim, K.J., Kil, M.J., Song, J.S., and Yoo, E.H. (2008). Efficiency of volatile formaldehyde removal by indoor plants: contribution of aerial plant parts versus the root zone. *Hortscience* 133: 521–526.
Klepeis, N.E., Nelson, W.C., Ott, W.R. et al. (2001). The National Human Activity Pattern Survey (NHAPS): a resource for assessing exposure to environmental pollutants. *Journal of Exposure Science and Environmental Epidemiology* 11 (3): 231.
Knief, C., Delmotte, N., Chaffron, S. et al. (2012). Metaproteogenomic analysis of microbial communities in the phyllosphere and rhizosphere of rice. *ISME Journal* 6: 1378–1390.
Korte, F., Kvesitadze, G., Ugrekhelidze, D. et al. (2000). Organic toxicants and plants. *Ecotoxicology and Environmental Safety* 47: 1–26.
Kumar, G.S. and Dubey, P.S. (1998). Differential response and detoxifying mechanism of *Cassia siamea* Lam. and *Dalbergia sissoo* Roxb. of different ages to SO_2 treatment. *Journal of Environmental Biology* 9 (3): 243–249.
LaFayette, P.R., Eriksson, K.E., and Dean, J.F. (1995). Nucleotide sequence of a cDNA clone encoding an acidic laccase from sycamore maple (*Acer pseudoplatanus* L.). *Plant Physiolology* 107: 667–668.
Langley, J.A. and Hungate, B.A. (2003). Mycorrhizal controls on belowground litter quality. *Ecology* 84: 2302–2312.
Larsson, S., Cassland, P., and Jonsson, L.J. (2001). Development of a *Saccharomyces cerevisiae* strain with enhanced resistance to phenolic fermentation inhibitors in lignocellulose hydrolysates by heterologous expression of laccase. *Applied and Environmental Microbiology* 67: 1163–1170.
Lawson, S.J., Galbally, I.E., Powell, J.C. et al. (2011). The effect of proximity to major roads on indoor air quality in typical Australian dwellings. *Atmospheric Environment* 45 (13): 2252–2259.
Lehmann, J. (2007). A handful of carbon. *Nature* 447: 143–144.
Lesaulnier, C., Papamichail, D., McCorkle, S. et al. (2008). Elevated atmospheric CO_2 affects soil microbial diversity associated with trembling aspen. *Environmental Microbiology* 10: 926–941.
Li, K., Xu, F., and Eriksson, K.E.L. (1999). Comparison of fungal laccases and redox mediators in oxidation of a nonphenolic lignin model compound. *Applied and Environmental Microbiology* 65: 2654–2660.
Lindow, S.E. and Brandl, M.T. (2003). Microbiology of the phyllosphere. *Applied Environmental Microbiology* 69: 1875–1883.

Lorenz, K. and Lal, R. (2014). Soil organic carbon sequestration in agroforestry systems: a review. *Agronomy for Sustainable Development* 34: 443–454.

McDonald, A.G., Bealey, W.J., Fowler, D. et al. (2007). Quantifying the effect of urban tree planting on concentrations and depositions of PM10 in two UK conurbations. *Atmospheric Environment* 41: 8455–8467.

McGuinness, M. and Dowling, D. (2009). Plant-associated bacterial degradation of toxic organic compounds in soil. *International Journal of Environmental Research and Public Health* 6: 2226–2247.

Mendes, A., Pereira, C., Mendes, D. et al. (2013). Indoor air quality and thermal comfort-results of a pilot study in eldery care centers in Portugal. *Journal of Toxicology and Environmental Health Part A* 76: 333–344.

Mikkonen, A., Li, T., Vesala, M. et al. (2018). Biofiltration of airborne VOCs with green wall systems - microbial and chemical dynamics. *International Journal of Indoor Environment and Health* 28 (5): 697–707.

Mohsenzadeh, F., Nasseri, S., Mesdaghinia, A. et al. (2010). Phytoremediation of petroleum-polluted soils: application of *Polygonum aviculare* and its root-associated (penetrated) fungal strains for bioremediation of petroleum-polluted soils. *Ecotoxicology and Environmental Safety* 73: 613–619.

Morikawa, H., Higaki, A., Nohno, M. et al. (1998). More than a 600-fold variation in nitrogen dioxide assimilation among 217 planta taxa. *Plant, Cell & Environment* 21: 180–190.

Morikawa, H. and Erkin, O.C. (2003). Basic processes in phytoremediation and some applications to air pollution control. *Chemosphere* 52: 1553–1558.

Morikawa, H., Takahashi, M., and Kawamura, Y. (2003). Air pollution clean-up using pollutant-philic plants-metabolism of nitrogen dioxide and genetic manipulation of related genes. In: *Phytoremediation: Transformation and Control of Contaminants* (eds. S.C. McCutcheon and J.L. Schnoor). Wiley.

Murthy, M.S.H., Raza, S.H., and Adeel, A. (1989). A new method in evaluation of SO_2 tolerance of certain trees. Air pollution & forest decline. *Proceedings in 14th International Meeting for Specialists in Air Pollution Effects on Forest. Ecosystem, IUFRO* P_2O_5, Interlaken, Switzerland (2–8 October). (ed. J.B. Bucher, I. Bucher, and N. Wall).Birmensdorf, pp. 486–488.

Myers, I. and Maynard, R.L. (2005). Polluted air—Outdoors and indoors. *Occupational Medicine* 55: 432–438.

National Research Council (NRC) (2010). *Advancing the Science of Climate Change.* Washington, DC: The National Academies Press.

Nawrot, T.S., Perez, L., Künzli, N. et al. (2011). Public health importance of triggers of myocardial infarction: a comparative risk assessment. *Lancet* 377: 732–740.

Newby, D.E., Mannucci, P.M., Tell, G.S. et al. (2015). Expert position paper on air pollution and cardiovascular disease. *European Heart Journal* 36: 83–93.

Newman, L., Gordon, M.P., Heilman, P. et al. (1999). Phytoremediation of MTBE at a California naval site. *Soil and Groundwater Cleanup* Feb–Mar: 42–45.

Nian, H.J., Meng, Q.C., Cheng, Q. et al. (2013). The effects of overexpression of formaldehyde dehydrogenase gene from *Brevibacillus brevis* on the physiological characteristics of tobacco under formaldehyde stress. *Russian Journal of Plant Physiology* 60: 764–769.

Normanly, J., Grisafi, P., Fink, G.R., and Bartel, B. (1997). *Arabidopsis* mutants resistant to the auxin effects of indole-3-acetonitrile are defective in the nitrilase encoded by the *NIT1* gene. *Plant Cell* 9: 1781–1790.

Nowak, D.J. (1994). Air pollution removal by Chicago's urban forest. http://www.nrs.fs.fed.us/pubs/gtr/gtr_ne186.pdf (accessed 3 January 2021).

Nowak, D.J., Civerloo, K.L., Rao, S.T. et al. (2000). A modeling study of the impact of urban trees on ozone. *Atmospheric Environment* 34: 1610–1613.

Nowak, D.J., Crane, D.E., and Stevens, J.C. (2006). Air pollution removal by urban trees and shrubs in the United States. *Urban Forestry and Urban Greening* 4: 115–123.

Oksanen, E., Häikiöl, E., Sober, J., and Karnosky, D.F. (2004). Ozone-induced H_2O_2 accumulation in field-grown aspen and birch is linked to foliar ultrastructure and peroxisomal activity. *New Phytologist* 161: 791–799.

Orwell, R.L., Wood, R.L., Terran, J. et al. (2004). Removal of benzene by the indoor plant/substrate microcosm and implication to air quality. *Water Air Soil Pollution* 157: 193–207.

Ostro, B.D., Tran, H., and Levy, J.I. (2006). The health benefits of reduced tropospheric ozone in California. *Journal of Air Waste Management Association* 56: 1007–1021.

Pandey, S.N. and Rao, D.N. (1980). Effect of coal smoke sulfur dioxide pollution on the accumulation of certain minerals and chlorophyll content of wheat plant. *Tropical Ecology* 19 (2): 155–162.

Papen, H., Gebler, A., Zumbusch, E., and Rennenberg, H. (2002). Chemolithoautotrophic nitrifiers in the phyllosphere of a spruce ecosystem receiving high atmospheric nitrogen input. *Current Microbiology* 44: 56–60.

Pawar, K. (1981). Pollution studies in Nagda area due to Birla Industrial Discharges. PhD thesis. School of Studies in Botany, Vikram University, Ujjain, M.P., India.

Pell, E.J., Schlagnhaufer, C.D., and Arteca, R.N. (1997). Ozone-induced oxidative stress: mechanisms of action and reaction. *Physiologia Plantarum* 100: 264–273.

Popek, R., Gawronska, H., Wrochna, M. et al. (2013). Paticulate matter on foliage of 13 woody species: deposition on surfaces and phytostabilization in waxes—a 3-year study. *International Journal of Phytoremediation* 15: 245–256.

Przybysz, A., Popek, R., Gawronska, H. et al. (2014). Efficiency of photosynthetic apparatus of plants grown in sites differing in level of particulate matter. *ACTA Scientiarum Polonorum Horticulture* 13: 17–30.

Rai, P.K., Panda, L.L.S., Chutia, B.M., and Singh, M.M. (2013). Comparative assessment of air pollution tolerance index (APTI) in the industrial (Rourkela) and non-industrial area (Aizawl) of India: an ecomanagement approach. *African Journal of Environmental Science and Technology* 7 (10): 944–948.

Rajkumar, M., Sandhya, S., Prasad, M.N.V., and Freitas, H. (2012). Perspectives of plant-associated microbes in heavy metal phytoremediation. *Biotechnology Advances* 6: 1562–1574.

Ramos, J.L., Molina, L., and Segura, A. (2009). Removal of organic toxic chemicals in the rhizosphere and phyllosphere of plants. *Microbial Biotechnology* 2: 144–146.

Raz, R., Roberts, A.L., Lyall, K. et al. (2015). Autism spectrum disorder and particulate matter air pollution before, during, and after pregnancy: a nested case-control analysis within the nurses' health study II cohort. *Environmental Health and Perspective* 123: 264–270.

Reddy, B.M. and Dubey, P.S. (2000). Scavenging potential of trees to SO_2 and NO_2 under experimental condition. *International Journal of Ecology and Environmental Sciences* 26: 99–106.

Rugh, C.L., Senecoff, J.F., Meagher, R.B., and Merkle, S.A. (1998). Development of transgenic yellow poplar for mercury phytoremediation. *Nature Biotechnology* 16: 925–928.

Saebo, A., Popek, R., Nawrot, B. et al. (2012). Plant species differences in particulate matter accumulation on leaf surfaces. *Science of Total Environment* 427–428: 347–354.

Samet, J.M., Dominici, F., Curriero, F.C. et al. (2000). Fine particulate air pollution and mortality in 20 US cities, 1987–1994. *The New England Journal of Medicine* 343: 1742–1749.

Samoli, E., Aga, E., Touloumi, G. et al. (2006). Short-term effects of nitrogen dioxide on mortality: an analysis within the APHEA project. *European Respiratory Journal* 27: 1129–1137.

Saravia, J., Lee, G.I., Lomnicki, S. et al. (2013). Particulate matter containing environmentally persistant free radicals and adverse infant respiratory health effects: a review. *Journal of Biochemical and Molecular Toxicology* 27: 56–68.

Saravia, J., You, D., Thevenot, P. et al. (2014). Early-life exposure to combustion-derived particulate matter causes pulmonary immunosuppression. *Mucosal Immunology* 7: 694–704.

Scheller, R.M., van Tuyl, S., Clark, K.L. et al. (2011). Carbon sequestration in New Jersey Pine Barrens under different scenarios of fire management. *Ecosystems* 14: 987–1004.

Schenzle, A., Lenke, H., Spain, J.C., and Knackmuss, H.-J. (1999). Chemoselective nitro group reduction and reductive dechlorination initiate degradation of 2-chloro-5-nitrophenol by *Ralstonia eutropha* JMP134. *Applied and Environmental Microbiology* 65: 2317–2323.

Schnoor, J.L., Licht, L.A., McCutcheon, S.C. et al. (1995). Phytoremediation of organic and nutrient contaminants. *Environmental Science and Technology* 29: 318A–323A.

Sedjo, R. and Sohngen, B. (2012). Carbon sequestration in forests and soils. *Annual Review of Resource Economics* 4: 127–144.

Segschneider, H., Wildt, J., and Forstel, H. (1995). Uptake of $^{15}NO_2$ by sunflower (*Helianthus anuus*) during exposures in light and darkness: quantities, relationship to stomatal aperture and incorporation into different nitrogen pools within the plant. *New Phytologist* 131: 109–119.

Setsungnern, A., Treesubsuntorn, C., and Thiravetyan, P. (2017). The influence of different light quality and benzene on gene expression and benzene degradation of *Chlorophytum comosum*. *Plant Physiology and Biochemistry* 120: 95–102.

Shetey, R.P. and Chephekar, S.B. (1978). Some estimations dust fall in the city of Bombay using plants. *Proceedings in Seminar on Recent Advances in Ecology*, New Delhi. pp. 61–70.

Singh, N., Yunus, M., Srivastava, K. et al. (1995). Monitoring of auto exhaust pollution by road side plants. *Environment Monitoring Assessment* 34: 13–25.

Sinha, R.K. and Singh, S. (2010). Plants combating air pollution. In: *Green Plants and Pollution: Nature's Technology for Abating and Combating Environmental Pollution (Air, Water and Soil Pollution Science and Technology)* (ed. R.K. Sinha). Palo Alto, CA: Nova Science Publishers.

Siswanto, D. and Thiravetyan, P. (2016). Improvement of trimethylamine uptake by *Euphorbia milii*: effect of inoculated bacteria. *Journal of Tropical Life Science* 6 (2): 123–130.

Smart, D.R. and Bloom, A.J. (2001). Wheat leaves emit nitrous oxide during nitrate assimilation. *Proceedings of the National Academy of Sciences of the United States of America* 98: 7875–7878.

Smith, W.H. (1990). *Air Pollution and Forests*. New York: Springer.

Soreanu, G., Dixon, M., and Darlington, A. (2013). Botanical biofiltration of indoor gaseous pollutants–a mini-review. *Chemical Engineering Journal* 229: 585–594.

Sriprapat, W. and Thiravertyan, P. (2013). Phytoremediation of BTEX from indoor air by *Zamioculcas zamiifolia*. *Water Air and Soil Pollution* 224: 1482.

Sriprapata, W. and Thiravetyan, P. (2016). Efficacy of ornamental plants for benzene removal from contaminated air and water: effect of plant associated bacteria. *International Biodeterioration & Biodegradation* 113: 262–268.

Su, Y.H. and Liang, Y.C. (2013). The foliar uptake and downward translocation of trichloroethylene and 1,2,3-trichlorobenzene in air-plant-water systems. *Journal of Hazardous Material* 252–253: 300–305.

Taha, H. (1996). Modeling impacts of increased urban vegetation on ozone air quality in the South Coast Air Basin. *Atmospheric Environment* 30: 3423–3430.

Tada, Y. and Kidu, Y. (2011). Glutathione-dependent formaldehyde dehydrogenase from golden pothos (*Epipremnum aureum*) and the production of formaldehyde detoxifying plants. *Plant Biotechnology* 28: 373–378.

Takada, S., Nakamura, M., Matsueda, T. et al. (1996). Degradation of polychlorinated dibenzo-*p*-dioxins and polychlorinated dibenzofurans by the white rot fungus *Phanerochaetesordida* YK-624. *Applied and Environmental Microbiology* 62: 4323–4328.

Takagi, M. and Gyokusen, K. (2004). Light and atmospheric pollution affect photosynthesis of street trees in urban environments. *Urban Forestry and Urban Greening* 2: 167–171.

Takahashi, M., Sasaki, Y., Ida, S., and Morikawa, H. (2001). Nitrite reductase gene enrichment improves assimilation of nitrogen dioxide in *Arabidopsis*. *Plant Physiology* 126: 731–741.

Takahashi, M., Kondo, K., and Morikawa, H. (2003). Assimilation of nitrogen dioxide in selected plant taxa. *Acta Biotechnology* 23: 241–247.

Tauber, M.M., Cavaco-Paulo, A., Robra, K.-H., and Gubitz, G.M. (2002). Nitrile hydratase and amidase from *Rhodococcus rhodochrous* hydrolyze acrylic fibers and granular polyacrylonitriles. *Applied Microbiology and Biotechnology* 66: 1634–1638.

Torpy, F.R., Irga, P.J., and Burchett, M.D. (2015). Reducing indoor air pollutants through biotechnology. In: *Biotechnologies and Biomimetics for Civil Engineering* (eds. F. Pacheco Torgal, J. Labrincha, M. Diamanti, et al.), 181–210. Cham: Springer.

Treesubsuntorn, C. and Thiravetyan, P. (2012). Removal of benzene from indoor air by *Dracaena sanderiana*: effect of wax and stomata. *Atmospheric Environment* 57: 317–321.

Treesubsuntorn, C., Boraphech, P., and Thiravetyan, P. (2017). Trimethylamine removal by plant capsule of *Sansevieria kirkii* in combination with *Bacillus cereus* EN1. *Environmental Science and Pollution Research* 24: 10139–10149.

Ugrekhelidze, D., Korte, F., and Kvesitadze, G. (1997). Uptake and transformation of benzene and toluene by plant leaves. *Ecotoxicology and Environmental Safety* 37: 24–29.

Vagaggini, B., Taccola, M., Cianchetti, S. et al. (2002). Ozone exposure increases eosinophilic airway response induced by previous allergen challenge. *American Journal of Respiratory Critical Care Medicine* 166 (8): 1073–1077.

Vallano, D. and Sparks, J. (2007). Foliar $\delta\ ^{15}N$ values as indicators of foliar uptake of atmospheric nitrogen pollution. In: *Stable Isotopes as Indicators of Ecological Change* (eds. T.E. Dawson and R.T.W. Siegwolf), 93–109. Amsterdam, The Netherlands: Elsevier Academic Press.

Varshney, C.K. and Mitra, I. (1993). Importance of hedges in improving urban air quality. *Landscape and Urban Planning* 25: 75–83.

Verma, A. (2003). Attenuation of automobile generated air pollution by higher plants. PhD thesis, University of Lucknow, Lucknow, India.

Viessman, W., Lewis, G.L., and Knapp, J.W. (1989). *Introduction to Hydrology*, 3e. New York, NY: Harper & Row Publishers.

Vora, A.B. and Bhatnagar, A.R. (1987). Comparative study of dust fall on the leaves in high pollution and low pollution area in Ahmedabad. V. Caused foliar injury. *Journal of Environmental Biology* 8 (4): 339–346.

Vorholt, J.A. (2012). Microbial life in the phyllosphere. *National Review of Microbiology* 10: 828–840.

Voriskova, J. and Baldrian, P. (2013). Fungal community on decomposing leaf litter undergoes rapid successional changes. *ISME Journal* 7: 477–486.

Weber, F., Kowarik, I., and Saumel, I. (2014). Herbaceous plants as filters: immobilization of particulates along urban street corridors. *Environmental Pollution* 186: 234–240.

Wellburn, A.R. (1990). Why are atmospheric oxides of nitrogen usually phytotoxic and not alternative fertilizers? *New Phytologist* 115: 395–429.

Wellburn, A. (1998). Atmospheric nitrogenous compounds and ozone—Is NO_x fixation by plants a possible solution? *New Phytology* 139: 5–9.

Weyens, N., Thijs, S., Popek, R. et al. (2015). The role of plant–microbe interactions and their exploitation for phytoremediation of air pollutants. *International Journal of Molecular Science* 16: 25576–25604.

Wisthaler, A. and Weschler, C.J. (2010). Reactions of ozone with human skin lipids: sources of carbonyls, dicarbonyls, and hydroxycarbonyls in indoor air. *Proceedings of the National Academy of Sciences* 107 (15): 6568–6575.

Wolkoff, P. (2003). Trends in Europe to reduce the indoor air pollution of VOCs. *Indoor Air* 13: 5–11.

Wolverton, B. and Wolverton, J.D. (1993). Plants and soil microorganisms: removal of formaldehyde, xylene, and ammonia from the indoor environment. *Journal of the Mississippi Academy of Sciences* 38 (2): 11–15.

Wolverton, B.C. (2008). *How to Grow Fresh Air. 50 Houseplants that Purify Your Home and Office*. Weidenfeld & Nicolson: London, UK.

Wood, R.A., Orwell, R.L., Tarran, J. et al. (2002). Potted-plant/growth media interactions and capacity for removal of volatiles from indoor air. *Journal of Hortical Science and Biotechnology* 77: 120–129.

Wood, R.A., Burchett, M.D., Alquezar, R. et al. (2006). The potted-plant microcosm substantially reduces indoor air VOC pollution: I. Office field-study. *Water, Air, & Soil Pollution* 175 (1): 163–180.

World Health Organization (WHO) (2010). *Guidelines for Indoor Air Quality: Selected Pollutants*. Geneva, Switzerland: World Health Organization, Regional Office for Europe, Copenhagen; World Health Organization (WHO).

Xiao, S.Q., Sun, Z., Wang, S.S. et al. (2011). Over expressions of dihydroxyacetone synthase and dihydroxyacetone kinase in chloroplasts install a novel photosynthetic HCHO- assimilation pathway in transgenic tobacco using modified gateway entry vectors. *Acta Physiologiae Plantarum* 34: 1975–1985.

Xu, Z.J., Qin, N., Wang, J.G., and Tong, H. (2010). Formaldehyde biofiltration as affected by spider plant. *Bioresource Technology* 101: 6930–6934.

Xu, A.J., Wu, M., and He, Y.Y. (2013). Toluene biofiltration enhanced by ryegrass. *Bulletin of Environmental Contamination and Toxicology* 90: 646–649.

Yang, J., McBride, J., Zhou, J., and Sun, Z. (2005). The urban forest in Beijing and its role in air pollution reduction. *Urban Forestry and Urban Greening* 3: 65–68.

Yin, S., Shen, Z., Zhou, P. et al. (2011). Quantifying air pollution attenuation within urban parks: an experimental approach in Shangai, China. *Environmental Pollution* 159: 2155–2163.

Yoo, M.H., Kwon, Y.J., Son, K.C., and Kays, S.J. (2006). Efficacy of indoor plants for the removal of single and mixed volatile organic pollutants and physiological effects of the volatiles on the plants. *Journal for American Society for Horticultural Science* 131: 452–458.

Zhou, S., Xiao, S., Xuan, X. et al. (2015). Simultaneous functions of the installed DAS/DAK formaldehyde-assimilation pathway and the original formaldehyde metabolic pathways enhance the ability of transgenic geranium to purify gaseous formaldehyde polluted environment. *Plant Physiology and Biochemistry* 89: 53–63.

Section 3

Biodiversity and Natural Resource Exploitation

9

Tree Benefits in Urban Environment and Incidences of Tree Vandalism: A Review for Potential Solutions

Krishna K. Chandra, Rajesh Kumar, and Gunja Baretha

Department of Forestry, Wildlife and Environmental Sciences, Guru Ghasidas Vishwavidyalaya (A Central University), Bilaspur, Chhattisgarh, India

9.1 Introduction

Currently, urbanisation is the main criterion of fast-growing nations and therefore cities are growing substantially in surface area (Avijit 2002; Chodak 2019). As of 2010, above 50% of the world's population live in urban areas (UNDP 2010) and develop as a centre of the economy (EEA 2012). It is estimated that in the tropical and subtropical regions, the population in urban areas would grow to four billion by 2025 (World Bank 2014). Even the most populated country like India accommodate nearly 34.47% of its current population to urban areas and contributes 63% of GDP (Sarbeswar et al. 2018; Neill 2021) which is expected to home 40% of India's population and contribute 75% of the country's GDP by 2030. Currently, in Europe, three-quarters of the population living in urban areas, and, if the current pace of urbanisation continues, an even larger population of the world will be living in urban areas in the future (Brune 2016). This will demand the development of physical, institutional, social, and economic infrastructure for urban residents to uplift their living standards and social connectedness. The development of Smart Cities is a forward step in that direction which increases the economy of the country as well as improves the living standards of people through advanced technology to create smart outcome for urban citizens. The smart city mission (SCM) intends to promote cities by developing infrastructure and provide a decent quality of life to inhabitants by creating a sustainable environment and smart solutions. The SCM efforts are to set a holistic model of smart city that can be replicated as such in various regions, parts of the country and adaptable for other countries throughout the world. SCM currently stands over INR 2000 Billion, consisting of 99 cities across the country (Anand et al. 2018).

Rapid urbanisation means more industries, a higher number of vehicles, big infrastructures, shopping malls, road connectivity, and many more which are the essential components of smart city and urban development. Unfortunately, these emit a higher amount of greenhouse gases (GHGs) such as CO_2, CO, SOx, NOx, volatile organic compound, etc., due

Urban Ecology and Global Climate Change, First Edition. Edited by Rahul Bhadouria, Shweta Upadhyay, Sachchidanand Tripathi, and Pardeep Singh.

to the preponderance of factories, vehicular emissions, particulate matters, etc. In addition, the clusters of buildings, concrete roads, and impermeable surfaces absorb the incoming solar radiation and release heat by longwave radiation causing urban heat islands (UHI) and acceleration in climate change. This makes the condition worse for living in urban areas and causes various types of health problems and thermal uncomfort to urban people. Doick and Hutchings (2013) and Kleerekoper et al. (2012) has also reported high temperatures, low humidity, reduced wind speed because of heightened buildings and high air pollutants in urban areas which hurt the health and well-being of urban residents. Thus the microclimate of urban areas is already more stressful than the rural areas, which are further expected to exacerbate the situation in the future due to changing climate and global warming (Gill et al. 2014). Therefore, adaptive measures are urgently needed side by side with urbanisation and it should be essential for all planners and agencies who are engaged in developing smart cities and urban areas to create a healthy living environment for the people (EEA 2012; Brune 2016).

A large number of studies have strongly highlighted the role and significance of urban trees and vegetation in ameliorating the urban microclimate through cooling the overheating urban environment and intercept air pollutions (Demuzere et al. 2014; Gill et al. 2014; Ratola and Jimenez-Guerrero 2017; Wolf et al. 2020, Ragula and Chandra 2020). Trees provide ample benefits to cities and their residents by creating a special microclimate and boosting the recreational and aesthetic values of the cities. Trees are the essential components of parks, gardens, green space, individual houses, parking, roadside plantations, and urban landscapes, and if trees will be healthy, that would provide long-term benefits to urban environment (Duinker et al. 2015). The pieces of evidence are available that how trees benefiting ecologically (Gillner et al. 2015), socially (Nowak and Dwyer 2007), and economically to the urban areas (Pandit et al. 2013). Ugle et al. (2010) has estimated the urban trees carbon storage potential in India and reported 23.8 million tons of stored carbon in urban trees of 7.79 million ha areas. Furthermore, they found that the urban trees are contributing to 2.21% of the carbon stored against the total carbon stock from forests. Trees have the highest benefits to urban society, and therefore its selection should be carefully done coupled with adequate management and planning for long-term sustainable development and to cope up with future climate change (Gillner et al. 2014). Trees endorse the nature-based solution both for visual and functional perspectives and are a key for green infrastructure and other urban ecosystems (Tiwary et al. 2016; Li et al. 2017; O'Brien et al. 2017; Pearlmutter et al. 2017). The ecological functions and services of urban trees structure, compositions have been studied extensively in few decades considering individual trees species, assemblages of trees, forested lands, gardens across public and private lands, streets, waterfronts, railways, etc. (Tyrvainen et al. 2005; Konijnendijk et al. 2006; Chen and Jim 2008; Roy et al. 2012; Davies et al. 2017; Wolf et al. 2020). All suggests trees as the simplest options for reducing the Green House Gases by sequestering carbon (Nowak et al. 2006, 2007; Chandra and Singh 2018; Ragula and Chandra 2020), decrease water runoff through interception and infiltration of higher quantity of rainwater (Xiao and McPherson 2002), reductions in the air temperatures. Greenery in the urban environment was also found to be beneficial in curing health problems for example in blood pressure, anxiety, asthma, cancer, etc. (Lovasi et al. 2008; Hirabayashi and Nowak 2016; Wang et al. 2016; Rao et al. 2017). Trees have been associated with the increased value of house and commercial properties, and offers an excellent return on investment, therefore the primary function of urban trees has now changed from a purely aesthetic and ornamental to environmental, economic, and social functions (Seamans 2013).

Undoubtedly, trees are the most important for urban areas to cope with climate change and make the environment healthy, livable, and joyful. Unfortunately, growing trees in urban areas are more challenging as the trees encounter various types of vandalism and grow in stressful environments than the rural areas and open landscape (Martin et al. 2012; Wolf et al. 2020). As there are increasing incidences of urban tree vandalism worldwide, the human tendency to damage and harm the trees cannot be overlooked. To solve this issue, integrated approaches are urgently required bringing urban administrator, civic society, and health professionals at a common platform. Interdisciplinary approaches are needed to establish a positive attitude in urban residents to trees and nature for a better urban environment (Mullaney et al. 2015). Through this chapter, efforts are made to highlight the benefits of urban trees for the overall wellbeing of the urban environment and to identify criteria of trees vandalism. Some potential solutions have also been suggested to minimise the incidences of vandalism in urban trees.

9.2 Benefits of Urban Trees

- *Reduction in urban heat islands.* A tree can cool the air temperature from 2 to 8 °C through the transpiration process and efficiently reduces the incidences of urban heat island. Trees also benefit in reducing the electricity bill and cooling cost of air conditioners by 30% during hot summer, and winter heating cost by 20–50%, if properly accommodated around buildings and house properties. Shashua-Bar et al. (2009) reported that these benefits can vary depending upon size, canopy, location, planting density of urban trees. Furthermore, it is established that roadside plantations can reduce the day temperatures between 5 and 20 °C (Burden 2006) and can reduce the energy cost between USD 2.6 and 77 per tree/per year. McPherson et al. (2005) concluded the saving of electricity by street trees by 95 kWh/tree/year equal to USD 15/tree/year.
- *Sequester carbon dioxide.* Trees act as carbon sinks by trapping the GHGs more specifically CO_2. Trees absorb CO_2 for photosynthesis and use it in growth and development (Ferrini and Fini 2010). Moore (2009) conducted a study to determine the carbon sequestration potential of the trees in Melbourne (Australia) and estimated one million tons of carbon in planted trees within and around city areas. A single tree can sequester 150 kg of CO_2 per year and contribute to climate change mitigation. A plantation of 0.4 ha area can trap CO_2 equal to the emission by a car to run for 41 800 km and release O_2 for 18 people every day (USDA 1990). A roadside street tree can sequester carbon to a value between USD 0.4 and 6 per year (Beecham and Lucke 2015). Forest in urban areas can offset 18.57% of the carbon emitted from the urban industries and can store carbon equal to 1.75 times of annual carbon emitted by industries (Zhao et al. 2010). In China, the urban trees of Shenyang city are storing 337 000 tons of carbon worth USD 13.88 million (Liu and Li 2011).

- *Improve air quality and filter pollutants.* Urban trees efficiently intercept dust particulates, traffic emissions, and other pollutants (Tallis et al. 2011) and save people's life against seasonal and other diseases. A roadside tree can capture nine times more pollutants compared to distant trees (Burden 2006). As the urban environment is characterised by a high concentration of air-borne pollutants, trees effectively reduce these elements and make the environment healthier to live in (Nowak et al. 2013). In economic terms, the value of the trees in reducing air pollution range from USD 0.34 to 42/tree/year.
- *Regulates water flow and reduces stormwater runoff.* Trees contribute to preventing floods and reducing the stormwater runoff in urban areas. USDA (2003) has estimated that a cluster of 100 mature trees can capture 379 m^3 of rainfall per year. Usually, during urbanisation and infrastructural development, soil surfaces are cemented which results in a reduction in rainwater infiltration into the deep layers of soil and increased runoff. Trees enable to improve the perviousness of the soil, and its water holding capacity and improve water balance by percolating large volume of surface water to below layers (Soares et al. 2011). Trees can reduce stormwater runoff between 3.2 and 11.3 kl per tree, which is valued at Australia US$ 3.4–58 per tree (Beecham and Lucke 2015).
- *Reduce noise pollution.* Trees absorb the high pitch of a sound and can reduce as much as 50% of the urban noise (USDA 2003).
- *Enhance property value and price.* Trees and out space landscape of commercial areas, offices, house properties increases property value by up to 20%, and enhance business in cities through attracting tourists and other buyers (Moskell and Broussard Allred 2013).
- *Contribute to human health.* It has been proved that trees reduce fatigue, mental stress, and anxiety in people in urban areas. The presence of trees augments livability and upgrade the quality of urban life by benefiting aesthetic and visual amenity. Looking at the tree continuously for 20 minutes was found to improve the restoration of perceived knowledge and happiness among students and people (Bogerd et al. 2018; Browning and Rigolon 2019). Trees can reduce mortality (Nowak et al. 2013), a decline in respiratory issues (Hirabayashi and Nowak 2016; Rao et al. 2017), decrease the incidence of lung cancer (Wang et al. 2016) and asthma in urban inhabitants (Lovasi et al. 2008)
- *Food and nutritional security.* It contributes to the increase of local food, and nutrients by providing fruit, nuts, and leaves for both human and animal consumption. The planting of trees in homestead gardens, near houses, parks, and footpaths are important in solving the issue of the poor society of the urban areas.
- *Enhance biodiversity.* Different trees have the potential for providing habitat and landscape connectivity for urban fauna (Rhodes et al. 2011). The planting of mixed species that the monoculture attracts avifauna and other animals and helps in the development of the ecosystem and enhances the biodiversity of the city's green environment.
- *Social benefits.* Trees were found to improve the overall urban environment by enhancing the opportunities for recreation, which promotes resident's social contact. Trees also contribute to stimulating social cohesion (Van Dillen et al. 2012), reduces urban crime, and increases public safety (Kuo and Sullivan 2001; Tarran 2009).

9.3 Selection Criteria for the Urban Trees

The presence of trees in urban areas presents a first impression of the urban environment therefore, the selection of trees is a very important consideration for long-term ecosystem services to society. As a tree takes time to mature and provides benefits for several decades, it should be planted from a long-term perspective rather than for the short term for achieving the highest possible outcome. Approaches such as 'right tree in the right place' are very important while choosing urban trees to current and future climatic conditions. Though several factors are considered selecting trees for urban areas such as drought tolerance, soils, coppicing ability, pest-disease resistance, and pollution tolerance (Doick and Hutchings 2013), the aesthetic value of the trees, shading nature, and trees toxicity and allergic potentials are also considered (Roloff 2013). Similarly, the size of the tree canopy, growth potential, and gestation period may also be kept in mind during tree species selection for different locations including, streets, gardens, parks, proximity to the individual house and residential colonies, waterfronts, and natural forest lands. Overall, the assessment of tree suitability for urban areas is complicated as the planner has to decide and ensure the traffic and pedestrian's safety, resident health, besides other factors. For few years scientists are regularly providing practical guidelines for urban tree selection based on stress tolerance, climate-species matrix, hardiness, and future climate (Bassuk et al. 2009; Roloff et al. 2009; GALK 2015). UK Forestry Commission (2015) has also given guidelines for selecting 'Right Tree in right place' to urban areas. Currently, there is a paradigm shift in the primary function of an urban tree to ecological interest and therefore the commercial interest of urban planners can be a conflict with maintaining diversity. Now focus is on higher tree diversity to make the urban environment more resilient to climate change (Brune 2016). Li et al. (2011) have suggested a 10–20–30 formula for selecting trees for urban areas illustrate maximum 10% of a single tree species, not more than 20% of a tree genus, and not more than 30% of a tree family to maintain higher floral diversity and control epidemic pest infestation.

Pauleit (2003) and Hemery (2007) strongly believe in non-native tree species to mitigate future climates while Ugle et al. (2010) and Ragula and Chandra (2020) have pointed out the benefits of indigenous tree species for the urban environment as the species has more adaptation to thrive in local environment than the exotic trees. Thus, native trees may be in central criteria during selection as they can protect biodiversity and the ecosystem in urban areas. However, studies demonstrate poor tree diversity in urban areas across Europe (Pauleit et al. 2002), whereas 2–3 common species dominate in the United States (Sanders 1981), and four species comprise 2/3 tree populations of Chicago (McPherson 2007). The issue of poor tree diversity is also a concern in major cities of Asian countries, where only a few species represent 49% of the urban forestry (Chacalo et al. 1994; Nowak et al. 2014). Depending on the requirement of shade and light in particular areas of the cities, a deciduous tree may be the best choice to manage passive solar design based on seasonal aspects. Similarly, large canopy trees live longer and provide a strong sense of wellbeing, capture higher CO_2, create ambience, and improve the physical health of urban people. Street trees should be planted in such a way to have spaced allowing light during winter and canopy shade in hot summer.

As it is difficult to predict the future climate and how the current trees will contribute to urban wellbeing in the future, trees with large genetic diversity and phenotypic plasticity may be the choice for urban areas. Has suggested adoption of tree species having a higher ability to adjust physiologically to changing climate. Still, the criteria for urban tree selection are unclear and selected solely based on local needs, silviculture of the species, costs, and perceptions and preferences of the urban planner. However, FAO (2016), Chinese Proverb (2016) has recommended some key elements to be considered for urban tree planting:

- Microclimate amelioration (temperature, drought, humidity for human thermal comfort)
- Passive solar design principles (sun/shade) for providing light during winter and shade on hot days
- Respect biodiversity (floral and faunal conservation, soil improvement, enrichment)
- Aesthetic and ornamental value (colour, textures, contrast, leaf shape and size, scent) of trees
- Architectural qualities of trees (canopy shape, foliage pattern, strong vertical presence)
- Human curiosity, surprise, fun, attractiveness
- Cultural heritage, traditional, ritual, ethnic references
- Mindfulness/calmness/health/wellbeing
- High potential of pollutant and CO_2 absorber
- Multiple ecosystem services

9.3.1 Sites for Urban Tree Planting

Buildings: Residential colonies, individual houses, offices, schools, universities, hospitals, commercial and industrial areas.

Outside spaces: Footpaths, walkways, parks, gardens, picnic areas, waterfront, cinema and malls, civic plazas, market places, sports stadiums, bus stops, railway station.

Street/Roadsides: Freeways, ramps, embankments, highways, city roads.

Parking areas: Large and small, civic, commercial, and private areas.

Homestead gardens: House backyards, nutritional gardens, animal, and pet houses.

9.4 Urban Trees Vandalism

Tree vandalism is applied for physical manifestations of undesirable behaviour and irresponsible attitude that cause damage to the trees (Black 1978; Nowak et al. 1990; Jim and Zhang 2013). Vandalism is a type of mechanical injury that starts from a point of tree damage and eventually leads to tree failure (Moore 2013). These phenomena impose tree disservices, resulting in unexpected economic losses, physical damage to the infrastructure, and poor health conditions (Lyytimaki 2017). The incidences of tree vandalism in urban areas occurred as carving and stripping the bark, girdling stem, breaking the stem, damage to branches and twigs, uprooting the younger trees, poisoning, burning, etc. (Sreetheran et al. 2006; Richardson and Shackleton 2014; Hamzah et al. 2018). All these vandalisms are anyhow related to the human and increase in anthropogenic pressure due to concentrated population in urban areas (Richardson and Shackleton 2014; Morgenroth et al. 2015). Vandalism may be intentional or accidental but the human factor is always in the centre (Nowak et al. 2004). Numerous reports depict the increasing trends of tree vandalism as in the Hong Kong about 10–15% of trees exhibit the sign of vandalism, it is 30% in European countries (Pauleit et al. 2002), and 42% in the Eastern Cape of South Africa (Richardson and Shackleton 2014). There are many examples of urban tree vandalism in India, particularly related to infrastructure development or other extension activities. Thus the incidence of tree vandalism cannot restrict with the boundary of countries and continents, and it is increasing annually across the world (Mullaney et al. 2015; Hamzah et al. 2018; Wolf et al. 2020). Hamzah et al. (2018) has conducted comprehensive studies on the types of tree vandalism based on tree location, value (utilitarian), and tree conditions. Interestingly, the result showed the highest vandalism of 61.53% in trees neighbouring individual houses than the other locations whereas higher utility of tree species has more chances of urban tree vandalism (Figure 9.1). Furthermore, young trees are more vulnerable to vandalism (50%), followed by mature trees (40.47%) compared with the tree conditions and age (Hamzah et al. 2018). As for as types of vandalism in urban trees are concerned, the chopping of tree branches and stem was the most frequent (65.71%), followed by breakage of branches and twigs (17.14%) either encounter intentionally for various uses or accidentally during the movement of vehicles and infrastructural development (Figure 9.2). Although urban trees are planted and managed with good strategies, the cases of urban tree vandalism remain unresolved and a question that arises to tree vandalism is why? To solve the issue, it is pertinent to identify the factors and criteria's that provoke tree vandalism as the acts cannot be overlooked. At present, the information on vandalism is scarce particularly on urban trees and community's vandalism, more information would guide urban horticulturists and tree managers to care more for the trees.

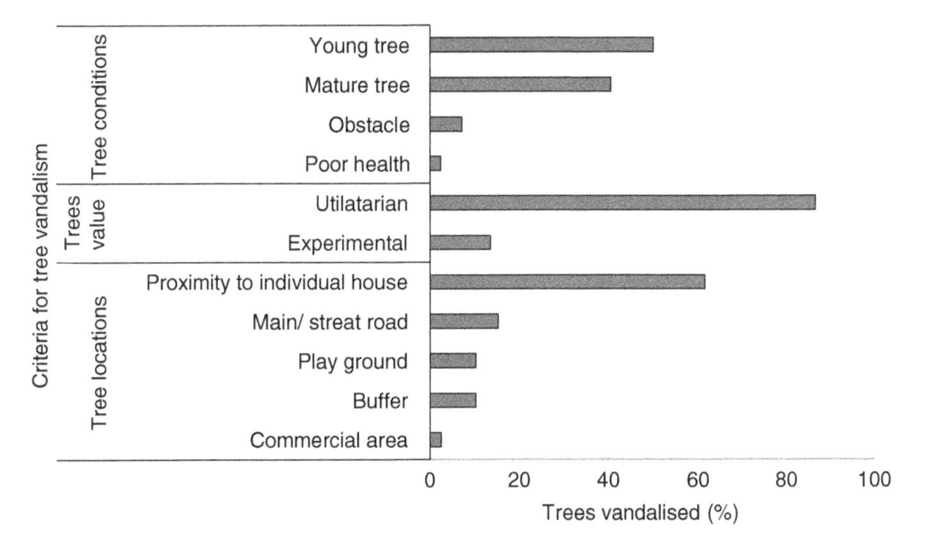

Figure 9.1 Incidence of urban tree vandalism and tree criteria's for vandalism. *Source:* Adapted from Hamzah et al. (2018).

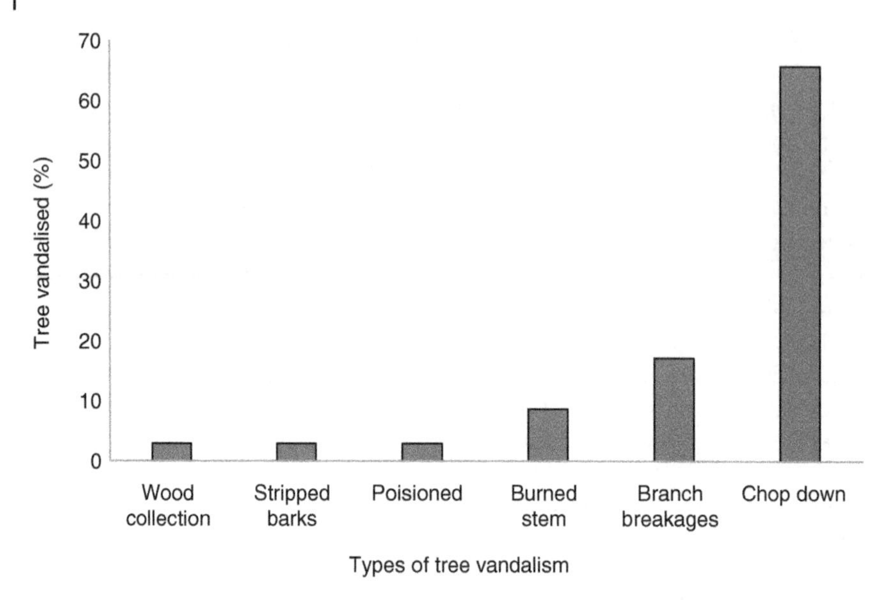

Figure 9.2 Types of tree vandalism in urban areas and its abundance. *Source:* Adapted from Hamzah et al. (2018).

9.4.1 Criteria Indicators for Tree Vandalism

Different criteria indicators have been identified causing tree vandalism in an urban environment; however, the few most important criteria of these notorious acts are considering in the study.

Public inconvenience and public preferences: Literatures indicate that the urban tree conditions reflect the public preferences (Abd Kadir and Othman 2012; Hamzah et al. 2020). When people reject trees, which comes in the form of vandalism. A study report demonstrates that the Kajang, Malaysia authority received 935 public applications for the removal of trees in 2016 (Hamzah et al. 2020) pointing out the fear related to public safety that would be inflicted by dangerous tree conditions. These circumstances create chances of urban tree vandalism and therefore trees should be selected based on public preferences and in the right location.

Human misconduct: The difference in people's perception and attitude towards trees and greenery is a major factor in tree vandalism. The population of urban areas commonly has a pessimistic attitude rather than an optimistic which accelerates the chances of tree vandalism. A high rate of urban tree vandalism is also reported when urban residents feel bore followed by the weak connectedness with nature (Richardson and Shackleton 2014). The misconduct of urban dwellers causes the widespread vandalism in urban trees as reported in California, England, South Africa, and Mexico (Nowak et al. 1990; Richardson and Shackleton 2014; Hernandez Zaragoza et al. 2015).

Urban stresses: As in urban areas, the level of stress is always high than the rural areas, due to pollution, fast movement of resident life, competition for livelihood, etc., the urban trees are also exposed higher to urban human stresses. These constraints affect the growth and performance of the trees adversely and also cause vandalism. The high anthropogenic activities of the urban area impact detrimentally to tree health also promote the incidences of tree vandalism (Vogt et al. 2015). Such stresses of urban inhabitants give rise to poor tree growth and disservices which influence urban tree populations and diversity (Delshammar et al. 2015). Moreover, the findings of Hamzah et al. (2020) classifies important criteria of urban tree vandalism under three broad categories such as specific motive action, ideology and practices, and victims of circumstances, so that the indicators can find out. Hamzah et al. (2020) have identified 32 criteria for urban tree vandalisms based on Delphi survey, and out of these criteria, top five criteria of tree vandalism have been highlighted in this chapter each for three broad categories (Table 9.1). It shows that the 'location of the tree' is the rank I criteria for vandalism, followed by the 'size of the tree' (rank II) under specific motives and actions category (Table 9.1). Similarly, ideologies and practice category the 'level of knowledge' is identified as rank I, while the rules and regulations are the second important criteria. The conflict of trees with other activities is scored rank I criteria under the victim of circumstances category. These have provided strong evidence for causes of urban tree vandalism and consideration of all these parameters would help reduce the incidences of tree vandalism.

9.4.2 Potential Solutions to Prevent Tree Vandalism

Worldwide, different strategies are being explored to check the incidences of tree vandalism and strengthen nature connectedness in the people. It is noticed that new generations are forgetting nature values, and usually not concern with nature compared to our parents

Table 9.1 Top five important criteria of tree vandalisation in urban areas.

Broad categories of tree vandalism	Criteria of tree vandalism	Ranking
Specific motive and action	Location of tree	I
	Size of tree	II
	Tree health condition	III
	Tree growth rates	IV
	Species of tree	V
Ideology and practices	Level of knowledge	I
	Rule and regulations	II
	Information on tree benefits	III
	Design and layout	IV
	Tree care monitoring	V
Victim of circumstances	Conflict with other activities	I
	Infrastructure upgrading/extension and urbanisation/development	II
	Priority of space usage	III
	Trees cause interference/obstruction	IV
	Tree for structure attachment	V

Source: Data from Hamzah et al. (2020).

and ancestors. This tendency of people raises the chances of tree vandalism in a common manner. To tackle the issue, a 'people awareness campaign' on tree benefits must be the foremost important activity followed by the initiatives which connect societies with nature (Hamzah et al. 2020). Such programmes often provide important information about 'what to look for, what to do' if tree vandalism is suspected. Tree vandalism solutions are just like with other vandalism gaiety, and urban communities can find effective ways to prevent such vandalism as damage of individual properties. Cities and small towns in Australia have seized a 'shaming policy', and now put a large display board next to trees that have been replaced because of vandalism, preventing house owners who are seeking a better view from enjoying it. The board is only removed when the replanted tree grows taller than the board. Such a policy can contribute more against tree vandalisation. The forest officers of Great Smoky Mountain, Tennessee, and North Carolina, offer to take photos of the culprits during vandalising trees and upload them onto social media, in hopes someone would recognise and turn him in. 'Reward to informer' maybe another potential solution to prevent urban tree vandalism. The sheriff's department of Eagan, MN is offering a reward of $5000 to anyone who informs tree vandalism. This is a kind of offence that leads to an arrest of a person involved in the act of vandalism. 'Adopt a Tree' programme is started in Baltimore County to encourage residents in tree care and protection. Participants voluntarily dedicate themselves to watering and care the tree extensively, in hopes of flushing out the poisoned trees and make them healthy. In India, the higher education department has already initiated a 'one student one tree' programme to enrich moral values among students by improving their nature connectedness, however, a lot has to be done for proper implementation of this programme in different universities and educational institutions. Similarly, a concept of 'memory trees' can also be initiated to increase greenery in urban environments involving common people of the areas. In this programme, on the occasion of birthday, marriage anniversary one can plant a tree or adopt trees and watch the growing tree with memory. This inculcates a sense of pride for the trees among the people of the area (Fox 2017).

As another measure, urban tree monitoring is a very important aspect of management. Urban tree monitoring protocols: Field Guide published by UTGL (2015) provides a long-term data collection procedure for different locations of urban trees includes site factors, tree species information, growth and tree condition data, management as well as socioeconomic information of community surrounding the tree (Hamzah et al. 2020). Also, the selection of mycotrophic (mycorrhizal) trees is a good choice for the urban area, to enable the tree to manage water and nutrient efficiently even in harsh conditions (Chandra 2013; Bhardwaj and Chandra 2018). Some unique ideas have also been experimented to control tree vandalism. The town administration of Lincoln, Neb has solved thieves' cases of urban trees with a cent percent success rate by using chemicals to deter the Grinches: fox urine (Fox 2017). A solution of glycerin and water is sprayed on the trunks of the trees, which has the chance of being thefts. After a spray of the solution, due to cold air urine smell is not too pungent initially, but once the tree warms up the odour is unbearable. If it is stolen and brought home, the high pungency of the tree trunk helps to seize trees and take action against thieves. To make it more effective signboards are also posted near trees indicating secret weapons, which results in no more cases of tree vandalism. In addition, strict rules and regulations are also essential tools to minimise tree vandalism, as people frequently visit near urban trees and cause damage while talking on mobile and spree with friends without fear of rules and policies. The incidences of tree vandalism in the urban area can be minimised by selecting an appropriate tree for different locations because any types of conflict between trees and properties emerge as a cause of vandalism. Therefore, the right tree in the right location is very important to make the urban environment healthier, happier, and sustainable.

9.5 Conclusions

Urban trees render ecological, economic, and social benefits that make the areas livable with quality standards. As different tree species show varied potentials and ecosystem services based on location, growth, and genetic diversity, a selection of the most suitable trees needs to suffice the requirements of urban areas. Even trees benefit a lot to mankind, these trees are exposed to intense urban stresses and become victims of vandalism. There is a large literature illustrating the increasing incidences of tree vandalism across the world. Urban tree vandalism is an act that adversely affects the growth and health of trees which hinders the ecosystem services of urban forestry. Inappropriate locations of the trees have more chances of vandalism therefore, the right trees should be planted at right places to avoid the common cause of tree vandalism. In addition, the high utilitarian of tree species also causes vandalism, particularly in public space. Numerous findings depict that all the trees whether native or exotic play an important role in modifying the urban environment, however, long-lived trees contribute for several decades and support the ecosystem. During selecting, the species for the urban environment, the consideration of future climate, and adjusting nature of trees with changing climate can be given preferences to fight against global warming. As there is a need for greater attention to preventing increasing trends of tree vandalism in urban areas, initiation of programmes such as adopt a tree, memory tree, awareness campaign on the benefits of trees to urban societies, regular monitoring of trees, care and protection would be helpful to reduce vandalism. Moreover, the provision of rewards for the person who informs vandalism and imposing strict rules and regulations against vandalism needs to be considered. Further research in the direction of tree vandalism associated with community and trees is required to strengthen the understanding of urban planners and selection of the most suitable trees for beautifying cities and creating a long-term microclimate with increasing sustainability for the future generation.

References

Abd Kadir, M.A. and Othman, N. (2012). Towards a better tomorrow: street trees and their values in urban areas. *Proceeding Social and Behavioral Sciences* 35: 267–274. https://doi.org/10.1016/j.sbspro.2012.02.088.

Anand, A., Sreevatsan, A., and Taraporevala, P. (2018). Policy brief – an overview of the smart cities mission in India. *Centre for Policy Research*: 1–17.

Avijit, G. (2002). Geo indicators for tropical urbanization. *Environmental Geology* 42: 736–742.

Bassuk, N., Curtis, D.F., Marranca, B.Z., and Neal, B. (2009). *Recommended Urban Trees: Site Assessment and Tree Selection for Stress Tolerance*, 122. Ithaca, NY: Urban Horticulture Institute, Cornell Universit.

Beecham, S. and Lucke, T. (2015). Street trees in paved urban environments – the benefits and challenges. https://treenet.org/resources/street-trees-in-paved-urban-environments-the-benefits-and-challenges (accessed 6 May 2021).

Bhardwaj, A.K. and Chandra, K.K. (2018). Soil moisture fluctuation influences AMF root colonization and spore population in tree species planted in degraded entisol soil. *International Journal of Biosciences* 13 (3): 229–243.

Black, M. (1978). Tree vandalism: some solutions. *Journal of Arboriculture* 4: 114–116.

Bogerd, N.V.D., Dijkstra, S.C., Seidell, J.C., and Maas, J. (2018). Greenery in the university environment: students' preferences and perceived restoration likelihood. *PloS One* 13 (2): e0192429.

Browning, M.H.E.M. and Rigolon, A. (2019). School green space and its impact on academic performance: a systematic literature review. *International Journal of Environmental Research and Public Health* 16: 429. https://doi.org/10.3390/ijerph16030429.

Brune, M. (2016). Urban trees under climate change. Potential impacts of dry spells and heat waves in three German regions in the 2050s. Report 24. Climate Service Center Germany, Hamburg.

Burden, D. (2006). 22 benefits of urban trees. http://www.michigan.gov/documents/dnr/22_benefits_208084_7.pdf (accessed 12 May 2021).

Chacalo, A., Aldama, A., and Grabinsky, J. (1994). Street tree inventory in Mexico city. *Journal of Arboriculture* 20: 222–226.

Chandra, K.K. (2013). Seedling quality, biomass production and nutrient uptake potential of AMF, *Azotobacter* and *Pseudomonas* in *Azadirachta indica* under nursery condition. *Journal of Biodiversity and Environmental Sciences* 3 (4): 39–45.

Chandra, K.K. and Singh, A.K. (2018). Carbon stock appraisal of naturally growing trees on farmlands in plains zone district of Chhattisgarh, India. *Tropical Ecology* 59 (4): 679–689.

Chen, W.Y. and Jim, C.Y. (2008). Assessment and valuation of the ecosystem services provided by urban forests. In: *Ecology, Planning, and Management of Urban Forests: International Perspectives* (eds. M.M. Carreiro, Y.-C. Song and J. Wu), 53–83. New York: Springer ISBN 978-0-387-71425-7.

Chinese Proverb (2016). Recommended urban trees: indigenous and exotic. The best time to plant a tree was 50 years ago. E19-5766-recommended-tree-for-urban-Environments.pdf. www.armadale.wa.gov.au. (accessed 8 May 2021).

Chodak, E.R. (2019). Pressures and threats to nature related to human activities in European urban and suburban forests. *Forests* 10: 765. https://doi.org/10.3390/f10090765.

Davies, H., Doick, K., Handley, P. et al. (2017). *Delivery of Ecosystem Services by Urban Forests*. Edinburgh, Scotland: Forestry Commission ISBN 978-0-85538-953-6.

Delshammar, T., Ostberg, J., and Oxell, C. (2015). Urban trees and ecosystem disservices – a pilot study using complaints records from three Swedish cities. *Arboriculture & Urban Forestry* 41 (4): 187–193.

Demuzere, M., Orru, K., Heidrich, O. et al. (2014). Mitigating and adapting to climate change: multi-functional and multi-scale assessment of green urban infrastructure. *Journal of Environmental Management* 146: 107–115.

Doick, K. and Hutchings, T. (2013). *Air Temperature Regulation by Urban Trees and Green Infrastructure*, 1–10. Farnham: Forestry Commission (UK) https://www.researchgate.net/publication/259889679_Air_temperature_regulation_by_urban_trees_and_green_infrastructure.

Duinker, P.N., Ordonez, C., Steenberg, J.W.N. et al. (2015). Trees in Canadian cities: indispensable life form for urban sustainability. *Sustainability* 7 (6): 7379–7396. https://doi.org/10.3390/su7067379.

EEA (2012). *Urban Adaptation to Climate Change in Europe: Challenges and Opportunities for Cities Together with Supportive National and European Policies*, 143. Copenhagen: European Environment Agency.

FAO (2016). *Guidelines on Urban and Peri-urban Forestry* (eds. F. Salbitano, S. Borelli, M. Conigliaro and Y. Chen). FAO Forestry Paper No. 178. Rome: Food and Agriculture Organization of the United Nations.

Ferrini, F. and Fini, A. (2010). Sustainable management techniques for trees in the urban areas. *Journal of Biodiversity and Ecological Sciences* 1: 1–20.

Fox, A. (2017). What cities are doing to fight tree Vandalism. Gov1. 1 August 2017. https://www.gov1.com/parks-recreation/articles/what-cities-are-doing-to-fight-tree-vandalism-jIos5KQ4elq2ma1D/ (accessed17 May 2021).

GALK (2015). GALK Stra-ßenbaumliste (online version). https://www.galk.de/arbeitskreise/stadtbaeume/themenuebersicht/strassenbaumliste.

Gill, S.E., Handley, J.F., Ennos, A.R., and Pauleit, S. (2014). Adapting the role cities for climate of the green change: infrastructure. *Built Environment* 33 (1): 115–133.

Gillner, S., Bräuning, A., and Roloff, A. (2014). Dendrochronological analysis of urban trees: climatic response and impact of drought on frequently used tree species. *Trees* 28 (4): 1079–1093.

Gillner, S., Vogt, J., Tharang, A. et al. (2015). Role of street trees in mitigating effects of heat and drought at highly sealed urban sites. *Landscape and Urban Planning* 143: 33–42. https://doi.org/10.1016/j.landurbplan.2015.06.005.

Hamzah, H., Othman, N., and Hussain, N.H.M. (2020). Setting the criteria for urban tree vandalism assessment. *Planning Malaysia* 18 (4): 12–32. https://doi.org/10.21837/pm.v18i14.815.

Hamzah, H., Othman, N., Hussain, N.H.M., and Simiset, M. (2018). The criteria of urban trees regarding the issues of tree vandalism. *IOP Conference Series: Earth Environment Science* 203: 012023. https://doi.org/10.1088/1755-1315/203/1/012023.

Hemery, G.E. (2007). Trees and climate change. a practical guide for woodland owners & managers, Nicholsons, p. 8. http://sylva.org.uk/forestryhorizons/documents/Nicholsos_ClimateChange_PracticalGuide.pdf (accessed 11 May 2021).

Hernandez Zaragoza, A.Y., Cetina Alcala, V.M., Lopez, M.A. et al. (2015). Identification of tree damages of three parks of Mexico city. *Mexican Journal of Forest Sciences* 6 (32): 63–82.

Hirabayashi, S. and Nowak, D.J. (2016). Comprehensive national database of tree effects on air quality and human health in the United States. *Environmental Pollution* 215: 48–57.

Jim, C.Y. and Zhang, H. (2013). Defect-disorder and risk assessment of heritage trees in urban Hong Kong. *Urban Forestry and Urban Greening* 12: 585–596.

Kleerekoper, L., van Esch, M., and Salcedo, T.B. (2012). How to make a city climate-proof, addressing the urban heat island effect. *Resources, Conservation and Recycling* 64: 30–38.

Konijnendijk, C.C., Ricard, R.M., Kenney, A., and Randrup, T.B. (2006). Defining urban forestry-a comparative perspective of North America and Europe. *Urban Forestry and Urban Greening* 4: 93–103.

Kuo, F.E. and Sullivan, W.C. (2001). Environment and crime in the inner city: does vegetation reduce crime? *Environment and Behaviour* 33: 343–367.

Li, F., Liu, X., Zhang, X. et al. (2017). Urban ecological infrastructure: an integrated network for ecosystem services and sustainable urban systems. *Journal of Clean Production* 163: S12–S18.

Li, Y.Y., Wang, X.R., and Huang, C.L. (2011). Key street tree species selection in urban areas. *African Journal of Agricultural Research* 6 (15): 3539–3550.

Liu, C. and Li, X. (2011). Carbon storage and sequestration by urban forests in Shenyang, China. *Urban Forestry and Urban Greening* 11 (2): 121–128.

Lovasi, G.S., Quinn, J.W., Neckerman, K.M. et al. (2008). Children living in areas with more street trees have lower prevalence of asthma. *Journal of Epidemiology and Community Health* 62: 647–649.

Lyytimaki, J. (2017). *Disservices of urban trees.* In: *Routledge Handbook of Urban Forestry* (eds. F. Francesco, C.C. Van Den Bosch and F. Alessio), 164–176. London: Routledge https://doi.org/10.4324/9781315627106.

Martin, N.A., Chappelka, A.H., Loewenstein, E.F., and Keever, G.J. (2012). Comparison of carbon storage, carbon sequestration, and air pollution removal by protected and maintained urban forests in Alabama, USA. *International Journal of Biodiversity Science, Ecosystem Services and Management* 8: 265–272.

McPherson, E.G. (2007). Benefit-based tree valuation. *Arboriculture and Urban Forestry* 33: 1–11.

McPherson, G.J.R., Simpson, P.J., Pepper, S.E.M., and Xiao, Q. (2005). Municipal forest benefits and costs in five US cities. *Journal of Forestry* 103 (8): 411–416.

Moore, G. (2009). *Urban trees: worth more than they cost.* In: *Proceedings of the 10th National Street Tree Symposium* (eds. D. Lawry, J. Gardner and M. Bridget), 7–14. Adelaide, South Australia: Adelaide University.

Moore, G.M. (2013). *Ring-barking and girdling: how much vascular connection do you need between roots and crown.* In: *The 14th National Street Tree Symposium* (ed. N. Wojcik), 87–96. Adelaide, Australia: University of Melbourne https://pdfs.semanticscholar.org/21d8/3db275574be4aedb9805071741eddf1afb6a.pdf?_ga=2.146494001.481915270.1570351625-2133761217.1557440011 (accessed 6 May 2021).

Morgenroth, J., Santos, B., and Cadwallader, B. (2015). Conflicts between landscape trees and lawn maintenance equipment – the first look at an urban epidemic. *Urban Forestry and Urban Greening* 14: 1054–1058.

Moskell, C. and Broussard Allred, S. (2013). Residents' beliefs about responsibility for the steward ship of park trees and street trees in New York city. *Landscape and Urban Planning* 120: 85–95.

Mullaney, J., Lukke, T., and Trueman, S.J. (2015). A review of benefits and challenges in growing street trees in paved urban environments. *Landscape and Urban Planning* 134: 157–166. https://doi.org/10.1016/j.landurbplan.2014.10.013.

Neill, A.O. (2021). Urbanization in India 2019. Statistics. https://www.statista.com/statistics/271312/urbanization-in-india (accessed 14 May 2021).

Nowak, D.J. and Dwyer, J.F. (2007). *Understanding the benefits and costs of urban forest ecosystems*. In: *Urban and Community Forestry in the Northeast* (ed. J. Kuser), 10–25. New York: Springer.

Nowak, D.J., Kuroda, M., and Crane, D.E. (2004). Tree mortality rates and tree population projections in Baltimore, Maryland, USA. *Urban Forestry and Urban Greening* 2: 139–147.

Nowak, D.J., McBride, J.R., and Beatty, R.A. (1990). Newly planted street tree growth and mortality. *Journal of Arboriculture* 16 (5): 124–129.

Nowak, D.J., Crane, D.E., and Stevens, J.C. (2006). Air pollution removal by urban trees and shrubs in the United States. *Urban Forestry and Urban Greening* 4: 115–123.

Nowak, D.J., Hirabayashi, S., Bodine, A., and Greenfield, E. (2014). Tree and forest effects on air quality and human health in the United States. *Environmental Pollution* 193: 119–129.

Nowak, D.J., Hirabayashi, S., Bodine, A., and Hoehn, R. (2013). Modeled PM 2.5 removal by trees in ten U.S. cities and associated health effects. *Environmental Pollution* 178: 395. https://doi.org/10.1016/j.envpol.2013.03.050.

Nowak, D.J., Hoehn, R., and Crane, D.E. (2007). Oxygen production by urban trees in the United States. *Arboriculture and Urban Forestry* 33: 220–226.

O'Brien, L., De Vreese, R., Kern, M. et al. (2017). Cultural ecosystem benefits of urban and Peri-urban green infrastructure across different European countries. *Urban Forestry Urban Greening* 24: 236–248.

Pandit, R., Polyakov, M., Tapsuwan, S., and Morand, T. (2013). The effect of street trees on property value in Perth, Western Australia. *Landscape and Urban Planning* 110: 134–142. https://doi.org/10.1016/j.landurbplan.2012.11.001.

Pauleit, S. (2003). Urban street tree plantings: identifying the key requirements. *Proceedings of the ICE-Municipal Engineer* 156 (1): 43–50.

Pauleit, S., Jones, N., Garcia-Martin, G. et al. (2002). Tree establishment practice in towns and cities – results from a European survey. *Urban Forestry and Urban Greening* 1: 83–96.

Pearlmutter, D., Calfapietra, C., Samson, R. et al. (2017). *The Urban Forest: Cultivating Green Infrastructure for People and the Environment*, 2017. Cham, Switzerland: Springer International Publishing AG ISBN 978-3-319-50279-3.

Ragula, A. and Chandra, K.K. (2020). Tree species suitable for roadside afforestation and carbon sequestration in Bilaspur, India. *Carbon Management* 11 (4): 369–380. https://doi.org/10.1080/17583004.2020.1790243.

Rao, M., George, L.A., Shandas, V., and Rosenstiel, T.N. (2017). Assessing the potential of land use modification to mitigate ambient NO_2 and its consequences for respiratory health. *International Journal of Environmental Research and Public Health* 14: 750.

Ratola, N. and Jimenez-Guerrero, P. (2017). Modelling benzoapyrene in air and vegetation for different land uses and assessment of increased health risk in the Iberian Peninsula. *Environmental Science and Pollution Research* 24: 11901–11910.

Rhodes, J.R., Ng, C.F., de Villiers, D.L. et al. (2011). Using integrated population modelling to quantify the implications of multiple threatening processes for a rapidly declining population. *Biological Conservation* 144: 1081–1088.

Richardson, E. and Shackleton, C.M. (2014). The extent and perceptions of vandalism as a cause of street tree damage in small towns in the Eastern Cape, South Africa. *Urban Forestry and Urban Greening* 13: 425–432. https://doi.org/10.1016/j.ufug.2014.04.003.

Roloff, A. (2013). *Baume in der Stadt: Besonderheiten – Funktion – Nutzen – Arten – Risiken*, 256. Stuttgart: Ulmer Eugen Verlag.

Roloff, A., Korn, S., and Gillner, S. (2009). The climate-species-matrix to select tree species for urban habitats considering climate change. *Urban Forestry and Urban Greening* 8 (4): 295–308.

Roy, S., Byrne, J., and Pickering, C. (2012). A systematic quantitative review of urban tree benefits, costs, and assessment methods across cities in different climatic zones. *Urban Forestry and Urban Greening* 11: 351–363.

Sanders, R.A. (1981). Diversity in the street trees of Syracuse, New York. *Urban Ecology* 5: 33–43.

Sarbeswar, P., Han, J.H., and Hawken, S. (2018). Urban innovation through policy integration: critical perspectives from 100 smart cities mission in India. *City, Culture and Society* 12: 35–43. https://doi.org/10.1016/j.ccs.2017.06.004.

Seamans, G.S. (2013). Mainstreaming the environmental benefits of street trees. *Urban Forestry and Urban Greening* 12: 2–11.

Shashua-Bar, L., Pearlmutter, D., and Erell, E. (2009). The cooling efficiency of urban landscape strategies in a hot dry climate. *Landscape and Urban Planning* 92: 179–186.

Soares, A.L., Rego, F.C., Mc Pherson, E.G. et al. (2011). Benefits and costs of street trees in Lisbon, Portugal. *Urban Forestry and Urban Greening* 10: 69–78.

Sreetheran, M., Philip, E., and Zakiah, S.M. (2006). A historical perspective of urban tree planting in Malaysia. *Unasylva* 57: 28–33.

Tallis, M., Taylor, G., Sinnett, D., and Freer-Smith, P. (2011). Estimating the removal of atmospheric particulate pollution by the urban tree canopy of London, under current and future environments. *Landscape and Urban Planning* 103: 129–138.

Tarran, J. (2009). *People and trees, providing benefits, overcoming impediments*. In: *Proceedings of the 10th National Street Tree Symposium* (eds. D. Lawry, J. Gardner and M. Bridget), 63–82. Adelaide, South Australia: Adelaide University.

Tiwary, A., Williams, I.D., Heidrich, O. et al. (2016). Development of multi-functional streetscape green infrastructure using a performance index approach. *Environmental Pollution* 208: 209–220.

Tyrvainen, L., Pauleit, S., Seeland, K., and de Vries, S. (2005). *Benefits and uses of urban forests and trees*. In: *Urban Forests and Trees*, vol. 2005 (eds. C. Konijnendijk, K. Nilsson, T. Randrup and J. Schipperijn), 81–114. Germany: Springer-Verlag: Berlin/Heidelberg ISBN 978-3-540-25126-2.

Ugle, P., Rao, S., and Ramachandra, T.V. (2010). Carbon sequestration potential of urban trees. *Lake 2010: Wetlands, Biodiversity and Climate Change 22nd–24th December 2010*, pp. 1–12.

UK Forestry Commission (2015). Right trees for changing climate database. http://www.righttrees4cc.org.uk/ (accessed 14 May 2021).

UNDP (United Nations Development Program) (2010). World urbanization prospects: the 2007 revision population database.

USDA (1990). United States Department of Agriculture. (1990). Benefits of Urban Trees. United States Forest Service Forestry. Report R8-FR 17.

USDA (2003). United States Department of Agriculture. (2003). Benefits of Urban Trees. United States Forest Service Forestry. Report R8-FR 71.

UTGL (2015). *Urban Tree Monitoring Protocols: Field Guide*. Gainesville, FL: Arboriculture Research & Education Academy of the International Society of Arboriculture www.urbantreegrowth.org (accessed 18 May 2021).

Van Dillen, S.M.E., De Vries, S., Groenewegen, P.P., and Spreeuwenberg, P. (2012). Green space in urban neighborhoods and residents health: adding quality to quantity. *Journal of Epidemiology and Community Health* 66: e8.

Vogt, J.M., Watkins, S.L., Mincey, S.K. et al. (2015). Explaining planted-tree survival and growth in urban neighborhoods: a social-ecological approach to studying recently-planted trees in Indianapolis. *Landscape and Urban Planning* 136: 130–143. https://doi.org/10.1016/j.landurbplan.2014.11.021.

Wang, L., Zhao, X., Xu, W. et al. (2016). Correlation analysis of lung cancer and urban spatial factor: based on survey in Shanghai. *Journal Thoracic Disease* 8: 2626–2637.

Wolf, K.L., Lam, S.T., McKeen, J.K. et al. (2020). Urban trees and human health: a scoping review. *International Journal of Environmental Research and Public Health* 17: 4371.

World Bank (2014). World development indicators: urbanization. http://wdi.worldbank.org/table/3.12 (accessed 16 May 2021).

Xiao, Q. and McPherson, E.G. (2002). Rainfall interception by Santa Monica's municipal urban forest. *Urban Ecosystem* 6: 291–302.

Zhao, M., Kong, Z., Escobedo, F.J., and Gao, J. (2010). Impacts of urban forests on offsetting carbon emissions from industrial energy use in Hangzhou, China. *Journal of Environmental Management* 91 (4): 806–813.

10

Environmental Status of Green Spaces in Bhaktapur District of Nepal – 2019

Samin Poudel[1,2], Shahnawaz[2], and Him L. Shrestha[1]

[1] *UNIGIS Kathmandu is a collaboration program of Kathmandu Forestry College affiliated with the University of Salzburg.*
[2] *Department of Geoinformatics - Z_GIS, University of Salzburg, Salzburg, Austria*

10.1 Introduction

Green spaces have been studied across multiple disciplines but the definition of green spaces varies across studies (Taylor and Hochuli 2017). USEPA (2017) defines green spaces as lands that are completely or partially covered with trees, grass, shrubs, and other vegetation. M'Ikiugu et al. (2012) define green spaces as outdoor places that have a significant amount of vegetation mainly semi-natural areas, parks, gardens, and other scattered vegetation.

Urban green spaces (UGS) are an integral part of the urban ecosystem and quantifying these spaces are of substantial importance in urban development (Liu et al. 2016). Jo (2002) suggests UGS makes only a partial contribution in the reduction of atmospheric carbon through UGS planning and management could be one of the more time-saving and cost-effective ways to slow the impact of climate change than to develop alternative energy sources. Previous studies (Liu et al. 2016; Loi et al. 2015; Gupta et al. 2012) have examined the green spaces focusing on the availability and distribution of green spaces. There have been studies by (Pokhrel 2019; Poudel 2012; Ishtiaque et al. 2017; Haack et al. 2002) regarding urban sprawl, suitability of green spaces in the study area but not looking into the environmental status of green spaces.

Asian cities have witnessed rapid urban transition characterised by uneven demographic densities, landscape changes, traffic and congestion, and various environmental challenges (Lahoti et al. 2019). Nepal is a landlocked country surrounded by China to the north and India to the East, West, and South. Cities in Nepal and mostly Kathmandu valley have become the educational and employment hub of the nation inducing large immigration in a rapid manner. Kathmandu Valley has experienced unprecedented urban growth of 4% per year since the 1970s, one of the fastest-growing metropolitan in South Asia (Muzzini and Aparicio 2013), with little intervention from the government to impose any restriction on land use (Ishtiaque et al. 2017).

Urban Ecology and Global Climate Change, First Edition. Edited by Rahul Bhadouria, Shweta Upadhyay, Sachchidanand Tripathi, and Pardeep Singh.

Recently this urban sprawl in the valley is mainly concentrated in the outskirts of the valley like areas in Bhaktapur district and other municipalities surrounding Kathmandu Metropolitan because of scarcity of land in Metropolitan as well as the availability of cheaper lands in the outskirts. In 2009 and 2016, the built-up areas of Madhyapur Thimi and Bhaktapur Municipal areas were among the fastest growing in Kathmandu Valley (Ishtiaque et al. 2017). This unmanaged and unplanned urban sprawl has compromised the available green spaces in the district. Community parks, playgrounds, and such green spaces are rare and not even available for many communities. With fewer green spaces, and rise in population, infrastructures, and vehicles the pollution level has increased significantly. Air pollution, water pollution, and other environmental problems are prominent in the region with serious impacts on people and other inhabitants.

This study aimed to identify, analyse the quality, assess the distribution, and determine the environmental status of green spaces in Bhaktapur district. This study follows the definition of green spaces as land covered with trees, shrubs, grass, and includes blue spaces in integration for their similar effects on nature and ecosystem. The study used identification, quantification, and scoring of green spaces using environmental parameters weights and ranking them as a representation of environmental status of green spaces. The research begins with land use land cover (LULC) classification based on Sentinel-2 imagery. The classes were vectorised and buffered to 100, 200, 300, 400, and 500 m distance each and integrated using arithmetic weighted overlay to derive the distance to green spaces in the district. The study followed (Loi et al. 2015) approach for analysing weighted urban green space index (WUGSI). WUGSI map was prepared based on weighted LULC, vegetation index, and distance to green spaces. Analysis of the environmental status of green spaces was carried out based on weighted UGS index and identified environmental parameters.

The study can be useful in green space planning. The availability and status of green spaces can help plan for evenly distributed green spaces for optimum environmental benefit. The identified green spaces of lower status can be upgraded to high quality while preserving green spaces of high environmental status. The identified areas with less quality green spaces can be can plan for new quality green spaces. This study can be beneficial to the local citizens concerned with the quality of the environment of the community, planners, land developers, local decision-makers, and public authorities responsible for urban development, environmental management, social affairs, and public health organisation.

10.2 Literature Review

10.2.1 Urban Development Overview

More than half of the world population are living in urban areas and it is projected to grow by 2.5 billion people between 2018 and 2050, with half of this growth concentrated in Asia and Africa (UN 2018). Research over several decades has shown that urbanisation has impacted atmospheric conditions, ecosystems, changes in carbon cycle throughout the globe possibly inducing climate change (Foley et al. 2005). Urban areas have been facing ever-increasing environmental problems (Apud et al. 2020). The changes as a result of built-up development have been increasingly recognised as a critical factor in relation to global change (Kumar et al. 2014).

In earlier stages of built-up development people tend to migrate inward to the city core but after a certain time urban development when unplanned usually moves outward of the city centre mostly because of overcrowding, traffic, and increased land prices (Rukhsana and Hasnine 2020). The uncontrolled changes in spatial structures due to unplanned physical development, population, and economic growth and rural–urban migration has serious implications on environmental and social factors which decreases the quality of life of residents (Nouri et al. 2014). The assessment of LULC transformation using remote sensing (RS) and geographic information system (GIS) analysis is of vital importance for the effective management of environmental and land resources (Shen et al. 2020).

10.2.2 Roles of Green Space

UGS are a vital contributor for sustainable development while the benefits of UGS stretches from economic, social, ecological, and environmental benefits to health and aesthetic (Haq 2011). There have been various studies that highlight the role of UGS in diminishing pollution, noise, and reduce other anthropogenic stresses on natural environment. Research has shown that Parks are able to filter up to 80% of air pollution, reduce lead content in air, and reduce noise by up to 12 dB, similarly, trees in avenues can filter up to 60% pollutants (Bernatzky 1982). Green spaces contribute significantly to improve microclimate and reduce pollution in a city (Makhelouf 2009).

UGS are deemed necessary as nature-based solutions for mitigating environmental problems caused by urbanisation ranging from reduction in noise and air pollution, and to reducing the impact of climate change (Andersson et al. 2017). The diversity and quality of UGS and human well-being has been linked to a wide range of beneficial ecosystem services of UGS, urban heat mitigation, and stormwater infiltration (Kopecká et al. 2017). Barton and Rogerson (2017) highlight the necessity of UGS for mental health and well-being of residents.

10.2.3 Green Spaces

Almanza et al. (2012) associated green spaces with vegetated parks, walkways, and playfields while describing vegetation levels ranging from sparsely vegetated streets, tree-lined walkways to playfields, parks, and forests. Tavernia and Reed (2009) described green spaces as areas of open lands, agricultural lands, forest, and wood perennials while associating green spaces including agriculture. Aydin and Çukur (2012) associated green spaces with the service to human needs and health while defining green spaces as a land use with prominent contribution to urban environment, ecology, aesthetics, and public health. Heinonen et al. (2014) also associated green spaces with human use and defined green spaces as recreational or undeveloped land. Gentin (2011) described green spaces as areas with significant green components while associating green spaces with considerable vegetation. Heckert (2013) similarly associated green space with vegetation and described as an area significantly covered with vegetation.

Gupta et al. (2012) define UGS in association with parks, gardens, informal vegetations, and riverfronts. Barton and Rogerson (2017) associated green spaces with diverse environmental areas such as nature reserves and wilderness environments. Gupta and Goyal (2014) categorised green spaces as parks, gardens, recreational areas, informal green spaces like riverfronts, playgrounds, and oxygen-producing greenery and pollution absorbent areas. Stanley et al. (2012) provide a multidisciplinary typology of open spaces while including green spaces as natural areas, semiwild areas, parks, gardens, cemeteries, orchards, agricultural fields, grazing areas, horticulture, and kitchen gardens. In contrast to the definition of vegetation, Kimengsi and Fogwe (2017) define green spaces as areas consisting entirely of trees.

Green spaces are most of the time defined in association with vegetation and a natural element though not at all the time (Taylor and Hochuli 2017). This consideration of vegetation as major factor is explained by Haq (2011) suggesting that ecologists, economists, social scientists, and planners agree on the definition of UGS as public or private open spaces in urban areas, covered primarily by vegetation, are directly (places of active/passive recreation) or indirectly (positive influence on the urban environment) available to people. The varying definition of green spaces across studies provides a challenge for researchers for developing a universal agreement on UGS but at the same time open a wide range of possibilities of defining green spaces as the requirement of the study.

10.2.4 Relevance of Green Space Study

The issues of land-use change and the environmental impact of urbanisation are increasingly gaining research attention (Zhou and Wang 2011). The bowl-shaped structure of Kathmandu valley restricts the passage of wind and the heavier pollutants remain in the air as such the valley area enclosing is particularly vulnerable to air pollution (Saud and Paudel 2018), which suggests availability of UGS is of significant importance. United Nations Development Plan (UNDP), Sustainable Development Goals (SDG) 2030 has highlighted "Sustainable Cities and Communities" as the 11th goal to achieve for 2030 which considers creating green spaces as one of the strategies for achieving this goal. UNDP (2016) target seven states with universal access to green and public space by 2030. WHO (2017) considers equality impacts of green spaces, availability, and accessibility as characteristic to impacts of UGS on health and well-being.

Similarly, the Planning Norms and Standard, 2015 prepared by the Ministry of Urban Development, Nepal has mentioned the preservation of the Natural Resource Promoted Area. Natural Resource Promoted Areas is defined as the urban area which is relatively undisturbed by humans but preserved in the natural form which includes all the natural and manmade greenery, water bodies, forest, environmentally sensitive area, and agriculture. The Land Use Policy, 2015 developed by the Ministry of Urban Development have made policies to tackle the challenge to conserve, develop and manage forests and greenbelts, open spaces, water, watershed or wetlands to mitigate the impact of climate changes and environmental protection.

10.2.5 Measurement of Green Spaces

Measurement of green spaces ensures the status of present availability and accessibility of available green spaces to urban inhabitants. Past research has shown that evaluation of green spaces is based on factors such as determining the quantity of UGS, determining the existing qualities; benefits and utilisation and lastly the functionality of UGS on the basis of location and accessibility (Haq 2011). There are two main methods to assess and measure the UGS subjective and objective method (Liu et al. 2016). The subjective methods use visual interpretation and self-reporting to evaluate UGS (Hoehner et al. 2005; Ellaway et al. 2005; Giles-Corti et al. 2005) while the objective methods use remote sensing techniques to measure the percentage of green space (Faryadi and Taheri 2009; Hur et al. 2010; Ruangrit and Sokhi 2004; Liu et al. 2016).

10.2.6 Indices for Measurement of Green Spaces

The urban neighbourhood green index (UNGI) has considered four parameters, percentage of green, built-up density, proximity to green, and height of the structure to compare the distribution of green spaces in neighbourhoods in terms and quantity and distribution (Gupta et al. 2012). The study used 100 m × 100 m grid cells as a proxy for neighbourhood comparison. Loi et al. (2015) developed and applied WUGSI using remote sensing and GIS techniques in Chandigarh, India, while considering the values of percentage of green, type of green, and proximity to green.

The UGS assessment was carried out sector wise in Chandigarh using comparative assessment of UGS. Liu et al. (2016) considered four parameters, green index (GI), building sparsity, building height, and proximity to green using multi-source high-resolution remote sensing data. The index compared the distribution of green spaces as well as configuration of green spaces in the buildings using leaf area index (LAI) based classification while the height of buildings has been calculated from light detection and ranging (LiDAR) data.

10.3 Study Area

The study area was Bhaktapur district lying in Kathmandu valley neighbouring Kathmandu, the capital city of Nepal, also the most populated city of the country. The proximity to the capital city and availability of undeveloped agricultural lands has resulted in rapid and unplanned infrastructural development in Bhaktapur. This haphazard urban development has compromised the green spaces in the district which contributes to the preservation of the natural environment. The environmental degradation with loss of green spaces could result in a wide range of environmental, health, and social problems. Kathmandu is one of the most polluted cities in the world, without any action for environmental preservation. Bhaktapur will face the same challenge in coming years.

Bhaktapur district is situated in the hilly region of Nepal as a part of Kathmandu Valley along with Kathmandu district comprising capital cities of Nepal and Lalitpur district. The district is a part of the Middle Mountain Physiographic region of Nepal. Bhaktapur lies between latitude, 27° 36′ N and 27° 44′ N and latitudes between 85° 21′ E and 85° 32′ E and covers an area of 119 km^2. The district is surrounded by Kavrepalanchok district in the east, Kathmandu and Lalitpur district in the west, Kathmandu and Kavrepalanchok district in the north and Lalitpur district in the south. The district includes four Municipalities namely Bhaktapur, Suryabinayak, Madhyapur Thimi, and Changunarayan.

Administratively the district is located in Bagmati Province of Nepal and is one of 77 districts of the country. The altitude of the district ranges from 1331 to 2191 m above sea level with regions of the valley and hills. The eastern region and nearly half of the northern and southern region of the district are covered with Mahabharat range hills. According to the census in 2001, the district had a population of 72 543 while according to the census in 2011 the district's population increased to 3 04 651 with a population growth rate of 2.96%. The temperature of the district falls within the warm temperate climate while average temperature ranges from 20 to 25 °C with monthly average temperature ranging from 2 to 35 °C. The average rainfall in the district is 56 mm per year. The summer season falls between April to September and winter between October and March. The location map of Bhaktapur district is presented in Figure 10.1.

10.4 Methods

The methodology of the study is presented in Figure 10.2.

10.4.1 Land Use Land Cover (LULC)

The research began by acquisition of Level 2A Sentinel-2 imagery (26 August 2019) of the study area. The downloaded bands were resampled in sentinel application platform (SNAP) to 10 m resolution using the reference band source product Band-2 tile with a resampling tool. The process of layer stacking was carried out where a single raster image was created from different bands of image. A single raster dataset was created from 11 multiple Sentinel-2 bands using composite bands, a data management tool in ArcGIS pro. The study area boundary was clipped using a clip tool to extract features from an overlayed study area boundary.

Some sections of the imagery were covered with clouds which were removed by using the Pixel editor in ArcGIS Pro by replacing the sections from subsequent images from 15 June 2019 and 20 September 2019. The LULC classification was performed using object-based image classification techniques using image segmentation and creating training samples of each class. The spectral detail value was set as 15.50, spatial detail as 15 and minimum segment size in pixels as 20. The segments of the image based on the spectral and spatial details were identified. A total of 1089 training samples were created for image classification: 230 forest samples, 7 shrubland samples, 14 barren samples, 226 agricultural, 7 grasslands, 17 water, and 588 developed. The samples were collected as evenly as possible throughout the study area so to decrease discrepancy in sampled pixels. Support vector machine was used for training dataset for classification. The segment attributes active chromaticity colour, mean digital number, standard deviation, count of pixels, and compactness were used for the classification. After using classification, tool adjustments were done to correct the misclassified classes using the Reclassify tool in ArcGIS Pro for a better classification result. The accuracy of the image was determined by using stratified random samples and confusion matrix.

The identified LULC classes were Forest, Shrubland, Grassland, Water, Agricultural, Barren, and Developed. The classes Forest, Shrubland, Grassland, and Water were considered green spaces for the study. The weights of each class were derived using Satty's pairwise comparison. The weight of each LULC class is given in Table 10.1.

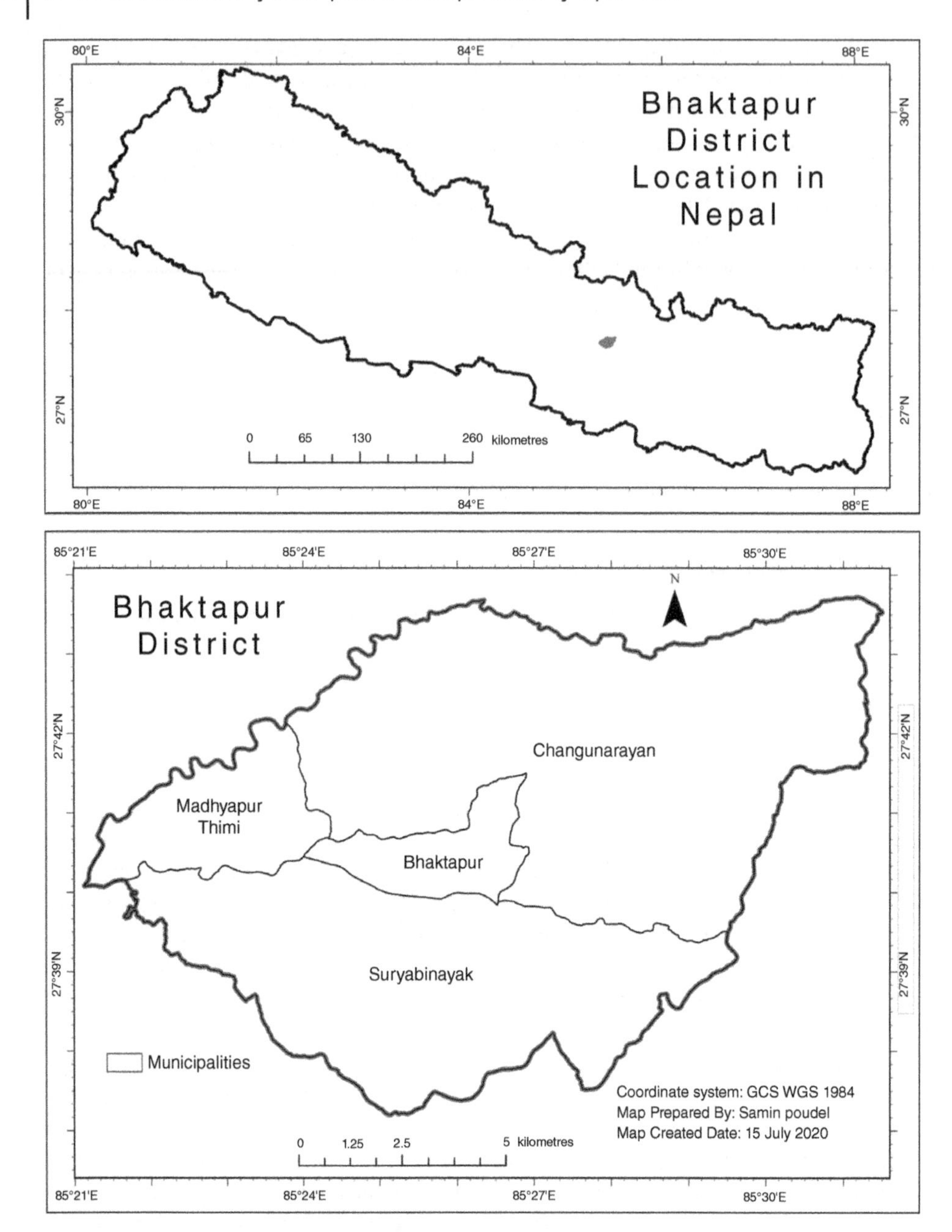

Figure 10.1 Location of Bhaktapur district.

10.4.2 Normalised Difference Vegetation Index (NDVI)

The processed image was used for deriving the normalised difference vegetation index (NDVI). The NDVI was measured using near-infrared (NIR) and Red band Sentinel-2 bands.

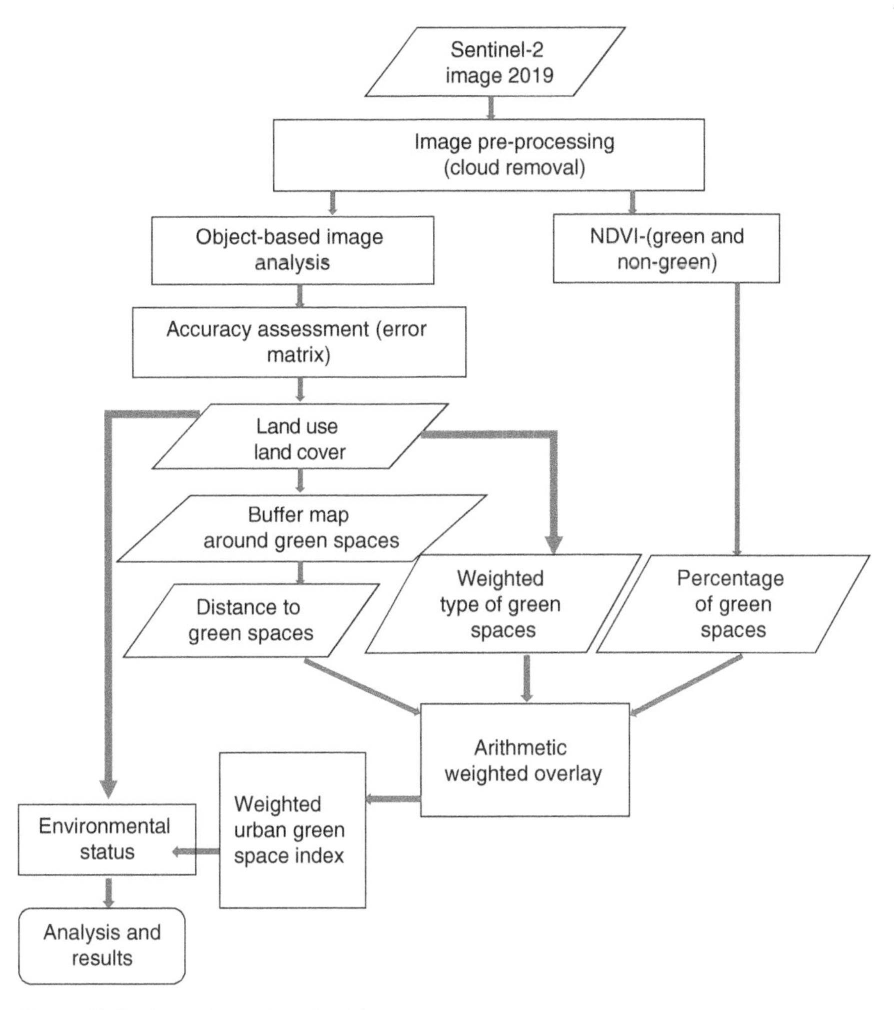

Figure 10.2 Flow chart of methodology.

Table 10.1 Weighted land use land cover (LULC).

Type of class	Weight value	Type of class	Weight value
Forest	10	Grassland	5
Shrubland	7	Water	3

$$NDVI = \frac{(NIR - RED)}{(NIR + RED)} \tag{1}$$

The NDVI value was classified into four classes. The weights assigned to NDVI categories are presented in Table 10.2.

Table 10.2 Weighted NDVI values of green spaces.

Parameter	NDVI value	Re-classified value	Weighted value	Quality classes
Percentage of green	≤0.199	0	0	Very low green
	0.2–0.49	1	3	Low green
	0.5–74	2	6	Moderate green
	0.75–1	3	10	High green

Table 10.3 Weight for each buffer distance to green spaces.

Distance from green spaces (m)	100	200	300	400	500
Forest	10	9	9	8	5
Shrubland	7	6	6	6	3
Grassland	5	4	4	4	2
Water	3	2	2	2	1

10.4.3 Distance to Green Spaces

The classified LULC was used to derive weighted distance to green spaces. Multi-ring buffer for each LULC class was created which was then reclassified with the weightage value. The weights were derived using Satty's pairwise comparison matrix. The weights for each buffer are given in Table 10.3.

10.4.4 Analytical Hierarchy Process (AHP)

The weighted values for LULC and multi-buffer for green spaces were based on analytical hierarchy process (AHP) multi-criteria decision-making tool developed by Professor Thomas L. Satty in the 1970s. The technique uses fundamental comparison scale to make the judgement of paired comparison on a scale of 1–9. To improve the consistency of the weights consistency index (CI) and consistency ratio (CR) was calculated.

$$\mathrm{CI} = \frac{(\lambda_{\max} - n)}{(n-1)} \tag{2}$$

$$\mathrm{CR} = \frac{\mathrm{CI}}{\mathrm{RI}(n)} \tag{3}$$

The acceptance limit of CR is 0.1 and not less than 0.1 (Saaty and Vargas 2012).

Random index (RI) depends on the number of pairs for comparison 'n' and is stated as in Table 10.4 (Kaewfak et al. 2021).

Table 10.4 Random index (RI) values number of pairs (Kaewfak et al. 2021; Saaty 1990).

n	3	4	5	6	7	8	9
RI	0.58	0.90	1.12	1.24	1.32	1.41	1.45

10.4.5 Weighted Urban Green Space Index (WUGSI)

WUGSI based on Loi et al. (2015) was calculated for each cell using arithmetic weighted overlay technique in ArcGIS Pro using the derived weight for LULC (Table 10.1), NDVI (Table 10.2), and distance from green spaces (Table 10.3), the weighted values as input layers were reclassified to this same scale values from 0 to 10. The Reclassify tool was used to assign weights for each layer in ArcGIS Pro.

10.4.6 Environmental Status Parameters

The environmental parameters and weightage were developed based on the environmental expert's consultation. Each parameter was set on a scale of 0–1. The area-wise weightage for water bodies was replaced for the class to sewage disposal as the river features were discontinuous during classification and an immense problem of sewerage disposal in rivers of Bhaktapur district. The weights were 1 for no sewer disposal and −4 for sewer disposal, the negative marking was due to the severity of the problem. The parameters included area, ownership, scale, access, maintenance, UGS class, protection status, waste/litter, wildlife/aquatic life, and WUGSI.

10.5 Results

Figure 10.3 shows the LULC map of Bhaktapur district in 2019. Most of the area in Bhaktapur district was found to have agricultural land use of 48.9% covering an area of 58.15 km^2. The second largest LULC is built-up with 27.3% and area 32.542 km^2 which is observed to be significant in the western region of the district which is closer to the Kathmandu metropolitan and also at the core area of Bhaktapur district which is the old city of Bhaktapur. The built-up expands radially throughout the district along roadways. The forest area of 22.9% with an area of 27.21 km^2 covers the north-eastern to the south-eastern boundary of the district which is also the edge of Kathmandu valley boundary of Bhaktapur district and also the high sloped hill areas.

The overall accuracy for the classified image was found to be 93% as shown in Tables 10.5 and 10.6. User's accuracy ranged from 96 to 70% with highest accuracy for forest and lowest accuracy for barren class, while producer's accuracy ranged from 100 to 44% with the highest accuracy for water and lowest accuracy for Shrubland class. The user's accuracy is considered a more reliable and relevant measure of classification to the user (Rwanga and Ndambuki 2017). The overall Kappa coefficient was 0.89 within the acceptable limit of accuracy. Figure 10.4 shows the green spaces identified in Bhaktapur district which includes Forest, Shrubland, Grassland, and Waterbodies. Suryabinayak and Changunaran

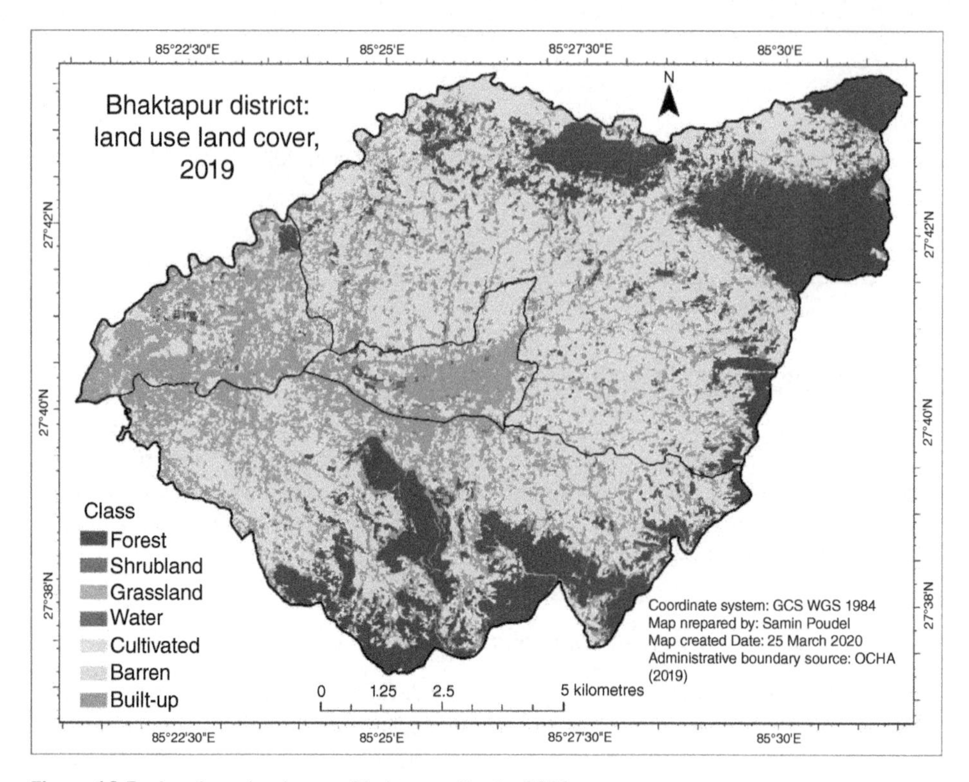

Figure 10.3 Land use land cover, Bhaktapur district 2019.

Table 10.5 Environmental parameters weightage.

	Value		**Value**
Area		UGS class	
To 1000	0.25	Vacant	0.1
1000–3000	0.5	Sports ground	0.25
3000–5000	0.75	Garden	0.5
5000+	1	Grassland/shrubland/riverbank/trees	0.75
Ownership		Park/forest	1
Private	0.25	Protection status	
Institutional	0.5	Not protected	0
Government/community	1	Protected	1
Scale		Waste/litter	
Neighbourhood	0.5	No	1
City	1	Yes	0
Access		Wildlife habitat	
Closed	0.1	No	0
Limited	0.25	Yes	1
Paid	0.5	WUGSI	
Open	1	Low	0
Maintenance		Moderate	0.25
Not maintained	0	High	0.5
Maintained	1	Very high	1

Table 10.6 Confusion matrix for accuracy assessment.

Class value	Forest	Shrubland	Grassland	Water	Cultivated	Barren	Built-up	Total	User's accuracy (%)	Kappa
Forest	109	4	0	0	1	0	0	114	96	0
Shrubland	0	8	0	0	1	0	1	10	80	0
Grassland	0	0	9	0	1	0	0	10	90	0
Water	0	0	0	8	0	0	2	10	80	0
Agricultural	4	3	1	0	230	0	6	244	94	0
Barren	0	0	0	0	1	7	2	10	70	0
Built-up	1	3	0	0	9	0	124	137	91	0
Total	114	18	10	8	243	7	135	535	0	0
Producer's accuracy (%)	96	44	90	100	95	100	92	0	93	0
Kappa	0	0	0	0	0	0	0	0	0	0.89

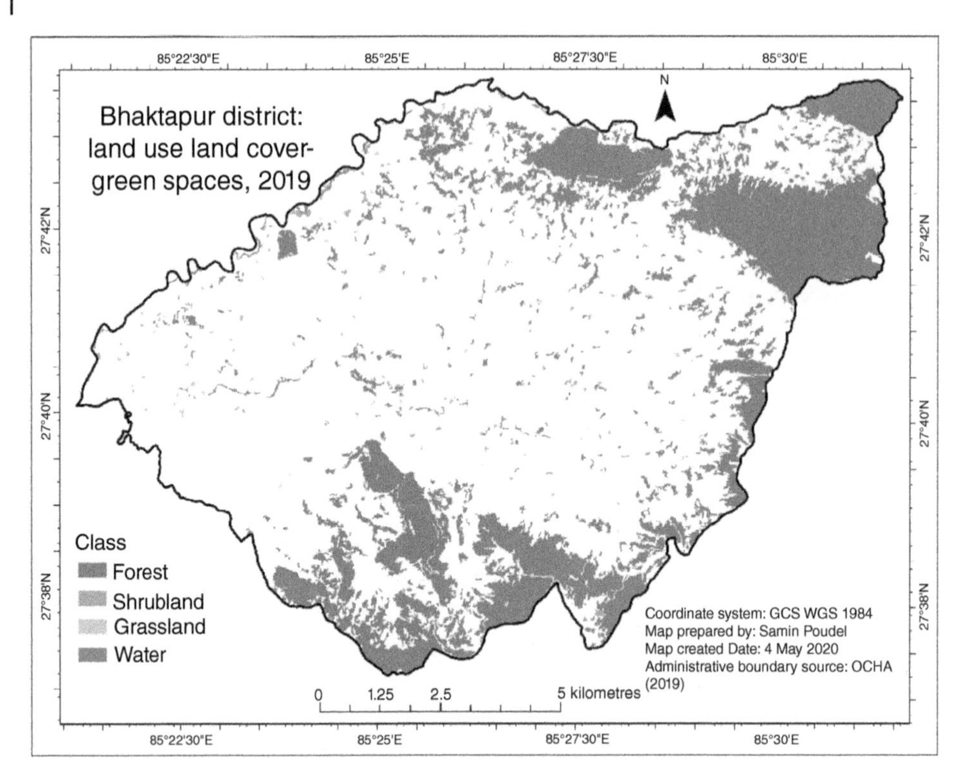

Figure 10.4 Green spaces in Bhaktapur district, 2019.

municipality had the highest 27.3% (11.58 km^2) and 25.4% (16 km^2) forest areas, respectively of the total municipal area.

Figure 10.5 shows the NDVI of the district classified according to Table 10.2. The map showed that the forest area had the highest quality green value in all municipalities of the district while shrublands were the least green. Changunarayan and Suryabinayak municipalities had the highest green value of 44 and 41% respectively.

The calculated value of arithmetic weighted overlay was classified into four classes based on geometric classification techniques. The result of WUGSI for the district of Bhaktapur is illustrated in Figure 10.6, which shows that green spaces are very sparsely distributed throughout the study area and most of the green spaces are farther from the settlement areas. Moderate quality WUGSI accounted for largest area of 28% with an area of 34.38 km^2 followed by 25% of high WUGSI with an area of 30.63 km^2, 24% very high WUGSI with an area of 30.01 km^2 and 23% poor WUGSI covering an area of 28.00 km^2. Changunarayan and Suryabinayak Municipalities have good green quality, i.e. very high-quality WUGSI of 27% covering an area of 16.68 km^2 and 28% covering an area of 11.72 km^2, respectively. Madhyapur Thimi Municipality has the lowest overall WUGSI (Figure 10.7).

Environmental status parameters were used to rank each identified green space with values and sum the values to weigh each identified green space and map them as points. The classification values are presented in Table 10.7.

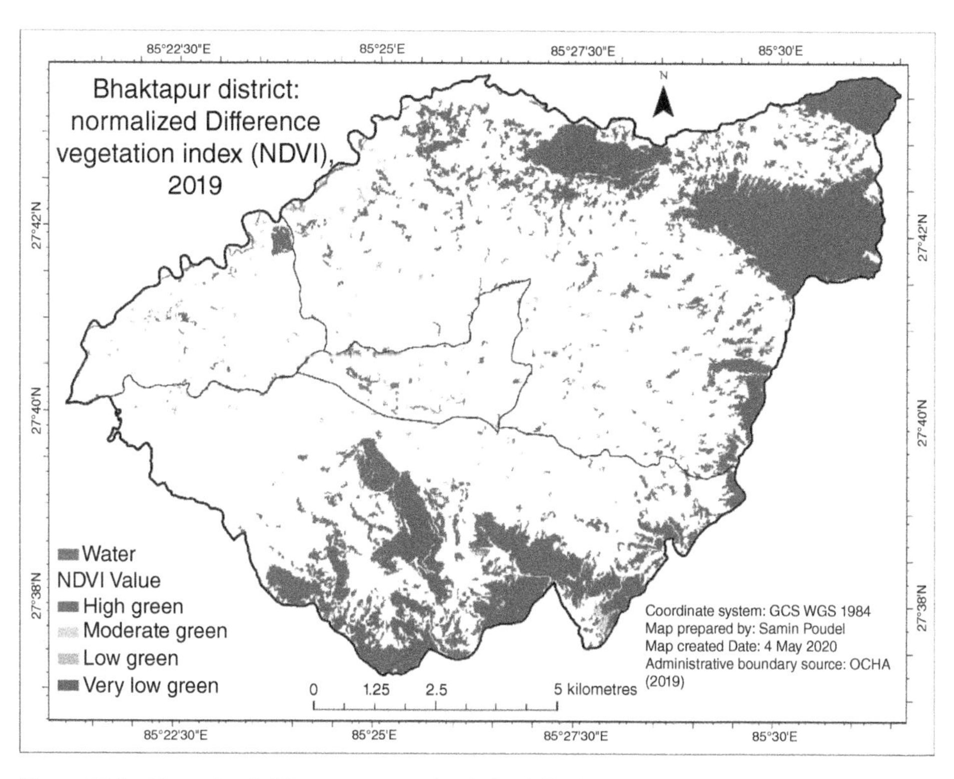

Figure 10.5 Normalised difference vegetation index (NDVI) in Bhaktapur district, 2019.

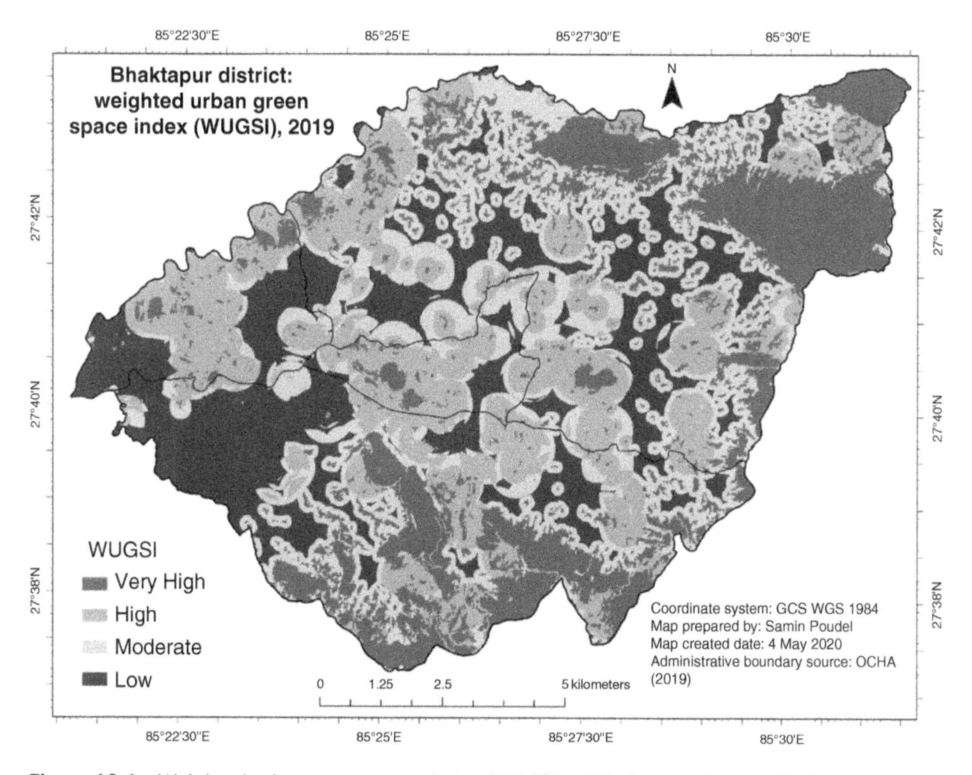

Figure 10.6 Weighted urban green space index (WUGSI) of Bhaktapur district, 2019.

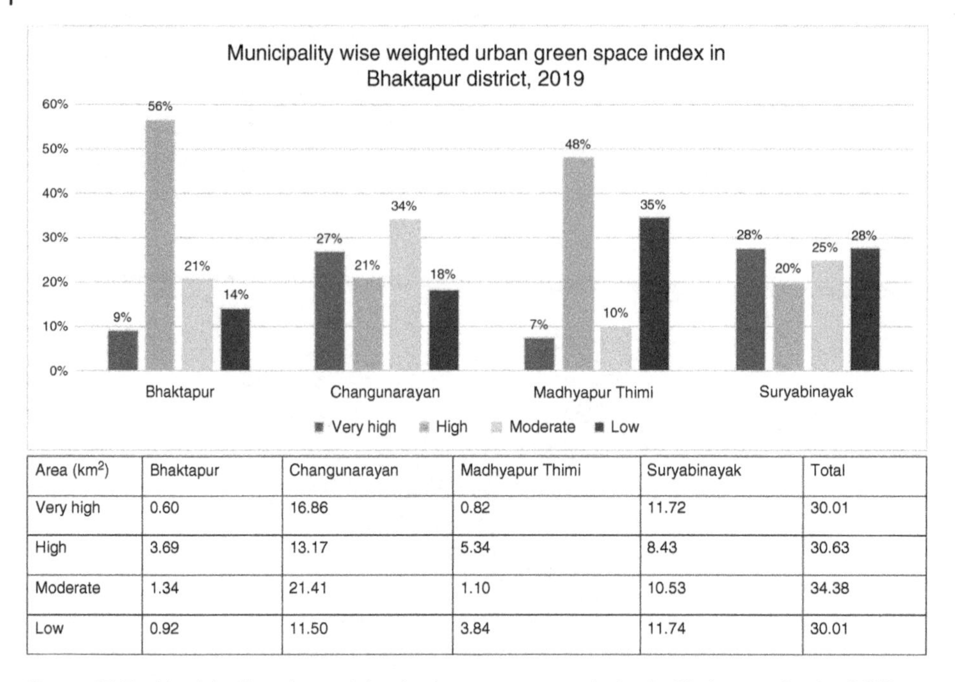

Area (km^2)	Bhaktapur	Changunarayan	Madhyapur Thimi	Suryabinayak	Total
Very high	0.60	16.86	0.82	11.72	30.01
High	3.69	13.17	5.34	8.43	30.63
Moderate	1.34	21.41	1.10	10.53	34.38
Low	0.92	11.50	3.84	11.74	30.01

Figure 10.7 Municipality wise weighted urban green space index in Bhaktapur district, 2019.

Table 10.7 Classification of environmental status values.

Value	Status	Value	Status
>8.5	Excellent	3.25–4.5	Poor
6.76–8.5	Good	<3.25	Very poor
4.6–6.75	Average		

Figure 10.8 represents the environmental status of the green spaces map of Bhaktapur district which includes forests, shrubland, grassland, and water bodies. The histograms of the environmental status of forests, water bodies, shrubland, and grassland are presented in Figure 10.9.

The number of identified forest spaces was 634 out of which 323 spaces had an environmental status less than 4.8. The mean for the status of forest spaces was higher than other spaces with 5.72 value. The number of identified water bodies were 45 out of which 31 spaces had an environmental status less than 2.4. Most of the water spaces were found to be in poor condition mainly because of sewerage disposal. The number of identified Shrubland spaces were 102 out of which 30 spaces had an environmental status less than 3.89. No spaces had a status value more than 8.5. Most of the shrubland spaces were found to be in poor condition with a mean of 4.5. The number of identified grassland spaces were 48 out of which 38 spaces had an environmental status less than 4.14. No spaces had a status value more than 8.25. Most of the grasslands were found to be in poor condition.

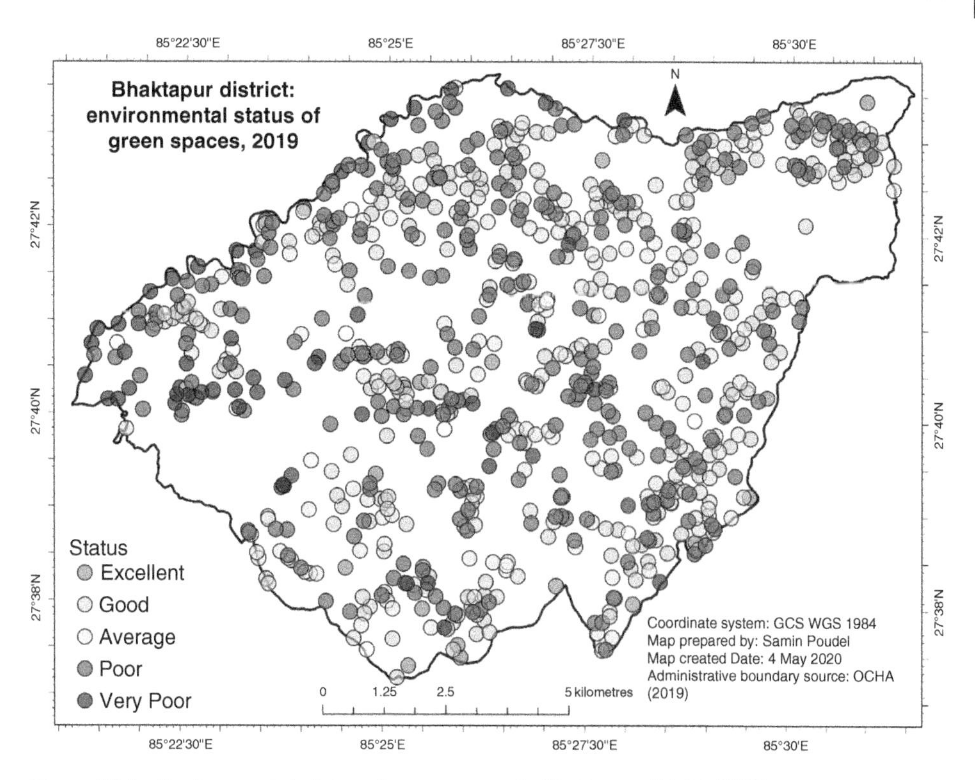

Figure 10.8 Environmental status of green spaces in Bhaktapur district, 2019.

10.6 Discussion

After the industrial revolution, there has been an unprecedented growth of urban residents while consequently leading to land-use change, land fragmentation, biodiversity loss, greenhouse gas emissions, heat islands, and contributing to climate change events (Páez et al. 2015). Resilience to expected climate change in cities has increasingly concerned planners and decision-makers and some of the viable solutions to achieving a liveable and sustainable future are quality UGS planning interventions that are climate effective while ensuring environmental justice (Emilsson and Sang 2017).

The environmental status study in integration with the index developed by Loi et al. (2015) provides a decision support system for effective green space planning. The study provides a basic measurement for identifying and rating green space environmental status. The results indicate that UGS in Bhaktapur were lower quality than Chandigarh, India which was expected as Chandigarh is one of the planned cities in India while urbanisation in Bhaktapur is haphazard. The moderate quality WUGSI accounted for largest area covering 34.38 km^2 (28%) of Bhaktapur district, in comparison the WUGSI in Chandigarh, India had largest high-quality green area with 46.09 km^2 (40.02%) (Loi et al. 2015). The concentration of built-up was found in Bhaktapur and Madhyapur Thimi are in line with the statement by Poudel (2012) with respect to the census of 2001. The urbanisation in Bhaktapur was observed as unplanned and haphazard in nature, randomly sprawling outward of core

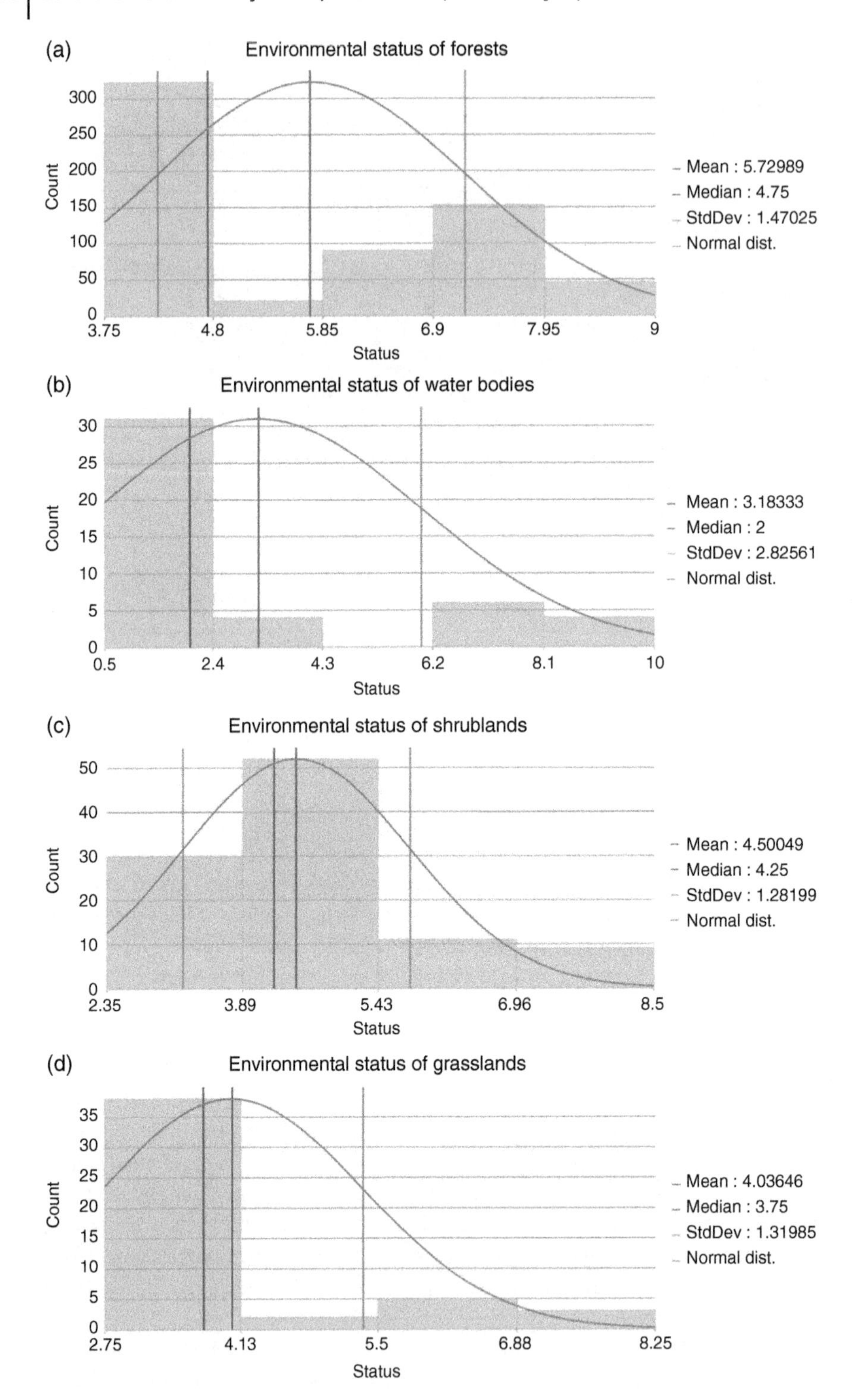

Figure 10.9. Histogram of environmental status of forests (a), water bodies (b), shrubland (c), and grassland (d) in Bhaktapur district, 2019.

town areas with insignificant intervention from the government in agreement with Ishtiaque et al. (2017) while also agreeing on the statement regarding fragmentation of forest habitats and biological corridor as observed during field visit but the actual implication needs to be studied in future.

The WUGSI showed that there was a 24% very high green quality of the total area of the district which is 30.01 km^2 while 23%, an area of 28.00 km^2 had a poor index value. The high-quality green was mostly contributed by forest areas in Suryabinayak and Changunarayan Municipality and cluster of small green spaces in Bhaktapur Municipality. The mean for the environmental status of green spaces in the district remained 5.8 (Average) on a scale of 1–10 for all green spaces mostly contributed by forest areas with an average environmental status while grassland, shrubland, and waterbodies had a mean poor status.

The forest showed better status which may be the result of protection status adopted by the government of Nepal for forests and trees while other green spaces are neglected. All four municipalities of the district had few parks and designated green spaces. The water bodies were facing major pollution due to sewage disposal. Though forests were conserved under government and community on paper the development of roadways were undergoing in forest areas which could create fragmentation in the natural habitat of wildlife. The forest areas accessible to the public were facing litter and waste disposal problems. Madhyapur Thimi and Bhaktapur municipalities had a mix of green spaces while Suryabinayak and Changunarayan mostly had forest areas only.

The study was conducted based on a medium resolution Sentinel image. The spatial resolution of the image of 10 m affected the classification process. In some cases, multiple LULCs were classified in a single pixel. Similarly, the environmental status parameters were only based on site observation and Google earth high-resolution imagery because of the unavailability of data for ownership of lands and wildlife habitat in the district.

10.7 Conclusion

Improving urban environments has become a concern for countries around the world with rapid urbanisation and consequent environmental degradation. Green spaces play a most important role in balancing the anthropogenic impacts on urban environments for sustainable urban development. The quality and distribution of green spaces should be ensured and should meet good environmental status for optimal benefit. Though developed countries have been incorporating greening concepts for their cities developing countries face a major challenge in awareness of green spaces and funding for green space development. The findings from the study showed that the distribution of green spaces is uneven throughout the district while the environmental status of green spaces was mostly in poor condition except for forest spaces with moderate environmental status. Further research is needed to analyse other parameters like chemical composition, biological richness, ecological parameters, and climatic effects relative to green space status. Similarly, socio-economic prospects, health prospects, and ecological prospects in relation to green spaces can be analysed.

References

Almanza, E., Jerrett, M., Dunton, G. et al. (2012). A study of community design, greenness, and physical activity in children using satellite, GPS and accelerometer data. *Health and Place* 18 (1): 46–54. https://doi.org/10.1016/j.healthplace.2011.09.003.

Andersson, E., Borgström, S., and Mcphearson, T. (2017). Double insurance in dealing with extremes: ecological and social factors for making nature-based solutions last. In: *Nature-Based Solutions to Climate Change Adaptation in Urban Areas. Theory and Practice of Urban Sustainability Transitions* (eds. N. Kabisch, H. Korn, J. Stadler and A. Bonn), 51–64. Cham: *Springer* https://doi.org/10.1007/978-3-319-56091-5.

Apud, A., Faggian, R., Sposito, V., and Martino, D. (2020). Suitability analysis and planning of green infrastructure in Montevideo, Uruguay. *Sustainability (Switzerland)* 12 (22): 1–18. https://doi.org/10.3390/su12229683.

Aydin, S.M.B. and Çukur, D. (2012). Maintaining the carbon-oxygen balance in residential areas: a method proposal for land use planning. *Urban Forestry and Urban Greening* 11 (1): 87–94. https://doi.org/10.1016/j.ufug.2011.09.008.

Barton, J. and Rogerson, M. (2017). Looking back to the future: the re-emergence of green care the importance of greenspace for mental health. *Bjpsych International* 14 (4): 79–81.

Bernatzky, A. (1982). The contribution of tress and green spaces to a town climate. *Energy and Buildings* 5 (1): 1–10. https://doi.org/10.1016/0378-7788(82)90022-6.

Ellaway, A., Macintyre, S., and Bonnefoy, X. (2005). Grafiti, greenery, and obesity in adults: secondary analysis of European cross sectional survey. *British Medical Journal* 331 (7517): 606–611. https://doi.org/10.1136/bmj.38583.728484.3A.

Emilsson, T. and Sang, A.O. (2017). Impacts of climate change on urban areas and nature-based solutions for adaptation. In: *Nature-Based Solutions to Climate Change Adaptation in Urban Areas. Theory and Practice of Urban Sustainability Transitions* (eds. N. Kabisch, H. Korn, J. Stadler and A. Bonn), 15–27. Cham: Springer https://doi.org/10.1007/978-3-319-56091-5_2.

Faryadi, S. and Taheri, S. (2009). Interconnections of urban green spaces and environmental quality of tehran. *International Journal of Environmental Research* 3 (2): 199–208.

Foley, J.A., DeFries, R., Asner, G.P. et al. (2005). Global consequences of land use. *Science* 309 (5734): 570–574. https://doi.org/10.1126/science.1111772.

Gentin, S. (2011). Outdoor recreation and ethnicity in Europe-a review. *Urban Forestry and Urban Greening* 10 (3): 153–161. https://doi.org/10.1016/j.ufug.2011.05.002.

Giles-Corti, B., Broomhall, M.H., Knuiman, M. et al. (2005). Increasing walking: how important is distance to, attractiveness, and size of public open space? *American Journal of Preventive Medicine* 28 (2 SUPPL. 2): 169–176. https://doi.org/10.1016/j.amepre.2004.10.018.

Gupta, K., Kumar, P., Pathan, S.K., and Sharma, K.P. (2012). Urban neighborhood green index - a measure of green spaces in urban areas. *Landscape and Urban Planning* 105 (3): 325–335. https://doi.org/10.1016/j.landurbplan.2012.01.003.

Gupta, P. and Goyal, S. (2014). Urban expansion and its impact on green spaces of Dehradun city, Uttarakhand, India. *International Journal of Environment* 3 (4): 57–73. https://doi.org/10.3126/ije.v3i4.11731.

Haack, B., Craven, D., Jampoler, S., and Solomon, E. (2002). Urban growth in Kathmandu, Nepal: mapping, analysis, and prediction. In: *Linking People, Place, and Policy* (ed. A. Mansoor), 263–282. Springer US https://doi.org/10.1007/978-1-4615-0985-1_12.

Haq, S.M.A. (2011). Urban green spaces and an integrative approach to sustainable environment. *Journal of Environmental Protection* 02 (05): 601–608. https://doi.org/10.4236/jep.2011.25069.

Heckert, M. (2013). Access and equity in greenspace provision: a comparison of methods to assess the impacts of greening vacant land. *Transactions in GIS* 17 (6): 808–827. https://doi.org/10.1111/tgis.12000.

Heinonen, J.B., Casanova, K., Richardson, A.S., and Larsen, P.G. (2014). Where can they play? Outdoor spaces and physical activity among adolescents in U.S. urbanized areas. *Bone* 23 (1): 1–7. https://doi.org/10.1016/j.ypmed.2010.07.013.Where.

Hoehner, C.M., Ramirez, L.K.B., Elliott, M.B. et al. (2005). Perceived and objective environmental measures and physical activity among urban adults. *American Journal of Preventive Medicine* 28 (2 SUPPL. 2): 105–116. https://doi.org/10.1016/j.amepre.2004.10.023.

Hur, M., Nasar, J.L., and Chun, B. (2010). Neighborhood satisfaction, physical and perceived naturalness and openness. *Journal of Environmental Psychology* 30 (1): 52–59. https://doi.org/10.1016/j.jenvp.2009.05.005.

Ishtiaque, A., Shrestha, M., and Chhetri, N. (2017). Rapid urban growth in the Kathmandu Valley, Nepal: monitoring land use land cover dynamics of a Himalayan City with landsat imageries. *Environments - MDPI* 4 (4): 1–16. https://doi.org/10.3390/environments4040072.

Jo, H.K. (2002). Impacts of urban greenspace on offsetting carbon emissions for middle Korea. *Journal of Environmental Management* 64 (2): 115–126. https://doi.org/10.1006/jema.2001.0491.

Kaewfak, K., Ammarapala, V., and Huynh, V.N. (2021). Multi-objective optimization of freight route choices in multimodal transportation. *International Journal of Computational Intelligence Systems* 14 (1): 794–807. https://doi.org/10.2991/ijcis.d.210126.001.

Kimengsi, J.N. and Fogwe, Z.N. (2017). Urban green development planning opportunities and challenges in sub-saharan Africa: lessons from Bamenda City, Cameroon. *International Journal of Global Sustainability* 1 (1): 1. https://doi.org/10.5296/ijgs.v1i1.11440.

Kopecká, M., Szatmári, D., and Rosina, K. (2017). Analysis of urban green spaces based on Sentinel-2A: case studies from Slovakia†. *Land* 6 (2): 25. https://doi.org/10.3390/land6020025.

Kumar, S., Radhakrishnan, N., and Mathew, S. (2014). Land use change modelling using a Markov model and remote sensing. *Geomatics, Natural Hazards and Risk* 5 (2): 145–156. https://doi.org/10.1080/19475705.2013.795502.

Lahoti, S., Kefi, M., Lahoti, A., and Saito, O. (2019). Mapping methodology of public urban green spaces using GIS: an example of Nagpur City, India. *Sustainability (Switzerland)* 11 (7): 1–23. https://doi.org/10.3390/su10022166.

Liu, Y., Meng, Q., Zhang, J. et al. (2016). An effective building neighborhood green index model for measuring urban green space. *International Journal of Digital Earth* 9 (4): 387–409. https://doi.org/10.1080/17538947.2015.1037870.

Loi, T.D., Tuan, P.A., and Gupta, K. (2015). Development of an index for assessment of urban green spacesat city level. *International Journal of Remote Sensing Applications* 5 (0): 78. https://doi.org/10.14355/ijrsa.2015.05.009.

M'Ikiugu, M.M., Kinoshita, I., and Tashiro, Y. (2012). Urban green space analysis and identification of its potential expansion areas. *Procedia - Social and Behavioral Sciences* 35 (December 2011): 449–458. https://doi.org/10.1016/j.sbspro.2012.02.110.

Makhelouf, A. (2009). The effect of green spaces on urban climate and pollution. *Iran Journal of Environment, Health, Science and Engineering* 6 (1): 35–40.

Muzzini, E. and Aparicio, G. (2013). *Urban Growth and Spatial Transition in Nepal: An Initial Assessment. Directors in Development*. Washington, DC: *World Bank* https://doi.org/10.1596/978-0-8213-9659-9.

Nouri, J., Gharagozlou, A., Arjmandi, R. et al. (2014). Predicting urban land use changes using a CA-Markov model. *Arabian Journal for Science and Engineering* 39 (7): 5565–5573. https://doi.org/10.1007/s13369-014-1119-2.

Páez, D., Rajabifard, A., and Gantiva, J.A.F. (2015). Methodological proposal for measuring and predicting urban green space per capita in a land-use cover change model: case study in Bogota. *Commission 7 Annual Meeting International Federation of Surveyors Article of the Month – January 2018*.

Pokhrel, S. (2019). Green space suitability evaluation for urban resilience: an analysis of Kathmandu Metropolitan City, Nepal. *Environmental Research Communications* 1 (10): 105003. https://doi.org/10.1088/2515-7620/ab4565.

Poudel, K.P. (2012). Urban sprawl and socioeconomic change in the Kathmandu Valley, Nepal. In: *Facets of Social Geography: Facets of Social Geography International and Indian Perspectives* (eds. A. Dutt, V. Wadhwa and F. Costa), 236–251. Foundation Books https://doi.org/10.1017/UPO9788175969360.015.

Ruangrit, V. and Sokhi, B.S. (2004). Remote sensing and gis for urban green space analysis – a case study of Jaipur City, Rajasthan. *Institute of Town Planners, India* 1 (2): 55–67.

Rukhsana and Hasnine, M. (2020). Modelling of potential sites for residential development at South East Peri-Urban of Kolkata. In: *Geoinformatics for Sustainable Development in Asian Cities. ICGGS 2018* (eds. S. Monprapussorn, Z. Lin, A. Sitthi and P. Wetchayont). Cham: Springer Geography. Springer https://doi.org/10.1007/978-3-030-33900-5_14.

Rwanga, S.S. and Ndambuki, J.M. (2017). Accuracy assessment of land use/land cover classification using remote sensing and GIS. *International Journal of Geosciences* 8 (4): 611–622. https://doi.org/10.4236/ijg.2017.84033.

Saaty, T.L. (1990). How to make a decision: the analytic hierarchy process. *European Journal of Operational Research* 48 (1): 9–26. https://doi.org/10.1016/0377-2217(90)90057-I.

Saaty, T.L. and Vargas, L. (2012). Models, methods, concepts & applications of the analytic hierarchy process. *International Series in Operations Research & Management Science 175*: 1–20. https://doi.org/10.1007/978-1-4614-3597-6.

Saud, B. and Paudel, G. (2018). The threat of ambient air pollution in Kathmandu, Nepal. *Journal of Environmental and Public Health* 2018: 1504591. https://doi.org/10.1155/2018/1504591.

Shen, L., Li, J.B., Wheate, R. et al. (2020). Multi-layer perceptron neural network and Markov chain based geospatial analysis of land use and land cover change. *Journal of Environmental Informatics Letters* 3 (1): 29–39. https://doi.org/10.3808/jeil.202000023.

Stanley, B., Stark, B., Johnston, K., and Smith, M. (2012). Urban open spaces in historical perspective: a transdisciplinary typology and analysis. *Urban Geography* 33 (8): 1089–1117. https://doi.org/10.2747/0272-3638.33.8.1089.

Tavernia, B.G. and Reed, J.M. (2009). Spatial extent and habitat context influence the nature and strength of relationships between urbanization measures. *Landscape and Urban Planning* 92 (1): 47–52. https://doi.org/10.1016/j.landurbplan.2009.02.003.

Taylor, L. and Hochuli, D.F. (2017). Defining greenspace: multiple uses across multiple disciplines. *Landscape and Urban Planning* 158: 25–38. https://doi.org/10.1016/j.landurbplan.2016.09.024.

UN (2018). World Urbanization Prospects: The 2018 Revision. Department of Economic and Social Affairs, Population Division. ST/ESA/SER. Vol. 12. https://population.un.org/wup/Publications/Files/WUP2018-Report.pdf

UNDP (2016). Goal 11 | Department of Economic and Social Affairs. *Department of Economic and Social Affairs Sustainable Development Goal 11: Sustainable Cities and Communities.* https://sdgs.un.org/goals/goal11

USEPA (2017). What is urban agriculture? Urban Environmental Program in New England. https://www3.epa.gov/region1/eco/uep/urbanagriculture.html

WHO (2017). Urban green spaces: a brief for action. Regional Office for Europe, 24. http://www.euro.who.int/__data/assets/pdf_file/0010/342289/Urban-Green-Spaces_EN_WHO_web.pdf?ua=1

Zhou, X. and Wang, Y.C. (2011). Spatial-temporal dynamics of urban green space in response to rapid urbanization and greening policies. *Landscape and Urban Planning* 100 (3): 268–277. https://doi.org/10.1016/j.landurbplan.2010.12.013.

11

Challenges and Opportunities of Establishing Jungle Flora Nursery in Urban Settlements

Deepti Sharma[1] *and Sujata Sinha*[2]

[1] *TerraNero Environmental Solutions Pvt. Ltd., Varun Garden, Manpada, Thane (W), Maharashtra, India*
[2] *Department of Botany, Deen Dayal Upadhyaya College (University of Delhi), New Delhi, India*

11.1 Introduction

Forest ecosystems are among the richest repositories of biodiversity on the earth. They provide habitat to 80% of the world's terrestrial species. And yet, forest ecosystems cover only 31% of the earth (FAO and UNEP 2020). Unfortunately, large tracts of forests have been lost beyond recovery (Weisse and Goldman 2017). Lloret and Batllori (2021) have reported the loss of forests due to climate change-induced extreme climate variability. Worryingly, this is an ongoing process that continues unabated despite the continued efforts of conservationists (Garcia et al. 2020). For instance, 27% of global forest loss can still be attributed to deforestation through permanent land-use change for commodity production (Curtis et al. 2018). As per reports, the yearly loss of tree-cover was as high as 29.7 million ha in 2016, which was 51% more than in 2015. In the year 2018, an area the size of the European nation Belgium (12 million ha) was lost (Weisse and Goldman 2017; Garcia et al. 2020). Forest degradation is defined as such changes brought about in a forest stand that lead to the long-term reduction of ecological attributes and functions, especially biodiversity, and change in the pattern of ecosystem goods and services being provided (Jenkins and Schaap 2018).

With increasing urbanisation and reducing forest cover, there is less and less land on which jungle flora species are flourishing. This is especially true for a developing country like India, which has conflictingly shown an appreciable increase in green cover (State of Forest Report 2019). Massive efforts behind increasing the green cover in India did bear fruit – a 2019 report from NASA confirmed that the global green leaf area had increased by 5% since the 2000s, and one-third of this could be attributed to India and China (Chen et al. 2019). Moreover, the study confirmed that this increased green leaf area could be attributed to massive tree plantation drives and intensive agriculture.

Urban Ecology and Global Climate Change, First Edition. Edited by Rahul Bhadouria, Shweta Upadhyay, Sachchidanand Tripathi, and Pardeep Singh.

However, the query is, if this rising green cover is ecologically beneficial? Have the right type of species been used for this purpose? The answer may be a disturbing 'no'! For instance, a survey conducted in the Thane district forest department nurseries revealed fewer than a dozen species per nursery and included several exotics and species of horticultural interest (Sharma et al. 2018).

Also, during several biodiversity surveys and site visits conducted by Sharma et al. (2015, 2017a, 2017b), it was revealed that exotic species had been planted, sometimes even within forest areas. Stands of Eucalyptus in Gadag district in Karnataka (2017b), of *Gliricidia sepium* and *Leucaena leucocephala* in several areas of Maharashtra, including Aurangabad and Pune district, that of *Acacia auriculiformis* near Matheran in Maharashtra (2015), and that of *Prosopis juliflora* (2017a) in Gujarat do create greenery, no doubt, due to their fast rate of growth. However, while *P. juliflora* is a noxious invasive weed (Tiwari 1999; Slate et al. 2020), Eucalyptus has been known to reduce the groundwater level in nearby areas (Mattos et al. 2019). *A. auriculiformis. L. leucocephala* and *G. sepium* do not support arboreal fauna biodiversity as much as native trees, creating what may be referred to as 'silent jungles' (Punalekar et al. 2010; Nulkar 2016). Such plantation of harmful exotic species is over and above the massive monoculture of teak plantations in Indian forests. This is in sharp contrast with well-established ecological tenets that assert that nature thrives on high diversity and exotic species plantation harms natural ecosystem balance. Hence, just like all that glitters is not gold, all that is green need not be the best-suited flora species.

Another ringing alarm bell about plantation drives is the unwitting plantation of trees in locales where the climatic and edaphic factors support grasslands or scrublands (Veldman et al. 2015). Indeed, the plantation of tree species here is akin to ecological violence.

Given the scale at which development projects are taking place in India – with real estate, infrastructure, and urbanisation claiming forest land by the acres, plantation drives, green belt development, avenue plantation, and gardening and landscaping must be adopted at a much larger scale. Not only that, these activities must be undertaken with greater emphasis on the ecological aspects of such activities.

In other words, any plantation activity, whether it is along the roadside area or in private gardens, in green-belts, or during plantation drives, must be viewed holistically as an eco-restoration activity. For instance, in the absence of the appropriate knowledge of ecology and biodiversity, in the aforementioned tree plantation activities, two major problems emerge – the unwitting plantation of exotic species and exclusion of those native species which has numerous benefits, such as providing habitat and food to other fauna species and forming ecological linkages with other flora species. These native species are not popularly known and are usually excluded from commercial or even government nurseries. It must be understood here that plant species do not exist in isolation – there is an extraordinarily intricate web of ecological linkages associated with each species. Be it symbiosis or predation – the healthy existence of a species and its population control within a community is determined by these ecological linkages (Mittelbach and McGill 2019; Heleno et al. 2020).

The population of exotic species may become impossible to control as they spread dangerously, occupying space that could have been utilised by a more ecologically useful native plant. Also, exotic species are unlikely to support biodiversities such as birds, butterflies, bees, or bats. Thus, they end up creating 'silent jungles'. Again, there is a loss of ecological benefits from native species.

Another issue is the bias in the species selected for plantation – we tend to plant species with pretty foliage or flowers, flowering all year, bearing edible fruits, growing quickly, or having religious or cultural worth. However, in this bias, we forget the ecological worth of those hundreds of other native species that may not bear edible fruits, may not bear attractive flowers or foliage, have no cultural significance, may be slow-growing – and in fact, are unattractive. However – each flora species has ecological importance – it is significant for the ecosystem.

Importantly, jungle flora species, and indeed most biodiversity, has been evaluated largely from the human benefit point of view – whether tangible or intangible. It is critical to observe the same through an ecological lens – if one were to conduct an ecological evaluation of such species, their support services in providing food and habitat to other fauna are critical. It acquires special significance when considering the importance of preserving the genetic stock of floral resources under supreme threat. The above thought is in conjunction with the working tenet of environmental ethics, which states that humans are not the most supreme creatures on earth but are rather as much a part of the earth's ecosystem as the other species; also, all natural resources need to be shared equitably between all. This relates closely with the concept of the intrinsic value of the environment, wherein it is stated that the environment has a value higher and beyond what we humans assign to it. It is understood as per environmental ethics that estimating the true worth of the environment will always be beyond human capacity. With this mindset, a whole new dimension has to be added to urban planning, where green infrastructure and ecological engineering ensure that human settlements and native biodiversity begin to co-exist.

With this background, the incorporation of jungle flora species in urban nurseries is a positive effort in the direction of genetic resource conservation and dependent faunal biodiversity augmentation in urban and semi-urban areas. In addition, the inclusion of jungle flora species in urban nurseries will help broaden the diversity in home gardens and city parks purely from an aesthetic viewpoint, with hitherto unseen, unknown flowers or foliage lending their natural beauty to asphalt-laden towns.

Hence, with the inclusion of jungle flora species in commercial urban nurseries, it will be possible to:

1) Maintain a collection of saplings/seedlings/seeds of wild species
2) Propagate the wild species to ensure their ready availability for plantation drives/avenue plantations/landscaping/eco-restoration activities
3) Maintain a databank of the edaphic and climatic requirements of each species, along with guidelines for their maintenance and growth
4) Study their various ecological linkages and maintain a database of the same

However, the incorporation of jungle flora species in urban ecosystems requires massive research and policy-level efforts, as very little data is available on the breeding and propagation techniques of a staggeringly high number of flora species. In this chapter, an effort has been made to collate data regarding common techniques for breeding forest trees. Also, the challenges likely to be encountered while incorporating jungle flora species in commercial urban nurseries and the potential ways out to overcome these challenges have been outlined in the subsequent sections.

11.2 Breeding Techniques: Jungle Flora Species

11.2.1 Plus and Elite Tree Selection

As per Kedharnath (1984), breeding of forest trees as a scientific practice started in Sweden. It largely involves the selection of superior parent trees (plus trees), assembling them as clones in seed orchards in such a way as to ensure more cross-pollination among the different clones and minimise inbreeding. Jo and Wilson (2005) consider plus tree selection as the very first step and a key element towards tree breeding. The purpose is to strike a balance between genetic diversity and the selection of desirable genetic qualities. Several candidate trees are short-listed from a natural stand of trees, based on their superior phenotypic qualities, after which they undergo grading or testing. Post this, plus trees are selected from among candidate trees and recommended for further propagation. The grading or testing on the basis of which this selection is made depends on the specific purpose of the breeding programme and can vary from good timber quality to high growth rate, pest resistance, or fruit-bearing capacity. From among the set of plus trees, elite trees are selected on the basis of progeny testing. The elite tree is used for mass seed production or vegetative propagation (Tewari 1994).

For instance, Singh et al. (2019) reported the propagation of *Prosopis cineraria* by the selection of plus trees from natural stands of the species on the basis of height, clear bole height, and girth at breast height. Arpiwi et al. (2018) worked to select plus trees for the species *Pongamia pinnata*, on the basis of criteria such as total height, clear bole height, diameter at breast height, canopy width, and oil content.

Some precautions to be observed during candidate tree selection are:

1) Not to select trees from a stand where logging has earlier been done – as the loggers are likely to have cut the best trees and left the inferior ones
2) Not to select trees too close to each other as these are likely to have low genetic diversity

11.2.2 Wild Seed Collection

As the name suggests, wild seed collection is the process of collecting seeds of wild plants for conservation. It is the most low-skilled and simplest approach and the first step towards the conservation of jungle flora species. Buckets, traps, pruners, and cloth or paper bags are necessary tools for wild seed collection. While hand-picking works best for large fruits, smaller fruits can be shaken loose from the tree by shaking. Pruning to obtain the seed clusters must be done with the utmost carefulness so as not to harm the plant.

Seeds so collected should either be propagated as seedlings – with or without pre-treatment – and create seed orchards (Hay and Probert 2013). Alternately, they may be cryopreserved in Seed Banks for the future. Cryopreservation of seeds is a highly recommended procedure for the long-term storage of seeds that maintains their viability. For instance, Faria et al. (2020) reported passion fruit cryopreservation using seeds of *Passiflora eichleriana*, *Passiflora crystallina*, and *Passiflora nitida* while Mursaliyeva et al. (2020) have discussed successful seed germination in *Allochrusa gypsophiloides* post cryopreservation and gibberellic acid treatment.

The reason why an entire year must be expended in this activity of wild seed collection is that the conservationists need to cover each season and ensure that different flora species with their differing phenology/flowering and seeding time are all included.

However, extreme care must be taken that:

1) the seed collection is not affecting the natural regeneration process of the forest – in most cases, we will collect ripened fruit containing the seed of the tree
2) there will be no collection from areas recovering from stress (such as forest fire), as here natural regeneration is of prime concern
3) the natural proportion of species among the seedlings/saplings is not disturbed if seeds are collected

Collected seeds may require some pre-treatment to break seed dormancy and hasten the germination process. Seed pre-treatment processes include air-drying or sun-drying of seeds, soaking of seeds in hot or cold water, acid treatment of seeds or irradiation of seeds.

11.2.3 Vegetative Propagation

Rooting of shoot cuttings or grafting are popularly used methods for vegetative propagation of plants. It overcomes the challenges of wild seed collection, seed dormancy, and low seed germination rates and viability remaining a low-cost, low-skill activity that can be undertaken easily without much investment.

11.2.4 Micro-propagation/Plant Tissue Culture (PTC)

Plant tissue culture (PTC) is the cultivation of plants *in vitro*, under sterile conditions, using suitable nutrient media and growth hormone combinations using cells, tissues, or organs of the parent plant. PTC techniques, despite their relatively higher costs, the requirement of dedicated laboratory, chemicals and skilled manpower, provide the option of large-scale propagation of species. The combination of nutrient media, plant hormones and growth conditions has to be optimised for each species and entails research effort to arrive at the best-suited factors. Both organogenesis and somatic embryogenesis pathways have been used to create plantlets (Muralidharan and Mascarenhas 1987; Martínez-Palacios et al. 2003; Bonga 2015).

Research institutes in India that are actively involved in forest tree species breeding are:

1) Kerala Forest Research Institute, Peechi, Kerala
2) Institute of Forest Genetics and Tree Breeding (IFGTB), Coimbatore, Tamil Nadu

Based on a literature survey of extant research works on various techniques for forest tree species propagation from India, the following data has been compiled in Table 11.1.

11.3 Challenges of Including Jungle Flora Species in Urban Nurseries

11.3.1 Lack of Awareness

In urban settlements, the plantation of trees and other floral species is conducted:

1) By private citizens, in their homes, balconies, and gardens
2) By the local municipality along avenues and in public parks and gardens

Table 11.1 Methods of propagation of a few jungle flora species.

Sr. No.	Species name	Propagation technique	Remarks	References
1	*Aquilaria agallocha*	Seed germination	Jun–Jul seed collection; air-drying fruits for 1–2 d in the shade to extract seeds; no seed pre-treatment needed; to be sown within 1–2 d of the collection; transplantation within 4–5 wk	SFRI Information Bulletin No. 9
2	*Amoora wallichii*		Jun–Jul seed collection; sun-drying fruits for 1–2 d to extract seeds; germination time 15–50 d	
3	*Chukrasia tabularis*		Jan–Mar seed collection; overnight soaking in cold water; germination time 7–35 d; transplantation after 30–45 d; sowing in 1 : 1 : 1 mixture of sand, soil, FYM	
4	*Morus laevigata*		Apr–May seed collection; no seed pre-treatment needed; germination time 10–45 d; transplantation after 60 d	
5	*Phoebe goalparensis*	Seed Germination	Oct–Nov seed collection; seeds need to be soaked overnight in tap-water, followed by removal of thin mesocarp of the seed and then immediate sowing after drying – in Nov; germination time 25–90 d; transplantation after 4 mo	
6	*Ailanthus grandis*		Feb–Mar seed collection; no seed pre-treatment needed; seeds are sown flat-horizontal after removal of wings; germination time 25–120 d	State Forest Research Institute, Dept of Environment and Forests, Govt. of Arunachal Pradesh, Itanagar, India
7	*Canarium strictum*		Nov–Jan seed collection; hot water treatment needed for seeds; germination time 26–140 d; high germination rate was observed in sand substratum; transplantation after seedlings attain the 3-leaf stage	

(*Continued*)

Table 11.1 (Continued)

Sr. No.	Species name	Propagation technique	Remarks	References
8	*Terminalia myriocarpa*		Dec–Jan seed collection; no seed pre-treatment needed; germination time 10–35 d; covering the bed with thatch enhances germination; transplantation after seedlings attain the 4-leaf stage	SFRI Information Bulletin No. 9 State Forest Research Institute, Dept of Environment and Forests, Govt. of Arunachal Pradesh, Itanagar, India
9	*Dipterocarpus macrocarpus*		Feb–Mar seed collection; no seed pre-treatment needed; direct sowing after removal of wings in a bed of sand; germination time 8–21 d; transplantation after germination	
10	*Altingia excelsa*		Dec–Feb seed collection; no seed pre-treatment needed; germination time 10–70 d; mulching is essential after sowing to maintain required humidity; transplantation after seedlings attain the 3-leaf stage	
11	*Anthocephalus chinensis*		Jan–Feb; Oct–Nov seed collection; overnight soaking of seeds needed; germination time 10–30 d; better results obtained in small boxes and trays than in seedbeds; transplantation after 60–75 d	
12	*Duabanga grandiflora*		Mar–May seed collection; no seed pre-treatment needed; germination time 10–40 d; transplantation after 60 d	
13	*Acrocarpus fraxinifolius*	Seed germination	Apr–May seed collection; overnight soaking in hot water or 2–3 min dunking in conc. sulphuric acid is needed; germination time 10–45 d in sand–soil–FYM mixture in 1 : 1 : 1 ratio; transplantation after 70 d	
14	*Shorea assamica*		Mar–Apr seed collection; no seed pre-treatment needed; germination time 10–25 d direct sowing in a bed of sand; transplantation immediately after germination	
15	*Mesua ferrea*		Aug–Sept seed collection; no seed pre-treatment needed; germination time 15–90 d; shade is necessary as species cannot tolerate drought conditions; transplantation of sprouted seeds	

16	*Michelia champaca*		Aug–Sept seed collection; no seed pre-treatment needed; germination time 10–45 d; red pulp/aril around seeds should be removed before sowing; transplantation after seedlings attain the 3-leaf stage	
17	*Abrus precatorius*	Seed germination	Germination time 30–40 d; no seed pre-treatment needed	Sharma et al. (2019)
18	*Hiptage benghalensis*	Seed germination	Germination rate 80%; germination time 15–20 d; no seed pre-treatment needed	
19	*Helicteres isora*	Rooting of stem cuttings	Germination rate 80%; germination time 15–20 d; no seed pre-treatment needed	
20	*Costus speciosus*	Seed and rooting of stem cuttings	Germination rate 70%; germination time 25–30 d; no seed pre-treatment needed	
21	*Acorus calamus*	Root suckers	Germination rate 60–70%; germination time 15–20 d	
22	*Sterculia villosa*	Seed germination	Vegetative propagation through grafting was attempted but did not succeed	
23	*Pongamia pinnata*		Germination was within a week. No pre-treatment needed	
24	*Madhuca longifolia*		Germination was within a week. No pre-treatment needed	
25	*Ziziphus jujuba*		Sapling growth rate was relatively slower	
26	*Sapindus emarginatus*	Seed germination	Large-sized seeds placed vertically at a depth of 1.0 cm in sand–soil–humus 1 : 1 : 1 mixture showed the most efficient germination	Venkatesh and Lakshmipathaiah (2009)
27	*Neolamarckia cadamba*		The best soil media for germinating these seeds was pure soil medium (100%); the best soil medium for height growth, diameter growth, and total dry weight of the seedlings was the mixed soil, cow-manure compost, and husk charcoal medium (3 : 1 : 1)	Irawan and Purwanto (2014)

(Continued)

Table 11.1 (Continued)

Sr. No.	Species name	Propagation technique	Remarks	References
28	*Melia dubia*	Stem cuttings	2000 mg l^{-1} IBA (liquid formulation). The sand was identified as the best rooting media for the multiplication of coppice. The stumps produced good coppice when cut at 120 cm above ground level with high survival percentage (60%) and increased sprout product	Geetha et al. (2019)
29	*Gmelina arborea*	Rooting of stem cuttings; seedlings; grafting		Romero (2004)
30	*Calophyllum inophyllum*	Seed germination	Moderately easy to propagate by seeds	Friday and Okano (2006)
31	*Ailanthus excelsa*	Grafting	On other plants like teak, *Dalbergia latifolia* etc.	Tomar et al. (2004)
32	*Sterculia urens*	Plant tissue culture	Murashige and Skoog's (MS) medium containing 2.0 mg l^{-1} 6-benzyl amino-purine (BAP) within 21 d of initial culture	Purohit and Dave (1996)
33	*Terminalia catappa*	Plant tissue culture	Nodal segments were cultured on MS medium supplemented with 6-benzyladenine (BA; 0.5–3.0 mg l^{-1}) or Kinetin (Kn; 0.5–3.0 mg l^{-1}) for bud breaking and multiple shoot induction. About 85% of the explant responded (2.8 ± 0.41 shoots per node with 2.7 ± 0.14 cm length) within 15 d of inoculation	Phulwaria et al. (2012)
34	*Pterocarpus marsupium*		The highest growth was obtained with growing nodal explants on Murashige and Skoog (MS) medium amended with 4.0 lM 6-benzyladenine (BA), 0.5 lM indole-3-acetic acid (IAA), and 20 lM adenine sulphate (AdS)	Husain et al. (2008)
35	*Tecomella undulata*		*Tecomella undulate* nodal explant on MS medium supplemented with indoleacetic acid (IAA) 0.05 mg l^{-1} and benzylaminopurine (BAP) 2.0 mg l^{-1}	Rathore et al. (1991)

36	*Prosopis cineraria*		*Prosopis cineraria* nodal shoot segment from pruned thorny adult trees on MS medium containing 0.1 $mg\,l^{-1}$ indole-3-acetic acid (IAA) + 2.5 $mg\,l^{-1}$ benzylaminopurine (BAP) + additives	Shekhawat et al. (1993)
37	*Ziziphus mauritiana*		*Z. mauritiana* nodal explant on MS + 7.5 $mg\,l^{-1}$ BAP + 0.1 $mg\,l^{-1}$ IAA, four to five shoots were produced	Rathore et al. (1992)
38	*Oroxylum indicum*		Cotyledonary node explant; best medium for proliferation was Murashige–Skoog (MS) medium with 6-benzyladenine (8.87 μM), indole-3-acetic acid (2.85 μM), and gibberellic acid (1.44 μM)	Dalal and Rai (2004)
39	*Bauhinia tomentosa*		Cotyledonary node and stem nodal segments showed maximum response on MS medium supplemented with 0.8 μM of thidiazuron	Naz et al. (2012)
40	*Morus alba*	Rooting of shoot or branch cuttings	Impact of auxin on rooting of shoot cuttings of various forest tree species	Nanda et al. (1968, 1970)
41	*Populus ciliata*			
42	*Dalbergia sissoo*			
43	*Lagerstroemia parviflora*			
44	*Bauhinia variegata*			
45	*Toona ciliata*			
46	*Aesculus indica*			
47	*Platanus orientalis*			
48	*Salix tetrasperma*		Nodal explants were grown on agar-solidified woody plant medium (WPM); shoot induction response was best on WPM supplemented with 6-benzyladenine (5.0 lM)	Khan et al. (2011)

(Continued)

Table 11.1 (Continued)

Sr. No.	Species name	Propagation technique	Remarks	References
49	*Tamarix aphylla*		Cuttings about 1.5 cm thick and about 40 cm long are planted about 30 cm deep in prepared nursery beds during February–March. Nursery beds are watered regularly until the cuttings root. Such rooted cuttings are then planted in the field, around August of the same year	Champion and Seth (1968)
50	*Aisandra butyracea*	Rooting and air-layering of cuttings	The cuttings obtained from the natural population and dipped in 500 ppm each IBA + NAA solution for 24 h prior to planting resulted in 79.7% rooting after 75 d of planting in summer. Only 30% of the air layers could root in summer with the aid of 1000 ppm NAA	Tewari and Dhar (1997)
51	*Excoecaria agallocha*	Cuttings and air-layering	Maximum rooting was recorded when the cuttings and air layers were treated with IBA alone up to 2500 ppm in all the three species. October was found to be best followed for the plantation of cuttings and initiation of air layers	Eganathan et al. (2000)
52	*Heritiera fomes*			
53	*Intsia bijuga*			
54	*Sonneretia apetala*	Air-layering	Rooting was observed only during monsoon and post-monsoon. The twigs were cut off after root formation, planted, and were found to establish themselves in the soil	Kathiresan and Ravikumar (1995)
55	*Xylocarpus granatum*			
56	*Bambusa balcooa*	Plant tissue culture	Multiple shoot formation (8–10) was observed from the excised tender node (12–18 mm in length) containing axillary bud isolated from secondary branches of 1½-yr-old culms when implanted on Murashige and Skoog (MS) medium containing 6-benzylaminopurine (BAP, 1.0 mg l^{-1}).	Mudoi and Borthakur (2009)
57	*Bambusa tulda*	Plant tissue culture	MS liquid medium enriched with 100 mM glutamine, 0.1 mM indole-3-acetic acid, and 12 mM 6-benzylaminopurine was the most effective	Mishra et al. (2008)

58	*Bambusa ventricosa*	Plant tissue culture	6-benzyladenine (6-BA) promoted bud sprouting, multiple bud induction, and proliferation while a-naphthaleneacetic acid (NAA) induced rooting in proliferated buds. The MS medium containing 22.2 lM 6-BA was optimal for bud initiation MS medium with 22.2lM 6-BA, 0.23lM Thidiazuron (TDZ: *N*-phenyl-*N*-[(1, 2, 3-thidiazol-5-yl) urea]), and 0.27lM NAA was effective for bud proliferation	Wei et al. (2015)
59	*Dendrocalamus strictus*	Rooting of planting stocks and plant tissue culture	*In vitro* regeneration through nodal culture; Murashige and Skoog's medium supplemented with 4 $mg\,l^{-1}$ BAP was found to be most effective	Goyal et al. (2015)
60	*Woodfordia fructosa*	Plant tissue culture	*In vitro* shoot tip culture from field-grown flowering plants and re-culture of the nodal segments of regenerated shoots in Schenk and Hildebrandt (1972) medium	Krishnan and Seeni (1994)
61	*Rhynchostylis retusa*		Half and quarter MS medium gave the earliest growth, while full MS medium supplemented with 10% coconut water helped shoot formation while full MS medium with fungal elicitor CVS4 aided root development	Oliya et al. (2021)
62	*Dendrobium densiflorum*		Seed culture on MS medium; half MS medium with 10% coconut water supported maximum seed generation. Full-strength MS medium with 15% coconut water aided highest number of shoot formation while maximum root development was on full-strength MS medium with 1.5 $mg\,l^{-1}$ IBA	Pant et al. (2021)
63	*Crotalaria longipes*		Nodal explants were cultured on MS medium with thidiazuron (TDZ) 1.0 $mg\,l^{-1}$ and NAA 0.5 $mg\,l^{-1}$ for maximum shoot induction	Seventhilingam et al. (2021)
64	*Rauwolfia serpentina*		Explant culture on full-strength MS medium	Ratnam (2021)

Whether individuals or municipalities, the source of saplings remains the private or government-owned plant nurseries; most of these nurseries give little or no thought to the ecological benefits of a species, relying more on its ornamental nature, faster growth rate or lower maintenance. Indeed, most of our nurseries are not equipped with the jungle flora species.

The challenge of inadequate awareness lies both on the supply and the demand sides, for most buyers are unbothered about the ecology of plants being purchased. Whether the species being sold are of native origin or exotic is thought of, if at all, much after the decision to purchase has been made based on concerns such as economic, aesthetic, medicinal, or socio-cultural.

11.3.2 Lack of Availability

Most of the commercial nurseries are mere retailers and not active plant-breeding centres. They purchase saplings germinated or micro-propagated elsewhere in horticultural research institutes or mass-cultivated in villages as other agricultural produce. On the other hand, the propagation of jungle flora species is the onus of forest nurseries and research centres, which work almost exclusively for reforestation and afforestation projects. Very few – if any – commercial nurseries are connected to such forestry research institutes. As a result, several native forest species of undeniable ornamental or medicinal benefits do not find their way in commercial nurseries.

Also, most of the research in forest species is centred on tree species, keeping in mind the perspectives of the lumber industry or agroforestry and silviculture concerns. Shrubs, herbs and climbers growing in the jungles have not been adequately researched at all. For instance, a Google Scholar search with keywords such as 'wild shrub species propagation', 'wild herb species propagation', 'forest shrub propagation', 'forest herb species propagation', did not yield relevant hits.

Thus, out of the jungle trees biodiversity repertoire, only a small fraction trickles into the nurseries, omitting hundreds of slow-growing, inedible fruit-bearing flora species with little-known benefits.

11.3.3 Lack of Research on Propagation Techniques

Some jungle flora species germinate slowly or require seed treatment prior to germination. The seed viability and germination rates can often be low. Hence, adequate species-specific research is required to understand how to break such seed dormancy periods or standardise other grafting and tissue culture techniques. Plant tissue culture techniques are costlier and ultimately produce clones of low genetic resource diversity. While nature thrives on diversity, any plant propagation done for commercial purposes will require uniformity of product. Hence, we keep coming back to the original dilemma of conservation of species for the sake of species and not human needs.

With more than 18 532 angiosperms, 81 gymnosperms, 2754 bryophytes, and 1293 pteridophytes (Singh 2020), the fact that research has been conducted on a mere handful of species stand out starkly.

11.3.4 Air Pollution Tolerance

Plantation along highways and roads with high traffic movement invariably exposes plants to the high content of dust, carbon, SOx (various sulphur oxides), and NOx (various nitrogen oxides), which can severely hamper plant growth and health. Thus, laboratory-based calculation of the air pollution tolerance index (APTI), which was designed by Singh et al. (1991) and has been used in several studies (Kaur and Nagpal 2017; Bharti et al. 2018; Yadav and Pandey 2020) may help select which jungle flora species are most suitable for avenue plantation. Briefly, this index takes into note the ascorbic acid, moisture, total chlorophyll content, and leaf extract pH to ascertain the air pollution tolerance of plant species and assigns values to them from 0 to 100. Plants scoring more than 30 in this index can tolerate air pollution.

Another index that uses the same 0–100 scale and scoring rules is the anticipated performance index (API), which adds the factor of the socio-economic importance of flora species during calculation (Panda et al. 2018). Several researchers working in the field of urban plantation have used API (Prajapati and Tripathi 2008; Pathak et al. 2011; Pandey et al. 2015).

A project undertaken by Sharma et al. (2019) for the Thane Municipal Corporation (Maharashtra) revealed that *Quisqualis indica, Passiflora* sp., *Thunbergia grandiflora, Operculina turpethum, Tinospora cordifolia*, and *Clitoria ternatea* showed reasonably high APTI values.

11.3.5 Over-exploitation Risk

The fear of over-exploitation of genetic resources from forests is real (Babu and Nautiyal 2015). Hence, activities of wild seed collection, micro-propagation, or grafting must be undertaken in a controlled manner, in cohesion with the forest department and their research institutes, working within the provisions of the existing legal framework in India.

11.4 Overcoming the Challenges

11.4.1 Creation of Native Forest Species Seed Banks

A seed bank is a repository of seeds and is hence a type of gene bank. A seed bank serves the purpose of preserving native genetic diversity and offers options of various research avenues from conservation to commercial. By preserving seeds of species that were once culturally popular, seed banks also serve a social purpose. Keeping in mind that seed species will have limited viability, seed banks alter the moisture and temperature conditions to ensure the longevity of the seeds.

11.4.2 Connecting the Dots Between Forest Species Research Institutes and Commercial Nurseries

Existing forest species research centres may initiate work in collaboration with commercial nurseries to conduct more copious research on jungle flora species that have potential as ornamentals. The volume of research required to bring more forest flora species to urban

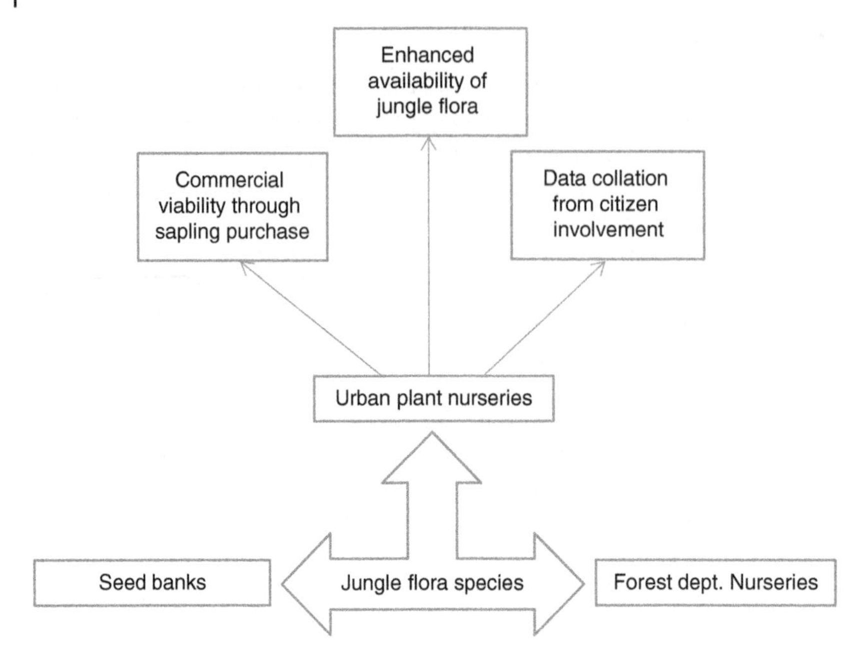

Figure 11.1 Socio-scientific importance of adding jungle flora species to urban plant nurseries.

home gardens is humongous, requiring manpower, and funds. If used judiciously, the involvement of commercial plant nurseries and home garden enthusiasts can convert this into a citizen science project with immense ecological benefits at lower costs.

11.4.3 Purchase of Wild Seeds

Wild seeds as a part of Non-Timber Forest Products can be harvested by local tribal populations in an ecologically sustainable manner. Purchase of such wild seeds by commercial nurseries working in conjunction with the forest department can be a positive step in this direction (Figure 11.1).

11.5 Conclusion

To conclude, the inclusion of jungle flora species in urban nurseries has high potential as a tool for conservation. However, this is a task ridden with challenges of massive investment in funds and skilled manpower for the required research. On the other hand, tackling these challenges with a few species at a time, a few case studies at a time can go a long way towards achieving the goal. The innovative approach of creating a bridge between forest research institutes and commercial nurseries and bringing in the aspects of citizen science in the project can help realise these aims. Future researchers may attempt to collect wild seeds or cuttings with help from the forest department and local tribal folk and try propagation using the best scientific practices.

References

Arpiwi, N.L., Wahyuni, I.G.A.S., Muksin, I.K., and Sutomo, S. (2018). Conservation and selection of plus trees of *Pongamia pinnata* in Bali, Indonesia. *Biodiversitas Journal of Biological Diversity* 19 (5): 1607–1614.

Babu, M.U. and Nautiyal, S. (2015). Conservation and management of forest resources in India: ancient and current perspectives. *Natural Resources* 6 (04): 256.

Bharti, S.K., Trivedi, A., and Kumar, N. (2018). Air pollution tolerance index of plants growing near an industrial site. *Urban Climate* 24: 820–829.

Bonga, J.M. (2015). A comparative evaluation of the application of somatic embryogenesis, rooting of cuttings, and organogenesis of conifers. *Canadian Journal of Forest Research* 45 (4): 379–383.

Champion, H.G. and Seth, S.K. (1968). General silviculture for India.

Chen, C., Park, T., Wang, X. et al. (2019). China and India lead in greening of the world through land-use management. *Nature Sustainability* 2 (2): 122–129.

Curtis, P.G., Slay, C.M., Harris, N.L. et al. (2018). Classifying drivers of global forest loss. *Science* 361 (6407): 1108–1111.

Dalal, N.V. and Rai, V.R. (2004). *In vitro* propagation of *Oroxylum indicum* Vent. a medicinally important forest tree. *Journal of Forest Research* 9 (1): 61–65.

Eganathan, P., Rao, C.S., and Anand, A. (2000). Vegetative propagation of three mangrove tree species by cuttings and air layering. *Wetlands Ecology and Management* 8 (4): 281–286.

Faria, A.D., Luz, P.B.D., Sobrino, S.D.P. et al. (2020). Efficacy of passion fruit cryopreservation using cryopotectant agents. *International Journal of Fruit Science* 20 (sup2): S627–S635.

FAO and UNEP (2020). The state of the world's forests 2020: forests, biodiversity and people.

Friday, J.B. and Okano, D. (2006). *Calophyllum inophyllum* (kamani). *Species Profiles for Pacific Island Agroforestry* 2 (1): 1–17.

Garcia, C.A., Savilaakso, S., Verburg, R.W. et al. (2020). The global forest transition as a human affair. *One Earth* 2 (5): 417–428.

Geetha, S., Venkatramanan, K.S., Warrier, K., and Warrier, R.R. (2019). Propagation protocols for enhancing conservation and utilization of *Melia dubia* cav. *Journal of Tree Sciences* 37 (2): 22–35.

Goyal, A.K., Pradhan, S., Basistha, B.C., and Sen, A. (2015). Micropropagation and assessment of genetic fidelity of *Dendrocalamus strictus* (Roxb.) nees using RAPD and ISSR markers. *3 Biotech* 5 (4): 473–482.

Hay, F.R. and Probert, R.J. (2013). Advances in seed conservation of wild plant species: a review of recent research. *Conservation Physiology* 1 (1): cot030.

Heleno, R.H., Ripple, W.J., and Traveset, A. (2020). Scientists' warning on endangered food webs. *Web Ecology* 20 (1): 1–10.

Husain, M.K., Anis, M., and Shahzad, A. (2008). *In vitro* propagation of a multipurpose leguminous tree (*Pterocarpus marsupium* Roxb.) using nodal explants. *Acta Physiologiae Plantarum* 30 (3): 353–359.

Irawan, U.S. and Purwanto, E. (2014). White jabon (*Anthocephalus cadamba*) and red jabon (*Anthocephalus macrophyllus*) for community land rehabilitation: improving local propagation efforts. *Agricultural Science* 2 (3): 36–45.

Jenkins, M. and Schaap, B. (2018). Forest ecosystem services. United Nations Forum on Forests.

Jo, C. and Wilson, E.R. (2005). The importance of plus-tree selection in the improvement of hardwoods. *Quarterly Journal of Forestry* 99: 45–50.

Kathiresan, K. and Ravikumar, S. (1995). Vegetative propagation through air-layering in two species of mangroves. *Aquatic Botany* 50 (1): 107–110.

Kaur, M. and Nagpal, A.K. (2017). Evaluation of air pollution tolerance index and anticipated performance index of plants and their application in development of green space along the urban areas. *Environmental Science and Pollution Research* 24 (23): 18881–18895.

Kedharnath, S. (1984). Forest tree improvement in India. *Proceedings: Plant Sciences 93* (3): 401–412.

Khan, M.I., Ahmad, N., and Anis, M. (2011). The role of cytokinins on *in vitro* shoot production in *Salix tetrasperma* Roxb.: a tree of ecological importance. *Trees* 25 (4): 577–584.

Krishnan, P.N. and Seeni, S. (1994). Rapid micropropagation of *Woodfordia fruticosa* (L.) Kurz (Lythraceae), a rare medicinal plant. *Plant Cell Reports* 14: 55–58. https://doi.org/10.1007/BF00233299.

Lloret, F. and Batllori, E. (2021). Climate-induced global forest shifts due to heatwave-drought. In: *Ecosystem Collapse and Climate Change* (eds. J.G. Canadell and R.B. Jackson), 155–186. Cham: Springer.

Martínez-Palacios, A., Ortega-Larrocea, M.P., Chávez, V.M., and Bye, R. (2003). Somatic embryogenesis and organogenesis of *Agave victoriae-reginae*: considerations for its conservation. *Plant Cell, Tissue and Organ Culture* 74 (2): 135–142.

Mattos, T.S., Oliveira, P.T.S.D., Lucas, M.C., and Wendland, E. (2019). Groundwater recharge decrease replacing pasture by Eucalyptus plantation. *Water* 11 (6): 1213.

Mishra, Y., Patel, P.K., Yadav, S. et al. (2008). A micropropagation system for cloning of *Bambusa tulda* Roxb. *Scientia Horticulturae* 115 (3): 315–318.

Mittelbach, G.G. and McGill, B.J. (2019). *Community Ecology*. Oxford University Press.

Mudoi, K.D. and Borthakur, M. (2009). *In vitro* micropropagation of *Bambusa balcooa* Roxb. through nodal explants from field-grown culms and scope for upscaling. *Current Science* 96: 962–966.

Muralidharan, E.M. and Mascarenhas, A.F. (1987). *In vitro* plantlet formation by organogenesis in *E. camaldulensis* and by somatic embryogenesis in *Eucalyptus citriodora*. *Plant Cell Reports* 6 (3): 256–259.

Mursaliyeva, V., Imanbayeva, A., and Parkhatova, R. (2020). Seed germination of *Allochrusa gypsophiloides* (Caryophyllaceae), an endemic species from Central Asia and Kazakhstan. *Seed Science and Technology* 48 (2): 289–295.

Nanda, K.K., Purohit, A.N., Bala, A., and Anand, V.K. (1968). Seasonal rooting response of stem cuttings of some forest tree species to auxins. *Indian Forester* 94 (2): 154–162.

Nanda, K.K., Anand, V.K., and Kumar, P. (1970). Some investigations of auxin effects on rooting of stem cuttings of forest plants. *Indian Forester* 96 (3): 171–187.

Naz, R., Anis, M., and Aref, I.M. (2012). Assessment of the potentiality of TDZ on multiple shoot induction in *Bauhinia tomentosa* L., a woody legume. *Acta Biologica Hungarica* 63 (4): 474–482.

Nulkar, G. (2016). Silent conflicts–human-wildlife interactions in urban spaces. *Journal of Ecological Society*: 34–43.

Oliya, B.K., Chand, K., Thakuri, L.S. et al. (2021). Assessment of genetic stability of micropropagated plants of *Rhynchostylis retusa* (L.) using RAPD markers. *Scientia Horticulturae* 281: 110008.

Panda, L.L., Aggarwal, R.K., and Bhardwaj, D.R. (2018). A review on air pollution tolerance index (APTI) and anticipated performance index (API). *Current World Environment* 13 (1): 55.

Pandey, A.K., Pandey, M., Mishra, A. et al. (2015). Air pollution tolerance index and anticipated performance index of some plant species for development of urban forest. *Urban Forestry & Urban Greening* 14 (4): 866–871.

Pant, B., Chand, K., Paudel, M.R. et al. (2021). Micropropagation, antioxidant and anticancer activity of pineapple orchid: *Dendrobium densiflorum* Lindl. *Journal of Plant Biochemistry and Biotechnology*: 1–11. https://link.springer.com/article/10.1007/s13562-021-00692-y#citeas.

Pathak, V., Tripathi, B.D., and Mishra, V.K. (2011). Evaluation of anticipated performance index of some tree species for green belt development to mitigate traffic generated noise. *Urban forestry & Urban Greening* 10 (1): 61–66.

Phulwaria, M., Ram, K., Gupta, A.K., and Shekhawat, N.S. (2012). Micropropagation of mature *Terminalia catappa* (Indian Almond), a medicinally important forest tree. *Journal of Forest Research* 17 (2): 202–207.

Prajapati, S.K. and Tripathi, B.D. (2008). Anticipated performance index of some tree species considered for green belt development in and around an urban area: a case study of Varanasi city, India. *Journal of Environmental Management* 88 (4): 1343–1349.

Punalekar, S., Mahajan, D.M., and Kulkarni, D.K. (2010). Vegetation of Vetal Hill, Pune. *Indian Journal of Forestry* 33 (4): 549–554.

Purohit, S.D. and Dave, A. (1996). Micropropagation of *Sterculia urens* Roxb.—an endangered tree species. *Plant Cell Reports* 15 (9): 704–706.

Rathore, T.S., Singh, R.P., and Shekhawat, N.S. (1991). Clonal propagation of desert teak (*Tecomella undulata*) through tissue culture. *Plant Science* 79 (2): 217–222.

Rathore, T.S., Singh, R.P., Deora, N.S., and Shekhawat, N.S. (1992). Clonal propagation of *Zizyphus* species through tissue culture. *Scientia Horticulturae* 51 (1–2): 165–168.

Ratnam, D.P.S.K. (2021). Studies on phytochemical, antimicrobial activity and micro propagation of medicinal plants from Eastern Ghats of Andhra Pradesh. *Asian Journal of Pharmacy and Technology* 11 (2): 111–115.

Romero, J.L. (2004). A review of propagation programs for *Gmelina arborea*. *New Forests* 28 (2): 245–254.

Schenk, R.U. and Hildebrandt, A.C. (1972). Medium and techniques for induction and growth of monocotyledonous and dicotyledonous plant cell cultures. *Canadian Journal of Botany* 50 (1): 199–204.

Seventhilingam, K., Selvam, H., and Kalaivanan, B.V. (2021). Micropropagation and clonal fidelity assessment of acclimatized plantlets of *Crotalaria longipes* Wight & Arn. using ISSR markers. *Vegetos* 34 (2): 325–331.

Sharma D., Walmiki, N., Jadhav, A., et al. (2015). India's first biodiversity race in Matheran. Unpublished raw data.

Sharma D., Walmiki, N. and Jadhav, A. (2017a). Ecology and biodiversity survey for three rivers Poorna, Auranga and Ambika in Gujarat for proposed tidal barrage projects. Unpublished raw data.

Sharma D., Walmiki, N., Jadhav, A., et al. (2017b). Ecology and biodiversity survey of Gadag district for a proposed gold mine project. Unpublished raw data.

Sharma, D., Jadhav, A., and Gosavi, S. (2018). Setting up a jungle flora nursery in Thane city. Unpublished raw data.

Sharma, D., Gosavi, S., Choudhary, P., and Jadhav, A. (2019). Erection of 11 green canopies covered with native climber species in Thane city. Unpublished raw data.

Shekhawat, N.S., Rathore, T.S., Singh, R.P. et al. (1993). Factors affecting *in vitro* clonal propagation of *Prosopis cineraria*. *Plant Growth Regulation* 12 (3): 273–280.

Singh, S.K., Rao, D.N., Agrawal, M. et al. (1991). Air pollution tolerance index of plants. *Journal of Environmental Management* 32 (1): 45–55.

Singh, P., Bangarwa, K.S., and Dhillon, R.S. (2019). Plus tree selection and progeny testing of Khejri (*Prosopis cineraria* (L.) Druce). *Journal of Pharmacognosy and Phytochemistry* 8 (5): 817–820.

Singh, P. (2020). Floristic diversity of India: an overview. In: *Biodiversity of the Himalaya: Jammu and Kashmir State* (eds. G.H. Dar and A.A. Khuroo), 41–69. Springer.

Slate, M.L., Tsombou, F.M., Callaway, R.M., and El-Keblawy, A.A. (2020). Exotic *Prosopis juliflora* suppresses understory diversity and promotes agricultural weeds more than a native congener. *Plant Ecology* 221: 659–669.

Tewari, D.N. (1994). *Biodiversity and Forest Genetic Resources*. International Book Distributors.

Tewari, A. and Dhar, U. (1997). Studies on the vegetative propagation of the Indian butter tree (*Aisandra butyracea* (Roxb.) Baehni). *Journal of Horticultural Science* 72 (1): 11–17.

Tiwari, J.W.K. (1999). Exotic weed *Prosopis juliflora* in Gujarat and Rajasthan, India-boon or bane? *Tigerpaper* 26 (3): 21–25.

Tomar, U.K., NEGI, U., Sharma, N., and Emmanuel, C.J.S.K. (2004). Successful grafting in *Ailanthus excelsa* ROXB.-a brief report. *My Forest* 40 (1): 35–37.

Veldman, J.W., Overbeck, G.E., Negreiros, D. et al. (2015). Where tree planting and forest expansion are bad for biodiversity and ecosystem services. *BioScience* 65 (10): 1011–1018.

Venkatesh, L. and Lakshmipathaiah, O.R. (2009). Effect of seed size on germination, viability and seedling biomass in *Sapindus emerginatus* (Linn). *Mysore Journal of Agricultural Sciences* 43 (4): 827–829.

Wei, Q., Cao, J., Qian, W. et al. (2015). Establishment of an efficient micropropagation and callus regeneration system from the axillary buds of *Bambusa ventricosa*. *Plant Cell, Tissue and Organ Culture (PCTOC)* 122 (1): 1–8.

Weisse, M. and Goldman, E.D. (2017). Global tree cover loss rose 51 percent in 2016. World Resources Institute blog.

Yadav, R. and Pandey, P. (2020). Assessment of air pollution tolerance index (APTI) and anticipated performance index (API) of roadside plants for the development of greenbelt in urban area of Bathinda City, Punjab, India. *Bulletin of Environmental Contamination and Toxicology* 105 (6): 906–914.

12

Effect of the Changing Climate and Urban Ecology on Spreading of Infectious Diseases Including SARS-CoV-2

Joy K. Dey[1,2], *Saiema Ahmedi*[3], *Nishi Jain*[4,5], *Sarika Bano*[5], *Nikhat Manzoor*[3], *and Sanjay Kumar Dey*[5]

[1] *Documentation and Publication Section, Central Council for Research in Homoeopathy, Ministry of AYUSH, Govt. of India, New Delhi, Delhi, India*
[2] *Dey Health Care and Research Foundation, Nalikul, West Bengal, India*
[3] *Department of Biosciences, Jamia Millia Islamia, New Delhi, Delhi, India*
[4] *Department of Biotechnology, Amity University, Noida, Uttar Pradesh, India*
[5] *Dr. B.R. Ambedkar Center for Biomedical Research, University of Delhi, New Delhi, Delhi, India*
Joy K. Dey, Saiema Ahmedi, and Nishi Jain have equally contributed for the current work. Sanjay Kumar Dey is the corresponding author for this chapter.

12.1 Introduction

A large number of factors are responsible for both climate change and urbanisation and thus urban ecology. This includes human activities including overexploitation of resources, overpopulation, deforestation, overuse of plastics, pollution, etc. causing a negative impact on the environment resulting into climate changes facilitated by natural calamities. Another risk factor responsible for increasing the worldwide health burden is urbanisation. Changes in human lifestyle with changes in urbanisation or globalisation have severely impacted on human health whether it is communicable (e.g. infectious diseases including fungal, bacterial, or viral infections) or non-communicable disorders (e.g. diabetes, cardiovascular disorders, cancer, etc.). Research findings indicate that epidemics can also coincide with urbanisation, geographic expansion, and migrant movement over time (Figure 12.1) (Tian et al. 2018). For example, the very first human Ebola outbreak occurred in Zaire (now DRC) in 1976, and since then approximately 28 known outbreaks of Ebola have occurred in Africa (Roca et al. 2015). Laboratory testing of reservoir competence shows that successful infection is possible in bats and rodents, but not in plants or arthropods (Peterson et al. 2004; Reiter et al. 1999; Swanepoel et al. 1996, 2007).

12.1.1 Urbanisation as a Factor to Increase Infectious Disorders

Urbanisation has been shown to alter the composition of wildlife communities and increase the number of species that thrive in urban areas, which has the potential to pose short- and

Urban Ecology and Global Climate Change, First Edition. Edited by Rahul Bhadouria, Shweta Upadhyay, Sachchidanand Tripathi, and Pardeep Singh.

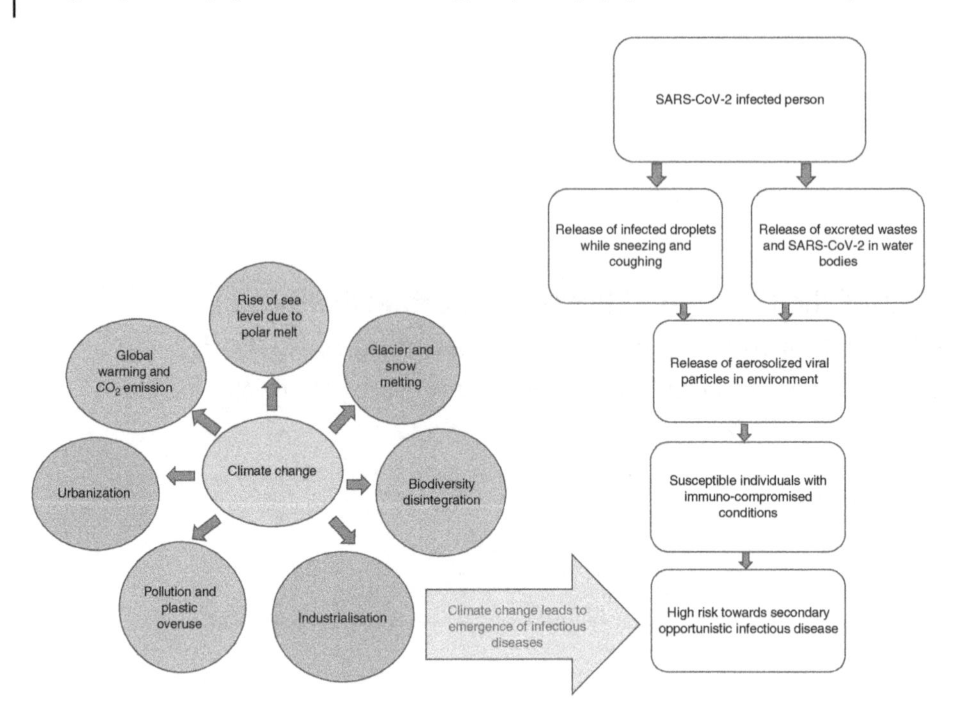

Figure 12.1 Flow diagram showing the basic outline of the effect of climate change and urban ecology on the spread of infectious diseases related to SARS-CoV-2.

long-term health problems and facilitate infectious disease emergence (Daszak et al. 2000; Gong et al. 2012; Hu et al. 2008; McKinney 2002). Many cities in the nineteenth century suffered an 'urban penalty' from infectious disease, such that deaths exceed births, and urban growth was only sustained by immigration subsidies from the countryside (Dye 2008). The early stage of urbanisation is characterised by mass rural–urban immigration alongside rapid urban expansion with poorly developed infrastructure (Normile 2008). New immigrants usually enter urban regions with poor housing and health care conditions, which are factors contributing toward a high risk of human infection. Due to altered food habits of modernised or urbanised humans, animal to human transmission may occur during hunting and consumption of the reservoir species or infected non-human primates or just increased eating habits of animal foods. The practice of butchering or eating bush meat or food contaminated with bat faeces (e.g. three species of tree roosting bats have been implicated as a reservoir) is also thought to contribute to this problem.

Globalisation in trade and travel, meanwhile, has facilitated the geographic expansion of the vector and has increased the possibility that travellers infected with the disease could come into contact with competent vectors (Semenza et al. 2016; Tatem et al. 2012).

In early epidemics, the re-use of non-sterile injections was responsible for many healthcare-associated transmissions (Breman et al. 2016). However, technical advancements have definitely played a key role in spreading of infection, globally. For example, travellers from affected areas, and laboratory scientists and others working with potentially infected materials and animals, are at high risk (Figure 12.1). Antibiotic resistance in the last decades has put an increasing pressure on human healthcare across the world (O'Neill 2014).

12.1.2 Changing Climate as Another Factor Responsible for Increasing the Infectious Disorders

Climate change leads to perturbations in the environmental conditions and disruption of ecological balance between nature and humans. Although the ecosystem is extremely dynamic and continuously adapts to these changes, human activities including overexploitation of resources, overpopulation, deforestation, overuse of plastics, and pollution, etc. cause a negative impact on the environment. Changes in the ecosystem and climate change has led to the extinction of several species and changed the epidemiology of several other species. Anthropogenic activities result in some serious problems like global warming that is slowly increasing the global temperature and as a result leading to the emergence of new fungal, bacterial, and viral diseases (Figure 12.1). Both field and laboratory experiments demonstrate that survival of vector species is affected by lower and upper temperature thresholds (Brady et al. 2013). Precipitation is another important factor influencing the availability of microhabitats for oviposition and larval development: heavy rainfalls – which are increasing in frequency due to climate change in some areas – have increased the abundance of the vector (Tjaden et al. 2017).

12.1.2.1 Deforestation That Affects Climate Indirectly

Deforestation leads to regional and global average temperature rise (Baker and Spracklen 2019; Singh et al. 2013; Prevedello et al. 2019). Loss of forests favours the emergence of fungal, bacterial, or viral infections through the following basic mechanisms.

Firstly, by climate-derived ecological disturbances interfering with the maintenance of pathogens in their natural environments and hosts; secondly, by favouring the presence, distribution, and proliferation of disease vectors in forest and urban areas, thirdly, by changes in temperature and rainfall patterns favouring pathogens' survival and reproduction and/or their ability to infect the human host. Changes in temperature also modify the ability of pathogens to infect vectors and to replicate in these animals (Baker and Spracklen 2019; Singh et al. 2013; Prevedello et al. 2019).

12.2 Spread and Emergence of Novel Fungal Infections with Changing Climate and Urban Ecology

Fungi play a very important role in regulating the ecosystem and hence are meaningful to human life as well. They are responsible for the decomposition of dead organic matter and hence help in recycling carbon and maintaining the environmental balance. Some fungi are an excellent food source like mushrooms and truffles. Their fermentation property is used for the production of bread, cheese, beer, wine, etc. Not only in the food industry, but fungi have also played a vital role in the pharmaceutical industry. Their secondary metabolites are used as scaffolds for the synthesis of life-saving medicines. For example, mycorrhiza is a symbiotic relationship between plants and fungi that enhances plant nutrition

Table 12.1 Some thermotolerant fungi and infections caused by them.

Fungal species	Temperature (°C)	Habitat/climate	Pathogenesis related activities	References
Aspergillus fumigatus	>55	Soil and decaying organic matter	Pulmonary Aspergillosis in humans	Bhabhra and Askew (2005), Couger et al. (2018)
Candida auris	>65	Tropical and subtropical climates	Candidiasis	Garcia-Solache and Casadevall (2010), Casadevall et al. (2019), Heaney et al. (2020)
Cryptococccus laurentii	>42	Found in soil	Skin infection, keratitis, endophthalmitis, lung abscess, peritonitis, meningitis, and fungemia	Johnston et al. (2016)
Paracoccidioides brasiliensis	>42	Endemic in Latin America	Mucous membrane ulceration of the mouth and nose	Longo et al. (2020)
Penicillium marneffei	>40	Soil, decaying vegetation, endemic in Southeast Asia	Systemic mycosis	Vanittanakom et al. (2006), Houbraken et al. (2014)
Cryptococcus gattii	>55	Tropical and subtropical climates	Pulmonary cryptococcosis in humans	Chen et al. (2013)

(Bleiker and Six 2014). Since fungi are simple eukaryotic organisms, they have been used as research model systems in the area of molecular biology. Studies on *Neurospora crassa* has resulted in significant advancements in modern genetics. Moreover, yeast is considered as a superior model system in recombinant DNA technology (Behie and Bidochka 2013). Unfortunately, due to unwanted changes in the natural environment by human activities, there is evolution of novel pathogenic species. Increase in global temperatures induces geographic expansion of fungi (Garcia-Solache and Casadevall 2010). Most fungal species survive in the temperature range of 12–30 °C, but now it has been observed that they start developing temperature tolerances (Robert and Casadevall 2009). Some species have been observed to grow at temperatures below −10 °C (Baxter and Illston 1980) or above 65 °C (Márquez et al. 2007). Table 12.1 summarises some thermotolerant fungi and their pathogenesis-related activities. Several factors that contribute to climate change include global warming, CO_2 emission and rise in sea level due to melting of polar ice and glaciers, industrialisation, urbanisation, disintegration of biodiversity, pollution, and excessive use of plastics (Hasnat and Rahman 2018). All these factors contribute to increase in overall global temperatures that lead to the emergence of thermoresistant pathogens which can cause various infectious diseases in immunosuppressed patients.

12.2.1 Emergence of Multidrug-resistant Strains of *Candida* and *Candidiasis*

12.2.1.1 Adaptation to Environmental Stress

Climate change and increase in global temperatures have resulted in emergence of thermally tolerant fungal species that can easily invade the mammalian thermal tolerant restriction zones (Casadevall et al. 2019). A thermal restriction zone is the difference between human basal temperatures and the environmental temperatures which provides protection to mammals from various microorganisms (Casadevall et al. 2019; Garcia-Solache and Casadevall 2010). *Candida* species are commensals and their growth is kept under control by several factors including the immune status and other protective barriers. Hence, they are not pathogenic at human physiological temperatures, i.e. 36.5–37.5 °C. However, *Candida auris*, a recently emerged *Candida* species which *can now easily* grow at higher temperatures like >40 °C (Jackson et al. 2019; Wang et al. 2018). Recent studies prove that emerged *C. auris* strains can survive even at extremely high temperatures in comparison to other *Candida* species, and its ability to grow at high temperatures, may have contributed to the evolution of this less prevalent species as a human pathogen (Casadevall et al. 2019). Another unique characteristic of *C. auris* is osmotolerance, which means having the ability to tolerate high salt concentrations (>10% NaCl, w/v) in comparison to other *Candida* species (Wang et al. 2018; Welsh et al. 2017). The yeast cells undergo morphogenesis to form pseudohyphae in response to high salt stress, suggesting that this species acquires the ability to undergo morphological switching under stressful conditions (Wang et al. 2018). Due to the emergence of heat tolerant and salt tolerant fungal species, these microbes have become a worldwide public health threat. Many studies have suggested that thermotolerance and osmotolerance are the key characteristics for survival over a long period of time on both biotic and abiotic surfaces (Biswal et al. 2017; Du et al. 2020).

12.2.1.2 Isolation and Epidemiology of *Candida auris*

C. auris was first isolated from the external part of the ear canal from a 70-year-old Japanese woman at Tokyo Metropolitan Geriatric Hospital in Japan in 2009 and thus named 'auris' (Satoh et al. 2009). The emergence of this multidrug-resistant fungal pathogen has been gaining a considerable attention due to its rapid expansion throughout the world over the past decade (Muñoz et al. 2018). *C. auris* was simultaneously found at several places including the Indian sub-continent, Venezuela, and South Africa during 2012–2015 with genetically varying clades (Casadevall et al. 2019). *Candida* species are now becoming the fourth major cause of all nosocomial infections (Perlroth et al. 2007). Genetically different clades of *C. auris* are related with differential resistance to conventionally used antifungal drugs suggesting that they will be able to continue phenotypically diverge in the future (Rhodes and Fisher 2019).

12.2.1.2.1 Major Risk Factors for C. auris Infections

The major risk factors for *C. auris* infections are mostly similar to those for other different *Candida* species. These are mainly opportunistic pathogens and are primarily involved with critically ill and immunosuppressed patients. The people at high risk include the elderly, diabetics, those undergoing surgery, presence of medical devices like catheters, immunocompromised conditions like cancer patients, patients suffering from chronic kidney diseases or undergoing hemodialysis, or the use of antibiotics or antifungal drugs (Du et al. 2020; Manzoor 2019; Sharma et al. 2020).

12.2.1.2.2 Major Virulence Traits of C. auris

12.2.1.2.2.1 Secretion of Hydrolytic Enzymes Secretion of extracellular hydrolytic enzymes plays a key contributing factor in virulence and pathogenicity of *Candida* species. Extracellular secretory enzymes majorly involves proteinases, lipases, phospholipases, hemolysins, which are virulence-associated enzymes (Du et al. 2020; Manzoor 2019; Sharma et al. 2020). These secretory enzymes help *Candida* in the invasion to host tissue and breaking their immune barrier.

12.2.1.2.2.2 Morphological Transitions Morphological transition is a common phenomenon used by polymorphic microorganisms to rapidly adapt the environmental changes (Jain et al. 2008; Justice et al. 2008). Most of the fungal species can easily undergo morphological switching under certain environmental conditions like increasing temperatures and high salinity. Other pathogenic *Candida* species like *Candida albicans* and *Candida tropicalis* can also undergo morphological transitions (Huang 2012). They can switch between cellular yeast form and long filamentous hyphal form in response to environmental stress. Pathogenesis involves two major morphological transitions – yeast to hyphae switching and the white to opaque transition (Biswas et al. 2007). Like other virulent *Candida* species, *C. auris* also having capacity to undergo several morphological phenotypic changes (Bentz et al. 2019; Yue et al. 2018), but the major regulatory mechanisms for this morphological switching in *C. auris* is still unknown. Most of the isolates of *C. auris* strain exist in the single-cell yeast form while many natural isolates of *C. auris* can form large clusters of long filaments of pseudohyphae-like cells (Borman et al. 2016). In comparison to *C. albicans*, phenotypic transition occurs more efficiently in *C. auris* and adapts more easily to the ever-changing environment (Du et al. 2020).

12.2.1.2.2.3 Biofilm Development Biofilms are comprised of microbial communities in which cells adhered or attaches to living or non-living surfaces (Nobile and Johnson 2015). Human tissues like teeth, mucosal lining of organs or an implanted medical device, can serve as a surface for biofilm development and infection can further spread to other parts of the body. Studies have suggested that *C. auris* are also capable to develop their biofilms on various surfaces but are relatively weakly attached to the surface in comparison to biofilms formed by *C. albicans* (Oh et al. 2011). Biofilm development ability varies among different *C. auris* isolates and even in genetically different clades (Singh et al. 2019). Several RNA-sequencing experiments were conducted and identified that genes encoding adhesion proteins, efflux pumps, and many virulence factors were up-regulated during *C. auris* biofilm formation (Kean et al. 2018). *C. auris* biofilms also play an important role in its pathogenicity, antifungal resistance, and survival in the environment and the host. Therefore, the development of novel therapeutic strategies is the demand of current situation to target biofilms of *C. auris* which is an important area for future research in discovering of new antifungals.

12.2.1.3 Antifungal Resistance and *Candidiasis*

C. auris is now considered a 'superbug' and becomes a major human health threat globally due to acquiring resistance to different classes of standard antifungal drugs (Spivak and Hanson 2018). Most isolates of *C. auris* are resistant to azole like fluconazole and showing high minimum inhibitory concentrations (MICs) values than that of other classes of antifungals like amphotericin B and echinocandins while some of the strains of *C. auris* are highly resistant to all classes of available antifungal drugs (Lockhart et al. 2017). Ergosterol is the major membrane sterol component of fungus and plays a very crucial role in maintaining cell integrity. It acts as a target site for azoles (e.g. fluconazole) and polyenes (e.g. amphotericin B) class of antifungals (Gray et al. 2012). The first-line antifungal drugs used in the clinics are azoles which inhibits ergosterol biosynthesis pathway by inhibiting fungal cytochrome P450 enzyme lanosterol demethylase which is important for the synthesis of ergosterol. The gene *ERG11* encodes this particular enzyme lanosterol demethylase in the *Candida* species. Studies suggested that three hotspot mutations occurred in Erg11 (Y132F, K143R, and F126L or VF125AL) in fluconazole-resistant strains of *C. auris* with different genetic clades (Lockhart et al. 2017).

Candida species are commensal microorganisms and are part of the normal human skin and gut microbiota in healthy individuals (Pappas et al. 2018). Defects in host defences provide them with an opportunity to become invasive and hence increase their colonisation resulting in opportunistic fungal infections. Invasive candidiasis is now not a single clinical entity, but it becomes a disorder with innumerable and multiple clinical manifestations that can have potential to affect any other organs, as each *Candida* species is having some unique traits and properties in relation to more invasive form, virulence, and antifungal susceptibility (Pappas et al. 2018). Hence, research exploring virulence-related traits and treatment of *C. auris* in more susceptible patients is currently needed. Also, new formulations of antifungals are the demand of current scenarios to use as combination therapies and development of new bioactive compounds might be useful for better therapeutic outcomes.

12.2.2 Mucormycosis in Immunocompromised Patients

Mucormycosis is also an opportunistic and life-threatening fungal infection that occurs mainly in immunosuppressed individuals. Conditions like cancer, organ, and bone marrow transplantation, AIDS, diabetic patients, etc. causes immunosuppression in the host. Due to changes in lifestyle and lack of physical activity results in increased incidence of diabetes mellitus, obesity, cancer, heart-related disorders, and organ transplantation (Kagaruki et al. 2015). This causes an increase in the number of patients at risk against this dreadful infection (Ibrahim et al. 2012).

12.2.2.1 Major Virulence Traits of Mucormycosis

Mucorales possess several pathogenic virulence properties that enable them to cause mucormycosis in immunocompromised patients. One such important virulence attribute is the ability to obtain the iron from the host cells. Iron is an essential element and performs many important roles like proper functioning of vital organs and maintenance of their equilibrium in body (Howard 1999). Many studies reported that availability of free or boundless iron in serum plays an important role in mucormycosis patients (Ibrahim et al. 2008). In mammalian hosts, iron is mainly found in bound form with some other carrier proteins like transferrin, lactoferrin, and ferritin (Kumar et al. 2017). Due to these iron sequestration mechanisms, the species *Rhizopus oryzae* reduces the availability of free iron and also reduces the toxic effects associated with free iron. The strategy of lowering down the free iron availability is a major universal host shielding and safeguarding mechanism against a variety of microbes including mucorales (Boelaert et al. 1993). Other pathogenic markers include the ability of *Rhizopus* to secrete hydrolytic enzymes mainly aspartic proteinases and lipases (Farley and Sullivan 1998). *Rhizopus* species are having an important active ketone reductase system, which enhances their growth in the acidic and glucose-rich environment like seen in ketoacidosis conditions and hence this mechanism plays very important role in their pathogenesis (Anand et al. 1992).

12.2.2.1.1 Adaptation to Environmental Stresses

Studies demonstrate that the association between weather patterns and climate changes are closely linked to mucormycosis (Al-Ajam et al. 2006; Shpitzer et al. 2005). Continuous alteration in seasonal pattern affects the incidence of mucormycosis and their fungal spores in the environment (Shpitzer et al. 2005). The sporangiospores from mucorales are small in structure and can easily be dispersed in the environment like an aerosol under suitable climatic conditions (Richardson 2009), and becomes more invasive in their most susceptible hosts. The continuous variation in weather patterns like low precipitation and high temperature is closely associated with high fungal spore counts. Studies hypothesise that extremely hot and dry summer conditions lead to their aerosolisation, spread and virulence of mucormycosis.

12.2.2.1.2 Antifungal Resistance and Mucormycosis

Mucormycosis is mainly caused by mucorales and this family comprises a large number of species. These fungi are also causing opportunistic fungal infection and possessing a high level of resistance to most of the currently using antifungal drugs. Animal model studies are of great importance to interrelate *in vitro* data and drugs that are already in human clinical trials. Currently available antifungal drugs are Amphotericin B and azoles like Posaconazole and Isavuconazole against mucorales, but new therapeutic strategies are needed to combat drug resistance (Dannaoui 2017). Different alternative strategies can be used like combination therapy which can be a better therapeutic approach to overcome resistance, but further studies are required to confirm its efficacy for patients.

12.2.3 Implication of Changing Climate and Urban Ecology on the Spread of Fungal Infections in Relation to SARS-CoV-2

Coronavirus disease 2019 (COVID-19) pandemic is the cause of serious public health concern of the century. Coronavirus or SARS-CoV-2 is a single-stranded RNA virus with a diameter ranging between 80 and 120 nm (Dey and Dey 2020; Dey et al. 2021). The outbreak of this disease was first reported from the Wuhan seafood market, China in December 2019, and within a short duration of time, it became a global health emergency (Chakraborty and Maity 2020; Dey and Dey 2020). Sources of viral transmission include human direct or indirect contact with respiratory aerosol droplets and contaminated objects. Human infected excreta are also reported to release viral particles which can ultimately enter the aquatic systems (Patel et al. 2020). SARS-CoV-2 causes lower respiratory infections. With diffuse alveolar damage and severe inflammatory exudation, COVID-19 patients are always immunosuppressed with decrease in CD4+ and CD8+ T lymphocytes (Yang et al. 2020a,b). Critically ill patients who were admitted to the intensive care units and required mechanical ventilation, or had prolonged hospital stays, even as long as 50 days, were more likely to develop secondary fungal co-infections. COVID 19 patients are developing varieties of fungal infections (e.g. black fungus, green fungus, etc.) during the middle and latter stages of the disease (Rizwan et al. 2021). This was seen to be more common in immunosuppressed patients (people suffering from cancer, diabetes, hypertension, or other chronic diseases). Increasing global temperatures give rise to emergence of more thermotolerant fungal species that are more resistant towards conventional antifungal drugs (Figure 12.1). These thermotolerant fungal species cause secondary fungal infections in SARS-CoV-2 patients (Yang et al. 2020a; Yang et al. 2020b).

12.3 Spread and Emergence of Newer Bacterial Infections with Change in Climate and Urban Ecology

Urbanisation has led to an increase in impermeable surfaces, the use of industrial energy, domestic heating and vehicle exhaust emissions, as well as a decrease in Greenland aquatic surfaces, and urban construction has been shown to alter the thermal conditions of the city in a variety of ways, thereby impacting local temperatures and increasing the incidence of bacterial infections, among others (Shao and Zeng 2012). Bacterial infections are usually regarded as more prevalent in the summer months (Shao and Zeng 2012). Indeed, an increase in the incidence of Gram-negative bacterial infections among hospitalised patients was reported with an increase in temperature. Higher temperatures can stimulate bacterial growth in the atmosphere and increase the virulence of Gram-negative bacteria, leading to an increase in the incidence of infection in warmer periods (Retailliau et al. 1979). In addition, seasonality may be related to lipid, a lipopolysaccharide moiety, which forms most Gram-negative bacteria outside the outer membrane and is controlled by environmental conditions.

12.3.1 Infection by *Acinetobacter baumannii*

Acinetobacter is a member of genus which is ubiquitous in nature. *A. baumannii* is one of them which are pathologically most significant and also a big source of nosocomial infection. It is a non-motile, anaerobic Gram-negative bacillus whose natural reservoir is still to be developed (Fournier et al. 2006). Nevertheless, it is present in many health care facilities and is a very powerful human coloniser in the hospitals. It is a successful nosocomial pathogen due to its environmental tolerance and its wide range of resistance determinants. Over the last 30 years, *Acinetobacter* has undergone significant taxonomic reform, showing in rising several global outbreaks, and resistance rates. For health care organisations,

A. baumanni creates global problem (Abbo et al. 2005). All existing antibiotics show resistivity against *A. baumannii* infection, indicating a sentinel case which the international health care community can respond promptly. The organism usually attacks the most helpless injured patients, those who are seriously ill with breaches of skin integrity and airway safety.

12.3.1.1 Major Infections Caused by *A. baumannii*

The hospital-acquired pneumonia infection is generally caused by *A. baumanni*. However, diseases influencing the central nervous system, skin and soft tissue, and bone have arisen as problematic for many species in more recent times (Al Shirawi et al. 2006). *A. baumannii* is a rapidly emerging global strain showing resistivity to all β lactams and also carbapenems.

12.3.1.2 Effect of Climate Change on *A. baumannii* Infection

Korea has already reported a sudden increase in *A. baumannii* infection in the community, causing complex pneumonia. Therefore, the development of complex community-related *A. baumannii* infections should not be neglected, particularly in light of Korea's climate change due to ongoing global warming (Fournier et al. 2006). In the 1970s, seasonal impacts were reported in *Acinetobacter calcoaceticus* isolation but it is uncertain in *A. baumanni*. No seasonality was observed with complex isolate *A. baumannii* multidrug-resistant (MDR) (Gootz and Marra 2008). A recent study in Korea found that fluctuation in *A. baumanni* temperature plays a significant role in their infection (De Silva et al. 2018; Chapartegui-González et al. 2018; Peleg et al. 2008).

12.3.2 Infection by *Mycobacterium tuberculosis*

While climate change is likely to be distally linked to tuberculosis (TB) incidence, it plays an important role in the seasonality and geographical variability of TB incidence. Similarly, while there are no clear favourable or unfavourable climate conditions for TB incidence, transmission could be helped by inadequate ventilation and overcrowding, according to published literature (Sumpter and Chandramohan 2013; McMichael et al. 2003).

There are more than 100 different species of mycobacteria and their identification is based on 16S rRNA sequence. Although this genus is well known for its effective human pathogens, *M. tuberculosis causes* TB, *Mycobacterium leprae* causes leprosy and *Mycobacterium ulcerans* causes ulcer. The bulk of its members are environmental bacteria that are not pathogenic. The World Health Organisation (WHO) declared *M. tuberculosis* as a global emergency in 1993. However, it is still one of the leading causes of human death due to bacterial infections today, with approximately two million deaths recorded every year (Glaziou et al. 2009). Reportedly, bacteria attack almost a third of the world's population. It is one of the most effective intracellular pathogens in the world. TB, which is caused by the bacterial pathogen mycobacterium tuberculosis (Mtb), is now one of the deadliest infectious diseases, having killed over a billion people over the last 200 years (Bussi and Gutierrez 2019; Paulson 2013). With 10.4 million new infections and close to 1.7 million deaths in 2017, TB killed more people than any other infectious disease (Bussi and Gutierrez 2019).

Pathogenicity of Mtb depends on its ability to survive and persist throughout infection within host macrophage cells. Its ability to remain in a dormant state until reactivation and to manifest itself in active TB is a remarkable trait of Mtb. Due to a variety of factors such as cost of care (Upadhyay et al. 2018) prolonged length of treatment, toxicity associated with drugs used for treatment, and rising antibiotic resistance, the treatment of the disease has been of huge concern. The dynamic design of the cell wall makes treatment difficult, which leads to the bacteria's persistence and survival.

Mtb was first discovered by Robert Koch in 1882. On the cell, it has an odd, waxy coating, rendering the cells impervious to Gram staining. While mycobacteria do not seem to match the Gram-positive category from an analytical point of view (i.e. they do not sustain the violet stain of the crystal), due to their lack of an outer cell membrane, they are categorised as acid-fast Gram-positive bacteria. In every 15–20 h, Mtb divides, which is relatively slow compared to other bacteria. Emerging drug resistance in Mtb is possibly caused by its peculiar cell wall, which is rich in lipids (mycolic acids), and is a key virulence factor. When Mtb is ingested by alveolar macrophages in the lungs, it prevents the phagosome from merging with the lysosome, preventing the bacteria from being absorbed. The early endosomal autoantigen 1 bridging molecule is blocked by Mtb, but this does not prevent nutrient-filled vesicle fusion. As a result, the bacteria grow unchecked within the macrophage (Upadhyay et al. 2018).

Mtb is mainly a mammalian respiratory system pathogen that infects the lungs, causing TB. It is highly aerobic and requires high oxygen levels for their survival. The genomic size of Mtb is 4.4 Mb (G-C content is ~65%), with 3959 genes; 40% of these genes are functional, and another 44% within the genome are non-functional. On the basis of DNA homology studies, the various species of the Mtb complex display a 95–100% DNA sequence similarity and the 16S rRNA gene sequence is exactly the same for all species. Chains of cells also form distinctive serpentine cords in smears made from colonies grown *in vitro*. This discovery was first made by Robert Koch, who connected virulent strains of the bacterium with the cord factor.

12.3.2.1 Multi-drug Resistant TB

Multi-drug resistant TB (MDR TB) is characterised as a disease caused by Mtb which is resistant to isoniazid and rifampicin. These are the most effective first line of drugs (Sharma and Mohan 2004). There are 300 000 new cases per year of MDR TB worldwide according to global TB management report of the WHO in 2004. For the treatment of TB, there are five drugs used: (i) Aminoglycosides, (ii) Fluoroquinolones, (iii) Polypeptides, (iv) *p*-amino salicylic, and (v) Thiamine.

12.3.2.2 Extensively Drug-resistant TB (XDR)

Extensively drug-resistant (XDR) TB is characterised by susceptibility to at least isoniazid and rifampicin, some fluoroquinolone and at least one of three second-line injectable drugs as an MTB-induced disease like amikacin, capreomycin, and kanamycin. The idea has relevant clinical relevance and has encouraged more uniform surveillance in different foreign environments. Recent surveillance data shows that the prevalence of drug resistance to TB has risen to the highest level ever recorded (McMichael et al. 2003; Aslam et al. 2018). The agar proportion method was the gold standard for drug-susceptibility testing. However, it takes several weeks for results to be measured for this technique. In resource-limited settings, more sensitive and specific diagnostic tests are still unavailable.

Although variable in different settings and among different strains, clinical manifestations have already shown that XDR TB is generally associated with higher morbidity and mortality than non-XDR TB. The treatment of XDR TB should include agents to which the organism is susceptible and should proceed for a period of 18–24 months. However, treatment in TB-endemic countries continues to be minimal, largely due to shortcomings in national health care models for TB. The ultimate drug-resistant TB control strategy is one that implements a systemic approach that incorporates political, social, economic, and scientific advancement.

The specific feature of mycobacteria is their cell wall they contain mycolic acids which form an effective permeability barrier from antimicrobial agent. While the chemical structures of mycolic acids are well known, much needs to be found about the role played by various annotated open reading frames (ORFs) for biosynthesis of mycolic acid and other lipids.

12.3.3 Implication of the Changing Climate and Urban Ecology on the Spread of Bacterial Infections in Relation to SARS-CoV-2

Carbapenem-resistant *Acinetobacter baumannii* (CRAB) is a significant public health problem. *A. baumannii*, an opportunistic pathogenic agent mainly associated with infection acquired in hospital ('Antibiotic resistance threats in the United States 2019. Atlanta, GA: US Department of Health and Human Services, CDC; 2019'). CRAB easily contaminates the patient care environment and the hands of health care professionals in health care facilities, lives on dry surfaces for extended periods, and can be spread by asymptomatically colonised people; these variables make it difficult to track CRAB outbreaks in acute care hospitals. In New Jersey, during COVID-19, cluster of CRAB infection was identified.

12.4 Spread and Emergence of Newer Viral Infections with Change in Climate and Urban Ecology

12.4.1 Ebola Viral Infection

One of the most virulent pathogens to affect humans is the Ebola virus (Worby et al. 2015). With nearly 20 outbreaks in the last 40 years have occurred, the first case of the known Ebola outbreak was in 1976 near the Ebola River in Zaire (Democratic Republic of Congo) (Curtis 2006; Hansen et al. 2015). 2300 cases and 1500 deaths were recorded before the 2014 epidemic itself (Curtis 2006; Hansen et al. 2015). Ebola virus infection is mostly called as Ebola hemorrhagic fever (Worby et al. 2015). The high mortality rate of this infection is dependent on the health facility available (Curtis 2006; Hansen et al. 2015).

At the onset of the disorder symptoms like high fever (temperature of up to 40 °C), malaise, fatigue, and body aches have been observed (Team 2014) (http://apps.who.int/iris/bitstream/10665/133833/1/roadmapsitrep4_eng.pdf?ua=1). It has been speculated that the virus was acquired by the humans due to interaction with the body fluids of infected animals (Bray et al. 2014).

12.4.1.1 Epidemiology and Clinical Analysis of Ebola Virus

Ebola is a single-stranded enveloped RNA virus. The first case was documented in Meliandou in Gueckedou in 2013 and later noted to have the high mortality rate of 86%. Patients are not considered infectious before the manifestation of symptoms with transmission to be reported by close contact with the body fluid of infected people. Incubation period may be generally for 5–6 days but may vary from 1 to 21 days in maximum number of cases (http://www.cdc.gov/vhf/ebola/outbreaks/2014-west-africa/case-counts.html).

12.4.1.2 Risk Factors for Ebola Viral Infection

Lab testing of supply skills shows that effective disease is feasible in bats and rodents, however not in plants or arthropods (Peterson et al. 2004; Reiter et al. 1999; Swanepoel et al. 1996, 2007). Creature to human transmission of Ebola may happen during chasing and utilisation of the repository species or contaminated non-human primates. The act of butchering or eating shrub meat or food polluted with bat excrement (three sorts of tree perching bats are involved as a supply) is furthermore thought to contribute. The early stage of urbanisation is characterised by mass rural–urban immigration alongside rapid urban expansion with poorly developed infrastructure (Jha and Tripathi 2020). New immigrants usually enter urban regions with poor housing and health care conditions, which are contributing toward creating an environment that is rather feasible for the breeding of several vectors (Jha and Tripathi 2020).

12.4.2 H1N1 Flu Infection

The spring of 2019 saw the emergence of a quick-spreading virus through Mexico primarily followed by a havoc throughout Europe. It was caused by an influenza strain derived from swine influenza strain (https://www.folkhalsomyndigheten.se/contentassets/1d7096c2b65d45b499c924d76333272c/influenza-in-sweden-2009-2010.pdf).

12.4.2.1 Risk Factors for H1N1 Flu

Meteorological factors associated with the rate of influenza transmission among individuals include precipitation, humidity, temperature, and solar radiation (Charland et al. 2009; Dalziel et al. 2018; Pica and Bouvier 2012). Not many socioeconomic variables affect the transmission throughout a country, possibly due to not much variation possible across that small landscape. The most drastically affected seems to be school children (Worby et al. 2015). There is a definitive and indirect effect of environmental factors in transmission of influenza. Temperature, for example, contributes right at the onset of the ailment (Hansen et al. 2015). Temperature seems to be the second most imperative risk factor and is estimated to have a large negative factor. Thus, a temperature drop is bound to have a great effect on the outbreak of influenza. It brings it to its peak. Humidity has also been documented to have some contribution as a risk factor (Hansen et al. 2015).

12.4.3 Encephalitis (A Viral Infection)

12.4.3.1 Epidemiology and Clinical Analysis

Chandipura virus (CHPV) was initially discovered during an acute febrile outbreak in Nagpur, Maharashtra state, India from two febrile cases (Bhatt and Rodrigues 1967;

Dhanda et al. 1970). It belongs to the genus *Vesiculovirus*, the family *Rhabdoviridae*. This virus has a single-stranded RNA genome with negative polarity and size of about 11 kb. Five structural proteins are coded by its genome: the nucleocapsid protein (N), the phosphoprotein (P), the matrix protein (M), the glycoprotein (G), and the large structural protein (L). These are produced in the form of five monocistronic mRNAs (Menghani et al. 2012). It is attributed that Children under 15 years of age are most susceptible to natural infection. Sandflies are the vectors for this virus while antibodies against this have been detected in a wide range of vertebrate animals (Mishra 2006). General clinical features include high-grade fever of short duration, vomiting, altered sensorium, generalised convulsions and decerebrate posture leading to grade IV coma, acute encephalitis-encephalopathy and death within a few to 48 h of hospitalisation (Sudeep et al. 2016). The available epidemiology analysis suggest that this disease mostly occurs in sporadic forms; however, it has potential to cause outbreaks. Recently, it was surveyed that the presence of this virus is recorded from the Indian subcontinent (India, Bhutan, Nepal, Sri Lanka) and Africa (Nigeria, Senegal) (Ba et al. 1999; Kemp 1975). The viral outbreaks and the prevalence of the virus in warmer parts of the world seem to indicate that this disease has a spike when the temperature goes up. This is also indicated by the fact that the outbreaks of this disease by other viruses of the same family seem to follow the same pattern (Rao et al. 2004; Korenberg 2000).

12.4.3.2 Mutations in Genomes

The mutations occurring in the genome of the CHPV virus has been a huge factor in the transmission of the disease as the resistance acquired is usually conferred harmful (Tsetsarkin and Weaver 2011).

12.4.4 Corona Viral Infection Including SARS-CoV-2

Mis-identified as cold when it first showed itself, the coronavirus was initially treated as simple non-fatal disease up until 2002 with estimated 500 cases found (Dey and Dey 2020). In 2003, 100 cases were noted with the virus clutching the United States of America, Hong Kong, Thailand, Vietnam, and Taiwan. COVID-19 was finally isolated in 2019 from bronchoalveolar lavage fluid in a patient from China (Dey et al. 2021).

12.4.4.1 Clinical Analysis and Epidemiology

COVID-19 or coronavirus is circular or pleomorphic, single abandoned, wrapped RNA, and covered with club moulded glycoprotein (Courouble et al. 2021). Corona has four subclasses namely, alpha, beta, gamma, and delta Covid (Rizwan et al. 2021). All of the subclasses of corona have numerous serotypes as well. Various of them were infecting humans while others infected creatures like pigs, birds, felines, mice, and canines (Table 12.2) (Buchholz et al. 2013; Gwaltney 1985; Mailles et al. 2013; Saif 2004). Humans can get the disease through close contact with an individual who has side effects from the infection incorporates hack and wheezing. Mostly, covid was spread by means of airborne zoonotic drops (Dey et al. 2020). Infection was reproduced in ciliated epithelium that caused cell harm and contamination at disease site (Courouble et al. 2021).

Table 12.2 Some viruses and the diseases caused by them.

Virus name	Genus, family	Host	Transmission	Disease
Chandipura virus	Vesiculovirus, Rhabdoviridae	Human, sandflies	Zoonosis, arthropod bite	Encephalitis
Ebolavirus	Ebolavirus, Filoviridae	Humans, monkeys, bats	Zoonosis, contact	Hemorrhagic fever
Human SARS coronavirus	Betacoronavirus, Coronaviridae	Human, bats, palm civet	Zoonosis	Respiratory
Zika virus	Flavivirus, Flaviviridae	Human, monkeys, mosquitoes	Zoonosis, arthropod bite	Fever, joint pain
Epstein-Barr virus	Lymphocryptovirus, Herpesviridae	Human contact and saliva	Zoonosis	Mononucleosis
Human astrovirus	Mamastrovirus, Astroviridae	Human	Huma fecal-oral	Gastroenteritis

Source: Modified from Human viruses and associated pathologies, Expasy. https://viralzone.expasy.org/678

12.4.4.2 Risk Factors Associated with Spread of Corona Infection

12.4.4.2.1 Urbanisation

SARS-CoV-2 (the infection answerable for COVID-19) has aggressively infected humans at a fish market in Wuhan. Wuhan is one of the biggest Chinese urban communities and a significant transportation hub with public and global connectivity (Connolly et al. 2020). Dramatic changes in demographic and social conditions, including an exponential increase in global transport, are responsible for much of this global emerging infectious disease problem (Dey et al. 2020). Urbanisation has posed growing challenges to global health governance due to the deeply political nature of health policies and planning decisions (Srivastava 2020). Staggered government grindings in wellbeing administration are adapted, moreover, by the current disintegration of general wellbeing foundation, accentuation on crisis reaction instead of anticipation, and related smugness inside certain nations are additionally answerable for such circumstance (Petersen 2010). It is likewise seen that the sickness is compounding through existing imbalances in the public arena along the lines of class and race inconsistencies, lopsided examples of versatility, admittance to disinfection foundations, and capacity to self-seclude. Rapid urbanisation has been demonstrated to be positive for rodents (*Rattus* spp.), population growth and related zoonoses (Menghani et al. 2012). The increase in the rainfall levels, furthermore, have a direct impact on the increasing number of insect vectors that directly relates to spread of the infectious diseases (Franklinos et al. 2019). The deforestation due to urbanisation has increased the number of potential animals to come in contact with the human population, especially transmitting diseases to us (Engering et al. 2013).

12.4.4.2.2 Climate Change

It refers to changes in weather conditions in the long term and patterns of extreme weather events. It may alter the levels of threat to human health, while further exaggerating existing health issues. The relative rodent density was found to be significantly correlated with monthly cumulative precipitation (Hansen et al. 2015) and the monthly mean temperatures. Factors like seasonality and changes in temperature, rainfall and humidity greatly influenced COVID-19 (Sajadi et al. 2020). It is widely understood that the process of fast-paced development and excessive human activities have resulted in an increase of $0.85 \pm 0.2\,°C$ in the global mean temperature during the period from 1880 to 2012 (Bai et al. 2013; Caminade et al. 2019). A recent report by the IPCC (https://www.ipcc.ch/site/assets/uploads/2018/05/SYR_AR5_FINAL_full_wcover.pdf) has forecasted an increase of 1.1–6.4 °C in the average global land and ocean surface temperature from 1990 to 2100. The report also highlighted the rising temperature's significant impact on transmission of infectious diseases (Bai et al. 2013). In China, the swift economic growth during the last few decades has been accompanied by environmental changes on an unprecedented level and a warming climate resulting in more frequent weather-related natural disasters (Change 2007).

Extreme weather events are a part of a daily happening for us nowadays, such as mudslides, landslides, floods in mountains as well in other regions, and especially droughts (Lin et al. 2014). Climate has been known to have a drastically direct impact on the lifecycles of various pathogens and viruses such as malaria, dengue, etc. Other diseases that seem to have imbibed a significant amount of change due to climate are rodent-borne diseases and the implications are clearly visible in the distribution, number of cases and ecology of the outbreaks and cases (Zhang et al. 2008). Our imprudent approach towards the environment has led to a drastic repercussion to it, especially to the temperature, humidity, and rainfall. The replicative cycle for most viruses becomes much shorter and a splurge in the cases is documented (Tabachnick 2010). Warmer climate is also associated with increase in number and activity of rodents which unambiguously impact number of hemorrhagic fevers with renal syndrome (HFRS) (Buchholz et al. 2013; Gwaltney Jr 1985; Saif 2004). On the other hand, rainfall provides better growth environment for vegetation providing food for the rodents and giving a boost to their population size and helping the infectious agents by providing more numbers of hosts (Bi et al. 2005; Xiao et al. 2014). Humidity similarly increases the rodent and viral reproductivity and thus incidents of HFRS.

12.5 Conclusion

Humans have directly or indirectly affected the environment severely. Several contributing factors result in change of climate and weather patterns. Perturbation of environmental temperatures results in the emergence of various infectious diseases mainly caused by bacteria, fungi, and viruses. Ongoing COVID-19 pandemic is a recent example of such increasing threat to humans, which has become a global health threat leading to patient hospitalisation and increase in nosocomial infections. Patients suffering from SARS-CoV-2 are more prone to secondary infections caused by fungi and bacteria. COVID-19 has caused disruption of immune protective barriers due to which patients become prone to opportunistic pathogens, some of which thermally tolerant survive at high temperatures. Increased temperatures as a result of global warming have made the treatment of these emerging thermally tolerant fungal strains extremely difficult as they show resistance to conventional antifungal drugs. Although climate change is the result of anthropogenic activities such as industrialisation and urbanisation, it is very important to take serious steps to protect our environment from unnecessary and unwanted interventions. The flora and fauna, be it in the environment or the human body, has to be preserved in the appropriate manner otherwise there will be disease, even worse than the recent pandemic of COVID-19.

It is still a mystery if we today are in any position to deal with the ardent situation of emerging infectious diseases as a result of the drastically changing climate and clarity being pursued. According to various Centre for Disease Control and Prevention (CDC, USA) researchers, there has been a prominent change in the transmission pattern of several air-borne, vector-borne and other transmutable diseases as the global climate is changing. In this report, a number of measures were suggested to transform and improve the health care system as well. Some critical measures are infrastructure, legislation, more coordination between officials, reforming the research, and increasing collaborative studies. Such studies need to be conducted in large numbers to get a better picture of the impending problems and the possible solutions.

Acknowledgements

All authors of the book chapter would like to acknowledge the Dr. B.R. Ambedkar Center for Biomedical Research (ACBR), University of Delhi, India for various help to complete the current work. Sanjay Kumar Dey acknowledges the University of Delhi, Institute of Eminence grant (IoE/2021/12/FRP).

References

Abbo, A., Navon-Venezia, S., Hammer-Muntz, O. et al. (2005). Multidrug-resistant *Acinetobacter baumannii*. *Emerging Infectious Diseases 11* (1): 22. https://www.ncbi.nlm.nih.gov/pmc/articles/PMC3294361/.

Al-Ajam, M., Bizri, A., Mokhbat, J. et al. (2006). Mucormycosis in the Eastern Mediterranean: a seasonal disease. *Epidemiology and Infection 134* (2): 341–346. https://www.cambridge.org/core/journals/epidemiology-and-infection/article/mucormycosis-in-the-eastern-mediterranean-a-seasonal-disease/F020E66E38D0CF94BBF4A97298327F90.

Al Shirawi, N., Memish, Z., Cherfan, A., and Al Shimemeri, A. (2006). Post-neurosurgical meningitis due to multidrug-resistant *Acinetobacter baumanii* treated with intrathecal colistin: case report and review of the literature. *Journal of Chemotherapy 18* (5): 554–558. https://doi.org/10.1179/joc.2006.18.5.554.

Anand, V.K., Alemar, G., and Griswold, J.A. Jr. (1992). Intracranial complications of mucormycosis: an experimental model and clinical review. *The Laryngoscope 102* (6): 656–662. https://doi.org/10.1288/00005537-199206000-00011.

Aslam, B., Wang, W., Arshad, M.I. et al. (2018). Antibiotic resistance: a rundown of a global crisis. *Infection and Drug Resistance 11*: 1645. https://www.ncbi.nlm.nih.gov/pmc/articles/PMC6188119/.

Ba, Y., Trouillet, J., Thonnon, J., and Fontenille, D. (1999). Phlébotomes du Sénégal: inventaire de la faune de la région de Kédougou. Isolements d'arbovirus. *Bulletin de la Société de Pathologie Exotique 92* (2): 131–135.

Bai, L., Morton, L.C., and Liu, Q. (2013). Climate change and mosquito-borne diseases in China: a review. *Globalization and Health 9* (1): 1–22. https://doi.org/10.1186/1744-8603-9-10.

Baker, J.C. and Spracklen, D.V. (2019). Climate benefits of intact Amazon forests and the biophysical consequences of disturbance. *Frontiers in Forests and Global Change* 2: 47. https://doi.org/10.3389/ffgc.2019.00047/full.

Baxter, M. and Illston, G. (1980). Temperature relationships of fungi isolated at low temperatures from soils and other substrates. *Mycopathologia 72* (1): 21–25. DOI: 10.1007%2FBF00443047.

Behie, S.W. and Bidochka, M.J. (2013). Potential agricultural benefits through biotechnological manipulation of plant fungal associations. *Bioessays 35* (4): 328–331. https://doi.org/10.1002/bies.201200147.

Bentz, M.L., Sexton, D.J., Welsh, R.M., and Litvintseva, A.P. (2019). Phenotypic switching in newly emerged multidrug-resistant pathogen *Candida auris*. *Medical Mycology 57* (5): 636–638. https://academic.oup.com/mmy/article/57/5/636/5133422?login=true.

Bhabhra, R. and Askew, D. (2005). Thermotolerance and virulence of *Aspergillus fumigatus*: role of the fungal nucleolus. *Medical Mycology* 43 (Supplement 1): S87–S93. https://academic.oup.com/mmy/article/43/Supplement_1/S87/1748298?login=true.

Bhatt, P.N. and Rodrigues, F. (1967). Chandipura: a new Arbovirus isolated in India from patients with febrile illness. *Indian Journal of Medical Research 55* (12): 1295–1305. https://www.cabdirect.org/cabdirect/abstract/19701000635.

Bi, P., Parton, K.A., and Tong, S. (2005). El Nino–southern oscillation and vector-borne diseases in Anhui, China. *Vector-Borne and Zoonotic Diseases 5* (2): 95–100. https://doi.org/10.1089/vbz.2005.5.95.

Biswal, M., Rudramurthy, S., Jain, N. et al. (2017). Controlling a possible outbreak of *Candida auris* infection: lessons learnt from multiple interventions. *Journal of Hospital Infection 97* (4): 363–370. https://www.sciencedirect.com/science/article/abs/pii/S0195670117305133.

Biswas, S., Van Dijck, P., and Datta, A. (2007). Environmental sensing and signal transduction pathways regulating morphopathogenic determinants of *Candida albicans. Microbiology and Molecular Biology Reviews 71* (2): 348–376. https://doi.org/10.1128/MMBR.00009-06.

Bleiker, K. and Six, D. (2014). Dietary benefits of fungal associates to an eruptive herbivore: potential implications of multiple associates on host population dynamics. *Environmental Entomology 36* (6): 1384–1396. https://academic.oup.com/ee/article-abstract/36/6/1384/503155.

Boelaert, J.R., de Locht, M., Van Cutsem, J. et al. (1993). Mucormycosis during deferoxamine therapy is a siderophore-mediated infection. *in vitro* and *in vivo* animal studies. *The Journal of Clinical Investigation 91* (5): 1979–1986. https://www.jci.org/articles/view/116419.

Borman, A.M., Szekely, A., and Johnson, E.M. (2016). Comparative pathogenicity of United Kingdom isolates of the emerging pathogen *Candida auris* and other key pathogenic Candida species. *MSphere 1* (4) https://doi.org/10.1128/mSphere.00189-16.

Brady, O.J., Johansson, M.A., Guerra, C.A. et al. (2013). Modelling adult *Aedes aegypti* and *Aedes albopictus* survival at different temperatures in laboratory and field settings. *Parasites and Vectors 6* (1): 1–12. https://doi.org/10.1186/1756-3305-6-351.

Bray, M., Hirsch, M., and Mitty, J. (2014). Epidemiology, pathogenesis, and clinical manifestations of Ebola and Marburg virus disease. *UpToDate 43*: 65–69.

Breman, J.G., Heymann, D.L., Lloyd, G. et al. (2016). Discovery and description of Ebola Zaire virus in 1976 and relevance to the West African epidemic during 2013–2016. *The Journal of Infectious Diseases 214* (suppl 3): S93–S101. https://academic.oup.com/jid/article/214/suppl_3/S93/2388104?login=true.

Buchholz, U., Müller, M.A., Nitsche, A. et al. (2013). Contact investigation of a case of human novel coronavirus infection treated in a German hospital, October-November 2012. *Eurosurveillance 18* (8): 20406. https://doi.org/10.2807/ese.18.08.20406-en.

Bussi, C. and Gutierrez, M.G. (2019). *Mycobacterium tuberculosis* infection of host cells in space and time. *FEMS Microbiology Reviews 43* (4): 341–361. https://academic.oup.com/femsre/article/43/4/341/5420823?login=true.

Caminade, C., McIntyre, K.M., and Jones, A.E. (2019). Impact of recent and future climate change on vector-borne diseases. *Annals of the New York Academy of Sciences 1436* (1): 157. https://www.ncbi.nlm.nih.gov/pmc/articles/PMC6378404/.

Casadevall, A., Kontoyiannis, D.P., and Robert, V. (2019). On the emergence of *Candida auris*: climate change, azoles, swamps, and birds. *MBio 10* (4) https://doi.org/10.1128/mBio.01397-19.

Chakraborty, I. and Maity, P. (2020). COVID-19 outbreak: migration, effects on society, global environment and prevention. *Science of the Total Environment 728*: 138882. https://www.sciencedirect.com/science/article/abs/pii/S0048969720323998.

Change, I.C. (2007). *The Physical Science Basis*. Cambridge University Press. https://journals.co.za/doi/10.10520/EJC93327.

Chapartegui-González, I., Lázaro-Díez, M., Bravo, Z. et al. (2018). *Acinetobacter baumannii* maintains its virulence after long-time starvation. *PLoS One 13* (8): e0201961. https://journals.plos.org/plosone/article?id=10.1371/journal.pone.0201961.

Charland, K., Buckeridge, D., Sturtevant, J. et al. (2009). Effect of environmental factors on the spatio-temporal patterns of influenza spread. *Epidemiology and Infection 137* (10): 1377–1387. https://www.cambridge.org/core/journals/epidemiology-and-infection/article/effect-of-environmental-factors-on-the-spatiotemporal-patterns-of-influenza-spread/F03CE61CA3F6C03AC6EEB0E0EA1830E8.

Chen, Y.-L., Lehman, V.N., Lewit, Y. et al. (2013). Calcineurin governs thermotolerance and virulence of *Cryptococcus gattii*. G3: genes, genomes. *Genetics 3* (3): 527–539. https://academic.oup.com/g3journal/article/3/3/527/6025703?login=true.

Connolly, C., Ali, S.H., and Keil, R. (2020). On the relationships between COVID-19 and extended urbanization. *Dialogues in Human Geography 10* (2): 213–216. https://doi.org/10.1177/2043820620934209.

Couger, B., Weirick, T., Damásio, A.R. et al. (2018). The genome of a thermo tolerant, pathogenic albino *Aspergillus fumigatus*. *Frontiers in Microbiology 9*: 1827. https://doi.org/10.3389/fmicb.2018.01827/full.

Courouble, V.V., Dey, S.K., Yadav, R. et al. (2021). Resolving the dynamic motions of SARS-CoV-2 nsp7 and nsp8 proteins using structural proteomics. *Journal of America Society for Mass Spectrometry.* 32 (7): 1618–1630. https://doi.org/10.1021/jasms.1c00086.

Curtis, N. (2006). Viral haemorrhagic fevers caused by Lassa, Ebola and Marburg viruses. *Hot Topics in Infection and Immunity in Children* III: 35–44. https://doi.org/10.1007/0-387-33026-7_4.

Dalziel, B.D., Kissler, S., Gog, J.R. et al. (2018). Urbanization and humidity shape the intensity of influenza epidemics in US cities. *Science 362* (6410): 75–79. https://science.sciencemag.org/content/362/6410/75.abstract.

Dannaoui, E. (2017). Antifungal resistance in mucorales. *International Journal of Antimicrobial Agents 50* (5): 617–621.

Daszak, P., Cunningham, A.A., and Hyatt, A.D. (2000). Emerging infectious diseases of wildlife--threats to biodiversity and human health. *Science 287* (5452): 443–449.

De Silva, P.M., Chong, P., Fernando, D.M. et al. (2018). Effect of incubation temperature on antibiotic resistance and virulence factors of *Acinetobacter baumannii* ATCC 17978. *Antimicrobial Agents and Chemotherapy 62* (1) https://doi.org/10.1128/AAC.01514-17.

Dey, J.K. and Dey, S.K. (2020). SARS-CoV-2 pandemic, COVID-19 case fatality rates and deaths per million population in India. *Journal of Bioinformatics, Computational and Systems Biology 2*:110.https://elynspublishing.com/index.php/journal/article/sars-cov-2-pandemic-covid-19-case-fatality-rates-and-deaths-per-million-populations-in-india.

Dey, J.K., Dey, S.K., and Sihag, H.J.H.L. (2020). Current insights into the novel Coronavirus disease 2019 (COVID-19) and its homoeopathic management. *Homœopathic Links 33* (03): 171–179. https://www.thieme-connect.com/products/ejournals/abstract/10.1055/s-0040-1715636.

Dey, S.K., Saini, M., Dhembla, C. et al. (2021). Suramin, Penciclovir and Anidulafungin bind nsp12, which governs the RNA-dependent-RNA polymerase activity of SARS-CoV-2, with higher interaction energy than Remdesivir, indicating potential in the treatment of Covid-19 infection. https://osf.io/preprints/urxwh/ (accessed 20 August 2021).

Dhanda, V., Rodrigues, F., and Ghosh, S. (1970). Isolation of Chandipura virus from sandflies in Aurangabad. *Indian Journal of Medical Research 58* (2): 179–180. https://www.cabdirect.org/cabdirect/abstract/19702703641.

Du, H., Bing, J., Hu, T. et al. (2020). *Candida auris*: epidemiology, biology, antifungal resistance, and virulence. *PLoS Pathogens 16* (10): e1008921. https://journals.plos.org/plospathogens/article?id=10.1371/journal.ppat.1008921.

Dye, C. (2008). Health and urban living. *Science* 319 (5864): 766–769. https://science.sciencemag.org/content/319/5864/766.abstract.

Engering, A., Hogerwerf, L., and Slingenbergh, J. (2013). Pathogen–host–environment interplay and disease emergence. *Emerging Microbes and Infections 2* (1): 1–7. https://doi.org/10.1038/emi.2013.5.

Farley, P.C. and Sullivan, P.A. (1998). The *Rhizopus oryzae* secreted aspartic proteinase gene family: an analysis of gene expression. *Microbiology 144* (8): 2355–2366. https://doi.org/10.1099/00221287-144-8-2355?crawler=true.

Fournier, P.E., Richet, H., and Weinstein, R.A. (2006). The epidemiology and control of *Acinetobacter baumannii* in health care facilities. *Clinical Infectious Diseases 42* (5): 692–699. https://academic.oup.com/cid/article/42/5/692/2052763?login=true.

Franklinos, L.H., Jones, K.E., Redding, D.W., and Abubakar, I. (2019). The effect of global change on mosquito-borne disease. *The Lancet Infectious Diseases 19* (9): e302–e312. https://www.sciencedirect.com/science/article/abs/pii/S1473309919301616.

Garcia-Solache, M.A. and Casadevall, A. (2010). Global warming will bring new fungal diseases for mammals. *MBio 1* (1) https://doi.org/10.1128/mBio.00061-10.

Gong, P., Liang, S., Carlton, E.J. et al. (2012). Urbanisation and health in China. *The Lancet 379* (9818): 843–852. https://www.sciencedirect.com/science/article/abs/pii/S0140673611618783.

Gootz, T.D. and Marra, A. (2008). *Acinetobacter baumannii*: an emerging multidrug-resistant threat. *Expert Review of Anti-infective Therapy 6* (3): 309–325. https://doi.org/10.1586/14787210.6.3.309.

Gray, K.C., Palacios, D.S., Dailey, I. et al. (2012). Amphotericin primarily kills yeast by simply binding ergosterol. *Proceedings of the National Academy of Sciences 109* (7): 2234–2239. https://www.pnas.org/content/109/7/2234.short.

Gwaltney, J. Jr. (1985). Virology and immunology of the common cold. *Rhinology 23* (4): 265–271. https://europepmc.org/article/med/3001912.

Hansen, A., Cameron, S., Liu, Q. et al. (2015). Transmission of haemorrhagic fever with renal syndrome in china and the role of climate factors: a review. *International Journal of Infectious Diseases 33*: 212–218. https://www.sciencedirect.com/science/article/pii/S1201971215000387.

Hasnat, M.A. and Rahman, M.A. (2018). A review paper on the hazardous effect of plastic debris on marine biodiversity with some possible remedies. *Asian Journal of Medical and Biological Research* 4 (3): 233–241. https://www.banglajol.info/index.php/AJMBR/article/view/38461.

Heaney, H., Laing, J., Paterson, L. et al. (2020). The environmental stress sensitivities of pathogenic Candida species, including *Candida auris*, and implications for their spread in the hospital setting. *Medical Mycology 58* (6): 744–755. https://academic.oup.com/mmy/article/58/6/744/5698107?login=true.

Houbraken, J., de Vries, R.P., and Samson, R.A. (2014). Modern taxonomy of biotechnologically important *Aspergillus* and *Penicillium* species. *Advances in Applied Microbiology 86*: 199–249. https://www.sciencedirect.com/science/article/pii/B9780128002629000044.

Howard, D.H. (1999). Acquisition, transport, and storage of iron by pathogenic fungi. *Clinical Microbiology Reviews 12* (3): 394–404. https://doi.org/10.1128/CMR.12.3.394.

Hu, X., Cook, S., and Salazar, M.A. (2008). Internal migration and health in China. *The Lancet 372* (9651): 1717–1719. https://www.thelancet.com/journals/lancet/article/PIIS0140673608613604/fulltext?isEOP=true.

Huang, G. (2012). Regulation of phenotypic transitions in the fungal pathogen *Candida albicans*. *Virulence 3* (3): 251–261. https://doi.org/10.4161/viru.20010.

Ibrahim, A., Spellberg, B., and Edwards, J. Jr. (2008). Iron acquisition: a novel prospective on mucormycosis pathogenesis and treatment. *Current Opinion in Infectious Diseases 21* (6): 620. https://academic.oup.com/cid/article/54/suppl_1/S16/284344?login=true.

Ibrahim, A.S., Spellberg, B., Walsh, T.J., and Kontoyiannis, D.P. (2012). Pathogenesis of mucormycosis. *Clinical Infectious Diseases* 54 (suppl_1): S16–S22. https://www.ncbi.nlm.nih.gov/pmc/articles/PMC2773686/.

Jackson, B.R., Chow, N., Forsberg, K. et al. (2019). On the origins of a species: what might explain the rise of *Candida auris*? *Journal of Fungi 5* (3): 58. https://www.mdpi.com/2309-608X/5/3/58.

Jain, N., Hasan, F., and Fries, B.C. (2008). Phenotypic switching in fungi. *Current Fungal Infection Reports 2* (3): 180–188. https://doi.org/10.1007/s12281-008-0026-y.

Jha, S.K. and Tripathi, V. (2020). Beginnings of urbanization in early historic India. *Studies in Humanities and Social Sciences 6* (2): 184–187.

Johnston, S.A., Voelz, K., and May, R.C. (2016). *Cryptococcus neoformans* thermotolerance to avian body temperature is sufficient for extracellular growth but not intracellular survival in macrophages. *Scientific Reports 6* (1): 1–9. https://www.nature.com/articles/srep20977.

Justice, S.S., Hunstad, D.A., Cegelski, L., and Hultgren, S.J. (2008). Morphological plasticity as a bacterial survival strategy. *Nature Reviews Microbiology 6* (2): 162–168. https://www.nature.com/articles/nrmicro1820.

Kagaruki, G.B., Kimaro, G.D., Mweya, C.N. et al. (2015). Prevalence and risk factors of metabolic syndrome among individuals living with HIV and receiving antiretroviral treatment in Tanzania. *Journal of Advances in Medicine and Medical Research*: 1317–1327. https://www.journaljammr.com/index.php/JAMMR/article/view/15714.

Kean, R., Delaney, C., Sherry, L. et al. (2018). Transcriptome assembly and profiling of *Candida auris* reveals novel insights into biofilm-mediated resistance. *MSphere 3* (4) https://doi.org/10.1128/mSphere.00334-18.

Kemp, G.E. (1975). Viruses other than arenaviruses from West African wild mammals: factors affecting transmission to man and domestic animals. *Bulletin of the World Health Organization 52* (4-6): 615. https://www.ncbi.nlm.nih.gov/pmc/articles/PMC2366648/.

Korenberg, E.I. (2000). Seasonal population dynamics of Ixodes ticks and tick-borne encephalitis virus. *Experimental and Applied Acarology 24* (9): 665–681. https://doi.org/10.1023/A:1010798518261.

Kumar, P., Nag, T.C., Jha, K.A. et al. (2017). Experimental oral iron administration: histological investigations and expressions of iron handling proteins in rat retina with aging. *Toxicology 392*: 22–31. https://www.sciencedirect.com/science/article/abs/pii/S0300483X17303086.

Lin, H., Zhang, Z., Lu, L. et al. (2014). Meteorological factors are associated with hemorrhagic fever with renal syndrome in Jiaonan County, China, 2006–2011. *International Journal of Biometeorology 58* (6): 1031–1037. 10.1007/s00484-013-0688-1.

Lockhart, S.R., Etienne, K.A., Vallabhaneni, S. et al. (2017). Simultaneous emergence of multidrug-resistant Candida auris on 3 continents confirmed by whole-genome sequencing and epidemiological analyses. *Clinical Infectious Diseases 64* (2): 134–140. https://academic.oup.com/cid/article/64/2/134/2706620?login=true.

Longo, L.V.G., Breyer, C.A., Novaes, G.M. et al. (2020). The human pathogen *Paracoccidioides brasiliensis* has a unique 1-cys peroxiredoxin that localizes both intracellularly and at the cell surface. *Frontiers in Cellular and Infection Microbiology 10*: 394. https://doi.org/10.3389/fcimb.2020.00394/full?report=reader.

Mailles, A., Blanckaert, K., Chaud, P. et al. (2013). First cases of Middle East Respiratory Syndrome Coronavirus (MERS-CoV) infections in France, investigations and implications for the prevention of human-to-human transmission, France, May 2013. *Eurosurveillance* 18 (24): 20502. https://doi.org/10.2807/ese.18.24.20502-en.

Manzoor, N. (2019). Candida pathogenicity and alternative therapeutic strategies. In: *Pathogenicity and Drug Resistance of Human Pathogens*, 135–146. Springer. https://link.springer.com/chapter/10.1007/978-981-32-9449-3_7.

Márquez, L.M., Redman, R.S., Rodriguez, R.J., and Roossinck, M.J. (2007). A virus in a fungus in a plant: three-way symbiosis required for thermal tolerance. *Science 315* (5811): 513–515. https://science.sciencemag.org/content/315/5811/513.abstract.

McKinney, M.L. (2002). Urbanization, biodiversity, and conservation: the impacts of urbanization on native species are poorly studied, but educating a highly urbanized human population about these impacts can greatly improve species conservation in all ecosystems. *Bioscience 52* (10): 883–890. https://academic.oup.com/bioscience/article/52/10/883/354714?login=true.

McMichael, A.J., Campbell-Lendrum, D.H., Corvalán, C.F. et al. (2003). *Climate Change and Human Health: Risks and Responses*. World Health Organization. https://www.google.co.in/books/edition/Climate_Change_and_Human_Health/tQFYJjDEwhIC?hl=en&gbpv=0.

Menghani, S., Chikhale, R., Raval, A. et al. (2012). Chandipura virus: an emerging tropical pathogen. *Acta Tropica 124* (1): 1–14. https://www.sciencedirect.com/science/article/abs/pii/S0001706X12002276.

Mishra, A.C. (2006). *Chandipura encephalitis*: a newly recognized disease of public health importance in India. *Emerging Infections 7*: 121–137. 10.1128/9781555815585.ch7.

Muñoz, J.F., Gade, L., Chow, N.A. et al. (2018). Genomic insights into multidrug-resistance, mating and virulence in *Candida auris* and related emerging species. *Nature Communications 9* (1): 1–13. https://www.nature.com/articles/s41467-018-07779-6.

Nobile, C.J. and Johnson, A.D. (2015). *Candida albicans* biofilms and human disease. *Annual Review of Microbiology 69*: 71–92. https://doi.org/10.1146/annurev-micro-091014-104330.

Normile, D. (2008). China's living laboratory in urbanization. *Science* 319: 740–743. https://science.sciencemag.org/content/319/5864/740.summary.

Neill, J.O. (2014). Antimicrobial resistance: tackling a crisis for the health and wealth of nations / the Review on Antimicrobial Resistance Chaired. In: *Review Paper-Tackling a Crisis for the Health and Wealth of Nations*, 1–20. HM Government Wellcome Trust. https://wellcomecollection.org/works/rdpck35v.

Oh, B.J., Shin, J.H., Kim, M.-N. et al. (2011). Biofilm formation and genotyping of *Candida haemulonii, Candida pseudohaemulonii*, and a proposed new species (*Candida auris*) isolates from Korea. *Medical Mycology 49* (1): 98–102. https://academic.oup.com/mmy/article/49/1/98/1392358?login=true.

Pappas, P.G., Lionakis, M.S., Arendrup, M.C. et al. (2018). Invasive candidiasis. *Nature Reviews Disease Primers 4* (1): 1–20. https://www.nature.com/articles/nrdp201826.

Patel, M., Chaubey, A.K., Pittman, C.U. Jr. et al. (2020). Coronavirus (SARS-CoV-2) in the environment: occurrence, persistence, analysis in aquatic systems and possible management. *Science of the Total Environment* 765: 142698. https://www.sciencedirect.com/science/article/abs/pii/S0048969720362276.

Paulson, T. (2013). Epidemiology: a mortal foe. *Nature 502* (7470): S2–S3. https://www.nature.com/articles/502S2a.

Peleg, A.Y., Seifert, H., and Paterson, D.L. (2008). *Acinetobacter baumannii*: emergence of a successful pathogen. *Clinical Microbiology Reviews 21* (3): 538–582. https://doi.org/10.1128/CMR.00058-07.

Petersen, J.L. (2010). *Praise, Blame, and Oracle: The Rhetorical Tropes of Political Economy*. Citeseer. https://www.proquest.com/openview/291cfc2bd68536f2ea105492c9728e55/1?pq-origsite=gscholar&cbl=18750.

Peterson, A.T., Bauer, J.T., and Mills, J.N. (2004). Ecologic and geographic distribution of filovirus disease. *Emerging Infectious Diseases 10* (1): 40. https://www.ncbi.nlm.nih.gov/pmc/articles/PMC3322747/.

Pica, N. and Bouvier, N.M. (2012). Environmental factors affecting the transmission of respiratory viruses. *Current Opinion in Virology 2* (1): 90–95. https://www.sciencedirect.com/science/article/abs/pii/S1879625711001891.

Prevedello, J.A., Winck, G.R., Weber, M.M. et al. (2019). Impacts of forestation and deforestation on local temperature across the globe. *PLoS One 14* (3): e0213368. https://journals.plos.org/plosone/article?id=10.1371/journal.pone.0213368.

Rao, B., Basu, A., Wairagkar, N.S. et al. (2004). A large outbreak of acute encephalitis with high fatality rate in children in Andhra Pradesh, India, in 2003, associated with Chandipura virus. *The Lancet 364* (9437): 869–874. https://www.sciencedirect.com/science/article/abs/pii/S0140673604169821.

Reiter, P., Turell, M., Coleman, R. et al. (1999). Field investigations of an outbreak of Ebola hemorrhagic fever, Kikwit, Democratic Republic of the Congo, 1995: arthropod studies. *The Journal of Infectious Diseases* 179 (Supplement_1): S148–S154. https://academic.oup.com/jid/article/179/Supplement_1/S148/880615?login=true.

Retailliau, H.F., Hightower, A.W., Dixon, R.E., and Allen, J.R. (1979). *Acinetobacter calcoaceticus*: a nosocomial pathogen with an unusual seasonal pattern. *The Journal of Infectious Diseases 139* (3): 371–375. https://www.jstor.org/stable/30110649.

Rhodes, J. and Fisher, M.C. (2019). Global epidemiology of emerging *Candida auris*. *Current Opinion in Microbiology 52*: 84–89. https://www.sciencedirect.com/science/article/abs/pii/S1369527419300177.

Richardson, M. (2009). The ecology of the Zygomycetes and its impact on environmental exposure. *Clinical Microbiology and Infection 15*: 2–9. https://doi.org/10.1111/j.1469-0691.2009.02972.x.

Rizwan, T., Kothidar, A., Meghwani, H. et al. (2021). Comparative analysis of SARS-CoV-2 envelope viroporin mutations from COVID-19 deceased and surviving patients revealed implications on its ion-channel activities and correlation with patient mortality. *Journal of Biomeolecular Structure and Dynamics*: 1–16. https://doi.org/10.1080/07391102.2021.1944319.

Robert, V.A. and Casadevall, A. (2009). Vertebrate endothermy restricts most fungi as potential pathogens. *The Journal of Infectious Diseases 200* (10): 1623–1626. https://academic.oup.com/jid/article/200/10/1623/881601?login=true.

Roca, A., Afolabi, M.O., Saidu, Y., and Kampmann, B. (2015). Ebola: a holistic approach is required to achieve effective management and control. *Journal of Allergy and Clinical Immunology 135* (4): 856–867. https://www.sciencedirect.com/science/article/pii/S0091674915002638.

Saif, L. (2004). Animal coronaviruses: what can they teach us about the severe acute respiratory syndrome? *Revue Scientifique et Technique-Office International des épizooties 23* (2): 643–660. https://europepmc.org/article/med/15702725.

Satoh, K., Makimura, K., Hasumi, Y. et al. (2009). *Candida auris* sp. nov., a novel ascomycetous yeast isolated from the external ear canal of an inpatient in a Japanese hospital. *Microbiology and Immunology 53* (1): 41–44. https://doi.org/10.1111/j.1348-0421.2008.00083.x.

Semenza, J.C., Lindgren, E., Balkanyi, L. et al. (2016). Determinants and drivers of infectious disease threat events in Europe. *Emerging Infectious Diseases 22* (4): 581. https://www.ncbi.nlm.nih.gov/pmc/articles/PMC4806948/.

Shao, P. and Zeng, X. (2012). Progress in the study of the effects of land use and land cover change on the climate system. *Climate and Environmental Research 1*: 103–111.

Sharma, Y., Rastogi, S.K., Perwez, A. et al. (2020). β-citronellol alters cell surface properties of *Candida albicans* to influence pathogenicity related traits. *Medical Mycology 58* (1): 93–106. https://academic.oup.com/mmy/article-abstract/58/1/93/5370456.

Shpitzer, T., Keller, N., Wolf, M. et al. (2005). Seasonal variations in rhino-cerebral Mucor infection. *Annals of Otology, Rhinology and Laryngology 114* (9): 695–698. https://doi.org/10.1177/000348940511400907.

Singh, O., Arya, P., and Chaudhary, B.S. (2013). Temperature trends at Dehradun in Doon valley of Uttarakhand. India. In: *Climatic Change & Himalayan Ecosystem-Indicator, Bio & Water Resources*, 37–56. Scientific Publisher India. https://www.google.co.in/books/edition/Climate_Change_Himalayan_Ecosystem_Indic/9qI4DwAAQBAJ?hl=en&gbpv=0.

Singh, R., Kaur, M., Chakrabarti, A. et al. (2019). Biofilm formation by *Candida auris* isolated from colonising sites and candidemia cases. *Mycoses 62* (8): 706–709. 10.1111/myc.12947.

Spivak, E.S. and Hanson, K.E. (2018). *Candida auris*: an emerging fungal pathogen. *Journal of Clinical Microbiology 56* (2) https://doi.org/10.1128/JCM.01588-17.

Srivastava, R.K. (2020). Urbanization-led neo risks and vulnerabilities: a new challenge. In: *Managing Urbanization, Climate Change and Disasters in South Asia*, 251–296. Springer. https://link.springer.com/chapter/10.1007/978-981-15-2410-3_6.

Sudeep, A., Gurav, Y., and Bondre, V. (2016). Changing clinical scenario in Chandipura virus infection. *The Indian Journal of Medical Research 143* (6): 712. https://www.ncbi.nlm.nih.gov/pmc/articles/PMC5094110/.

Sumpter, C. and Chandramohan, D. (2013). Systematic review and meta-analysis of the associations between indoor air pollution and tuberculosis. *Tropical Medicine and International Health 18* (1): 101–108. 10.1111/tmi.12013.

Swanepoel, R., Leman, P.A., Burt, F.J. et al. (1996). Experimental inoculation of plants and animals with Ebola virus. *Emerging Infectious Diseases 2* (4): 321. https://www.ncbi.nlm.nih.gov/pmc/articles/PMC2639914/.

Swanepoel, R., Smit, S.B., Rollin, P.E. et al. (2007). Studies of reservoir hosts for Marburg virus. *Emerging Infectious Diseases 13* (12): 1847. https://www.ncbi.nlm.nih.gov/pmc/articles/PMC2876776/.

Tabachnick, W. (2010). Challenges in predicting climate and environmental effects on vector-borne disease episystems in a changing world. *Journal of Experimental Biology 213* (6): 946–954. https://journals.biologists.com/jeb/article/213/6/946/10167/Challenges-in-predicting-climate-and-environmental.

Tatem, A., Huang, Z., Das, A. et al. (2012). Air travel and vector-borne disease movement. *Parasitology-Cambridge 139* (14): 1816. https://www.cambridge.org/core/journals/parasitology/article/abs/air-travel-and-vectorborne-disease-movement/FD43E8CB468C98652870203EA62E7FBD.

Team, W.E.R. (2014). Ebola virus disease in West Africa—the first 9 months of the epidemic and forward projections. *New England Journal of Medicine 371* (16): 1481–1495. https://doi.org/10.1056/NEJMoa1411100.

Tian, H., Hu, S., Cazelles, B. et al. (2018). Urbanization prolongs hantavirus epidemics in cities. *Proceedings of the National Academy of Sciences United States of America 115* (18): 4707–4712. https://www.pnas.org/content/115/18/4707.short.

Tjaden, N.B., Suk, J.E., Fischer, D. et al. (2017). Modelling the effects of global climate change on Chikungunya transmission in the 21st century. *Scientific Reports 7* (1): 1–11. https://www.nature.com/articles/s41598-017-03566-3.

Tsetsarkin, K.A. and Weaver, S.C. (2011). Sequential adaptive mutations enhance efficient vector switching by Chikungunya virus and its epidemic emergence. *PLoS Pathogen 7* (12): e1002412. https://journals.plos.org/plospathogens/article?id=10.1371/journal.ppat.1002412.

Upadhyay, S., Mittal, E., and Philips, J. (2018). Tuberculosis and the art of macrophage manipulation. *Pathogens and Disease* 76 (4): fty037. https://academic.oup.com/femspd/article/76/4/fty037/4970761?login=true.

Vanittanakom, N., Cooper, C.R., Fisher, M.C., and Sirisanthana, T. (2006). *Penicillium marneffei* infection and recent advances in the epidemiology and molecular biology aspects. *Clinical Microbiology Reviews 19* (1): 95–110. https://doi.org/10.1128/CMR.19.1.95-110.2006.

Wang, X., Bing, J., Zheng, Q. et al. (2018). The first isolate of *Candida auris* in China: clinical and biological aspects. *Emerging Microbes and Infections 7* (1): 1–9. https://doi.org/10.1038/s41426-018-0095-0.

Welsh, R.M., Bentz, M.L., Shams, A. et al. (2017). Survival, persistence, and isolation of the emerging multidrug-resistant pathogenic yeast *Candida auris* on a plastic health care surface. *Journal of Clinical Microbiology 55* (10): 2996–3005. https://doi.org/10.1128/JCM.00921-17.

Worby, C.J., Chaves, S.S., Wallinga, J. et al. (2015). On the relative role of different age groups in influenza epidemics. *Epidemics 13*: 10–16. https://www.sciencedirect.com/science/article/pii/S1755436515000511.

Xiao, H., Tian, H.-Y., Gao, L.-D. et al. (2014). Animal reservoir, natural and socioeconomic variations and the transmission of hemorrhagic fever with renal syndrome in Chenzhou, China, 2006–2010. *PLoS Neglected Tropical Diseases 8* (1): e2615. https://journals.plos.org/plosntds/article?id=10.1371/journal.pntd.0002615.

Yang, W., Cao, Q., Qin, L. et al. (2020a). Clinical characteristics and imaging manifestations of the 2019 novel coronavirus disease (COVID-19): a multi-center study in Wenzhou city, Zhejiang, China. *Journal of Infection 80* (4): 388–393. https://www.sciencedirect.com/science/article/abs/pii/S0163445320300992.

Yang, X., Yu, Y., Xu, J. et al. (2020b). Clinical course and outcomes of critically ill patients with SARS-CoV-2 pneumonia in Wuhan, China: a single-centered, retrospective, observational study. *The Lancet Respiratory Medicine 8* (5): 475–481. https://www.sciencedirect.com/science/article/pii/S2213260020300795.

Yue, H., Bing, J., Zheng, Q. et al. (2018). Filamentation in *Candida auris*, an emerging fungal pathogen of humans: passage through the mammalian body induces a heritable phenotypic switch. *Emerging Microbes and Infections 7* (1): 1–13. https://doi.org/10.1038/s41426-018-0187-x.

Zhang, Y., Bi, P., and Hiller, J.E. (2008). Climate change and the transmission of vector-borne diseases: a review. *Asia Pacific Journal of Public Health 20* (1): 64–76. https://doi.org/10.1177/1010539507308385.

Perlroth, J., Choi, B., and Spellberg, B. (2007). Nosocomial fungal infections: epidemiology, diagnosis, and treatment. *Medical Mycology* 45 (4): 321–346.

Sharma, S.K. and Mohan, A. (2004). Multidrug-resistant tuberculosis. *Indian Journal of Medical Research* 120 (4): 354–376.

Glaziou, P., Floyd, K., and Raviglione, M. (2009). Global burden and epidemiology of tuberculosis. *Clinics in Chest Medicine* 30 (4): 621–636.

Sajadi, M.M., Habibzadeh, P., Vintzileos, A. et al. (2020). Temperature, humidity, and latitude analysis to estimate potential spread and seasonality of coronavirus disease 2019 (COVID-19). *JAMA Network Open* 3 (6): e2011834–e2011834.

13

Human–Wildlife Conflict in the Mumbai Metropolitan Region – An Empirical Study

Deepti Sharma[1] *and Prachi Sinha*[2]

[1] *TerraNero Environmental Solutions Pvt. Ltd., Thane (W), Maharashtra, India*
[2] *Ministry of Home Affairs, Government of India, National Investigation Agency, New Delhi, India*

13.1 Introduction

Human–wildlife conflict may be defined as interactions between human beings and wild fauna species, leading to a negative impact on human beings and wild animals or both – this impact may be serious enough to entail the loss of human or animal life or can cause damage to physical assets or wildlife habitats. It can also cause morbidity and mortality through zoonotic infections (Richardson et al. 2020) and emotional and psychological discomfort because of sociocultural conditioning (Penteriani et al. 2016).

With rising human encroachment into wildlife habitats and environmental changes caused by the climate, human–wildlife conflicts in the urban areas are on the rise, both in terms of frequency and intensity (Donihue and Lambert 2014; Johnson and Munshi-South 2017). Incessant urbanisation has caused humans to usurp wildlife habitats by the acres, causing loss of feeding and breeding grounds for faunal species. Wild animals move from one natural area to another for various reasons such as food, shelter, community, dispersion, mating, and childbirth. Urban areas lure wild fauna species with their ready food sources such as piles of garbage and cultivated plants, and most importantly, freshwater (Cox and Gaston 2018). However, wild animals face problems due to habitat fragmentation while crossing a road/railway line or other manmade infrastructure. Noise, moving vehicles, wide roads, and roads between habitats can lead to confusion or trauma to the animals, making it difficult for them to cross over. Ultimately, they become a victim of roadkill or a railway track victim (Nielsen et al. 2003). Also, several fauna species get trapped in human households, getting injured unintentionally or intentionally.

For instance, Beisner et al. (2015) discuss the conflict with rhesus macaques in Dehradun, India. While Priston and McLennan (2013) point out that several deaths have ensued in India due to conflict with the rhesus macaques, Sharma et al. (2011) record the man–monkey conflicts in the Jodhpur city of the Rajasthan state in western India. Similarly,

Urban Ecology and Global Climate Change, First Edition. Edited by Rahul Bhadouria, Shweta Upadhyay, Sachchidanand Tripathi, and Pardeep Singh.

several researchers have reported human–leopard conflicts (Hathaway et al. 2017; Ranade et al. 2015; Sen and Pattanaik 2015), especially in the Mumbai region and elsewhere (Odden et al. 2014, across Maharashtra and Himachal Pradesh; Sidhu et al. 2017, Anamalai hills). Elephant–human conflict in India's northeastern cities has been documented (Choudhury 2004; Das et al. 2012). Snakes have been commonly reported to be in conflict with human beings (Longkumer et al. 2016, North Bihar; Jadesh et al. 2014; Gulbarga). Various bat species are also involved in conflict situations (Nulkar 2016); for instance, residents from Mumbai and Pune region insisted on expelling bat colonies from their neighbourhood areas. Hatkar et al. (2019) have described the rescue and rehabilitation of loggerhead sea turtles from the Dahanu coast in northern Maharashtra.

Apart from the loss of life, risk of morbidity, and psychological or emotional discomfort, human–wildlife conflicts cause substantial economic losses. Manral et al. (2016) discuss the economic losses that ensue as an outcome of human–wildlife conflict. As per Ravenelle and Nyhus (2017), about USD 230 million has been paid as compensation since 1950 in 50 countries.

There is a pressing need to manage human–wildlife conflicts in urban areas. The risk of injury or loss of life due to wild animals is one that no citizen would like to face. Simultaneously, several citizens recognise the value of biodiversity in urban areas and are willing to do their bit for the conservation of species. Also, there is rising awareness regarding the legal implications of killing wild animals.

Mumbai, the financial capital of India, is a highly urbanised city. However, it also has rich biodiversity thriving in the Sanjay Gandhi National Park (SGNP) in its Borivali locality, Powai, Vihar and Tulsi Lakes, and amidst the mangrove patches at Bhandup and Vikhroli localities along the Thane Creek and elsewhere in the city. The larger exurban area around Mumbai is covered under the Mumbai Metropolitan Region (MMR) is richer in mangroves and other ecosystems.

However, this richness of biodiversity often finds itself at odds with the thriving city, creating incidents of human–wildlife conflicts. Be it the Indian Marsh crocodiles of the Powai Lake (Sen and Nagendra 2019) or the leopards of SGNP (Bhatia et al. 2013), golden jackals of the Bhandup Pumping Station or rare birds from the IIT Bombay Campus and Godrej-owned mangroves at Vikhroli, and there have been numerous records of such species either getting injured by traffic, getting trapped amidst wires or caught inside homes, or attacking human beings in panic. It is critical to study such incidents in detail, with the ultimate aim of reducing human–wildlife conflict and ensuring peaceful coexistence of biodiversity in urban ecosystems.

The Maharashtra State Forest Department, along with several voluntary Non-Government Offices (NGOs), is involved in rescuing wild animals (mainly birds, mammals, and reptiles) from near-fatal situations such as being trapped in electric wires/trees/towers/drains/homes/offices and burn injuries from fire or electrocution, dehydration, road-accident injuries, human-inflicted injuries, illegal trade, and failed poaching attempts.

With this background, the present study was conducted by collating and analysing wild animal rescue data, as old as available, from the MMR, to address the following research questions:

1) Which wild fauna species in a given locality of the MMR are involved in the human–wildlife conflict?

2) In which specific locations are human–wildlife conflict most common?
3) Is there a seasonal peak in the human–wildlife conflicts?

Hence, the objectives of the present study are to:

1) Collate wild animal rescue/roadkill/poaching/smuggling data from the Maharashtra State Forest Department and its partner NGOs from 2012 to 2018
2) Analyse the data with respect to species involved, seasonal peaks in the conflicts and locations where conflicts occurred
3) Map the locations where conflicts were reported

The outcome of this study is likely to help organise effective training programmes for local citizens to reduce human–animal conflicts. Also, planning an effective animal transit programme across the MMR is crucial, lowering the chances of roadkills. This study is the first attempt to collect long-term data regarding wild animal rescue operations in the MMR and subjecting it to analysis to ensure that meaningful conclusions are drawn.

Using keywords such as 'human–wildlife conflict', 'human–animal conflict', 'human–leopard conflict Mumbai', and research publications were collated from Google Scholar.

Table 13.1 gives a brief literature review of research papers addressing the issues of human–wildlife conflicts, reasons for its existence, contributors in conservation and mitigation through the use of various technologies.

13.2 Methodology

13.2.1 Study Area

The project was carried out in the complete MMR. The MMR has the following municipal corporations – Greater Mumbai, Thane, Navi Mumbai, Panvel, Ulhasnagar, Vasai-Virar, Kalyan-Dombivali, Mira-Bhayander, and Bhiwandi-Nizampur, and municipal councils – Palghar, Ambernath, Kulgaon-Badlapur, Pen, Alibag, Uran, Matheran, and Karjat. Among these, Greater Mumbai, Thane, and Navi Mumbai are among the most urbanised, followed by Panvel, Ulhasnagar, Vasai-Virar, Kalyan-Dombivali, Mira-Bhayander, and Bhiwandi-Nizampur. The municipal councils are relatively less urbanised.

The forest divisions in this area include – Mangrove Cell, SGNP, Tungareshwar Wildlife Sanctuary, and Forest Divisions at Thane, Badlapur, Uran, Pen, Alibaug, and Kalyan-Dombivali.

As per Bajaru et al. (2020), the MMR terrestrial biodiversity is rich, including 213 species (25 mammals, 135 birds, 16 amphibians, and 36 reptiles). This same study identified seven different habitats in the MMR – semi-evergreen forest, moist deciduous forest, scrub forest, mangrove, grassland, agriculture, and settlements (Figure 13.1).

13.2.2 Data Collection and Analysis

13.2.2.1 Sources of Data

There were three major sources of data – the forest department, volunteers of NGOs recognised by the forest department and individual volunteers recognised by the forest department.

Table 13.1 Literature review.

Sr. No.	Year and citation	Findings
1	Trombulak and Frissell (2000); Underhill and Angold (1999)	Roads are linear obstructions that interfere with animal movement, especially those of slow-moving species. Physical harm and injuries caused to wild animal species due to speeding traffic are often fatal or near-fatal. The alteration caused to the physical and chemical environment by road construction causes changes in animal behaviour. With newly constructed roads, human disturbance can reach hitherto untouched areas. Plantations along roads are largely ornamental exotics that are at odds with local flora
2	Santos et al. (2017), p. 11	Mortality of smaller taxons like reptiles and amphibians in railway track killings is often ignored or overshadowed by the deaths of large mammals. Since Mumbai is a city with local trains and simultaneously thrives with a rich biodiversity, this is an important aspect of human–wildlife conflict in the study area and must be given research and administrative attention
3	Seiler and Olsson (2017), p. 277	Deterrents and warning signals to repel wildlife away from railway tracks have been discussed by these authors
4	Russo and Ancillotto (2015)	Even the so-called synurbic species that are more tolerant of anthropogenic activities are at risk in urban ecosystems
5	Chace and Walsh (2006)	Infectious diseases, predator assemblage, and food – apart from the risk of collision with artificial objects such as glass facades are factors impacting the survival of bird species in urban areas
6	Freitas et al. (2013)	River proximity and herbaceous vegetation cover were associated with the most road-killed vertebrate groups. For all species and for mammals separately, roadkill was associated with river proximity, whereas for large and arboreal mammals, reptiles and owls, roadkills were associated with higher herbaceous vegetation cover
7	Shine and Koenig (2001)	Wild animal rescue activities (i) impact the mortality pattern of wildlife species abundance at the lower geographical level, and (ii) maintain records that yield an overview of local biodiversity patterns
8	Vyas (2013)	The rescue-related records maintained by NGOs and volunteers serve as a record of local biodiversity
9	Li et al. (2005).	The road greenway network can act as an important corridor for people and wildlife. Therefore, they are an important component of the green network in the built-up area
10	Wu and Hobbs (2002)	Network of patches and corridors can provide connectivity of natural elements and help preserve linkage between different ecosystems
11	Clevenger et al. (2002)	A GIS environment, regional or landscape-level connectivity models can facilitate the identification and delineation of barriers and corridors for animal movement
12	Nielsen et al. (2003)	Techniques such as GIS, remote sensing, and multivariate statistics to analyse the roadkill occurrence and rural landscape were used to reduce roadkill issues
13	Rudd et al. (2002)	It is critical that natural habitats within a city are linked through patches of vegetation such as backyard habitats and planted boulevards to permit safe transit to urban wildlife
14	McKinney (2002)	Awareness raising among the urbanised human population, regarding the importance of conservation of native species and reducing the ill-effects of urbanisation on native species

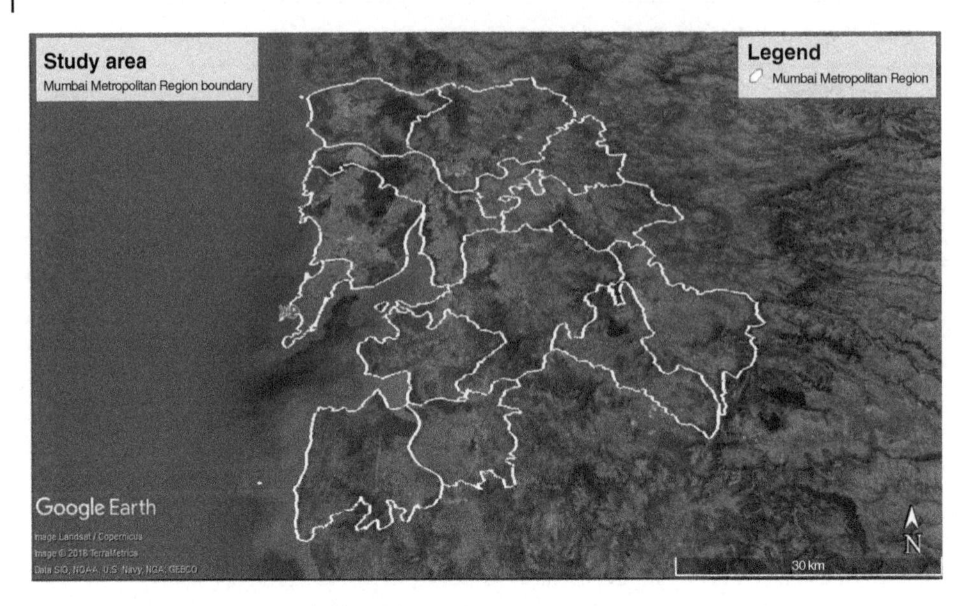

Figure 13.1 Google Earth imagery of the study area.

Each volunteer – whether from recognised NGOs or working individually – is supposed to maintain a written record of all the rescue activities they have performed. Forest department officers involved in the rescue operations also maintain this data in their files. This data are filed every year and maintained by the forest department with the help of its personnel as well as its partner NGOs. Few such NGOs are – Resqink Association for Wildlife Welfare (RAWW), Eco-Echo, Trust for Rescue Afforestation Conservation and Knowledge (TRACK), Sarp Mitra, Plant and Animals Welfare Society (PAWS), and Wildlife Welfare Association (WWA). In addition, the Thane Society for the Prevention of Cruelty to Animals (SPCA) is also active in providing medical aid to the injured animals.

A list of NGOs and individual rescuers contacted is given in Table 13.2 while Table 13.3 shows the details of the individual forest department offices. A total of 13 NGOs and 5 individual rescuers/volunteers were contacted.

13.2.2.2 Data Collection

Data were collected by personally visiting each forest office mentioned in Table 13.3, and after submitting a formal request, authentic photocopies of their hand-filled forms containing rescue data were obtained. A list of recognised local NGOs and individual volunteers was also obtained from the forest office.

Later, each NGO and individual was contacted and personally visited. If data were available in the soft copy, it was obtained through electronic transfer. Hard copies of data were photocopied to generate authentic facsimiles.

In some cases, formal entry of data had not been carried out, and due to this quantitative information could not be obtained. In such cases, a format for data entry, which was provided by the forest department, was forwarded to the NGOs and they were asked to submit future data in the given format.

Table 13.2 List of NGOs and individuals contacted.

Sr. No.	Area	Name of organisation	Contact person	Data format	Response
NGOs					
1	Mumbai	Resqink Association for Wildlife Welfare (RAWW)	Pawan Sharma	Soft copy in different format	Data received from 2014 till present
		Spreading Awareness on Reptile and Rehabilitation Programme (SAARP)	Santosh Shinde Saket	Soft copy in different format	Data of 2017 received and about to receive more
		Plant & Animals Welfare Society (PAWS - Mumbai)	Sunish Subramania Kunj	Data not received	Data not received
2	Thane	Trust for Rescue Afforestation Conservation and Knowledge (TRACK)	Nitesh Pancholi Chandrakant	Data in hard and soft copy	Data not received
		Wildlife Welfare Association (WWA)	Aditya Patil Rohit Mohite	Soft copy in different format	Data of 2017 received
3	Kalyan	Kalyan Saarp Seva	Suhas Pawar Datta Bombe	Hard copy of release data of Kalyan, Dombivali, Titwala, Shahpur, and Ambernath areas	Obtained data through orientation/personal interview
		Plant & Animals Welfare Society (PAWS – Dombivli)	Nilesh	Soft copy in different format	Data not received

(*Continued*)

Table 13.2 (Continued)

Sr. No.	Area	Name of organisation	Contact person	Data format	Response
4	Raigad	Owls	Kunal Salunkhe	No data available	Obtained data through orientation/personal interview
		Friends of Nature, Uran	Anuj Patil Jaywant Thakur	No data available	Obtained data through orientation/personal interview
		Snake and Reptile Protection (SARP), Raigad	Pradeep Kulkarni Aniruddha Joshi	Data available in hard copy	Data not received
		Sarp Vishva, Karjat	Nandkumar Tandel	Data available in hard copy	Data of 2017 received and about to receive more
		Care of Nature Samjik Satha Veshvi, Uran	Raju Mumbaikar	Data available in hard copy	Obtained data through orientation/personal interview
		Punarvasu	Shashank Padale	Active since last eight years but has registered as an NGO and started maintaining data from last two months	Obtained data through orientation/personal interview
Individual rescuers					
1	Raigad	Kapil Dev and Family (individual) Pen	Kapil Dev	Hardcopies of last 1–1.5 yr available	Data received on 16 October 2018
		Mandar Gadkari (Individual) Alibaug	Mandar Gadkari	No data available	Obtained data through orientation/personal interview
		Rohan Nimbalkar (Individual. Turbe)	Rohan	Data available in hard copy	Obtained data through orientation/personal interview
2	Vasai	Suraj Pandy (Individual)	Suraj Pandy	Data of last three months available. Active since last two year	2017 data received
		Ashutosh Randive (Individual)	Ashutosh	No data available	Data not received

Table 13.3 Data obtained from forest range offices.

Sr. No.	Name of organisation	Contact person	Data format	Response
1	Office of Deputy Conservator Forests, Thane Forest Division, Lal Bahadur Shashtri Mark, Marathor Circle, Naupada Thane – 400062	Dr. Jitendar Ramgaokar (Deputy Conservator of Forests, Thane)	Hardcopy in files	Data from 2016 to 2017 received
2	Forest Department Office, Mahatma Phule Chowk, Murbad Road, Kalyan, Thane, Mumbai 421301	Mr. Jadhav (Forester) Shri. Waghire (Range Forest officer, Kalyan)	Hardcopy in files	Data from 2017 to 2018 received
3	Office of the Deputy Conservator of Forests (Territorial) Alibag, Near Collector Office, At. Po. Tal. Alibag Dist. Raigad Pin 402 201	Maneesh Kumar (Deputy Conservator of Forests)	Hardcopy in files (gave references of local rescuers from their office)	Partial data from 2017 received (Only of 1926 Helpline data, no species name or any other data available.)
4	Forest Colony, Takka Village, Panvel, Navi Mumbai, Maharashtra 410206	Shri. Kupte Nandkumar Nanasaheb (Assistant Conservator of Forests Territorial and Campa)	—	Data not received
5	Township Hall, JNPT township, Uran 400707	Shri. Shashank Kadam (Range Forest officer, Uran)	—	Data not received
6	Near Pen, Tahsildar office	Shri. Gaikwad (Range Forest officer, Pen)	—	Data not received
7	Badlapur	Shri. Chandrakant Shelke (Range Forest officer, Badlapur)	—	Data not received

Also, short focus group discussion (FGD) sessions were conducted with such volunteers to obtain the trends in their rescue activities with respect to number, frequency, seasonality, geography, and species identity. The qualitative data so received were used to buttress the quantitative data obtained.

13.2.2.3 Data Entry and Analysis

Most of the data obtained were in the form of field diaries or forms filled up by the individual rescuers in the NGOs'/Forest Department's pre-fixed format. Hence, the authors created the digital copy of these data by entering them in MS Excel 2010. Some of the data received in soft copy but in MS Word format were converted to MS Excel 2010 files.

Further, data analysis was also done using Microsoft Excel 2010. The workflow has been provided in Figure 13.2 below:

13.3 Results and Discussion

13.3.1 Overview of Findings

The details of formally recorded data, which was collated in the present study, have been provided in Table 13.4.

A general observation was that more data were obtained from highly urbanised areas of the MMR, viz., Mumbai and Thane – RAWW and WWA are Mumbai- and Thane-based NGOs, respectively. This leads to the possible conclusion that the higher the degree of urbanisation,

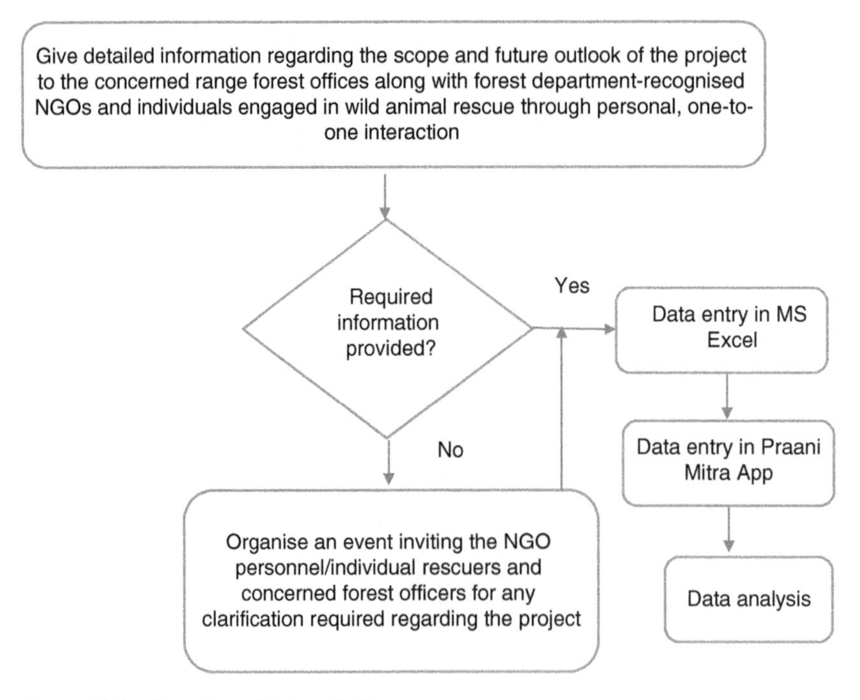

Figure 13.2 Overview of data collation process.

Table 13.4 Rescue data obtained from various NGOs and forest department.

Year	RAWW	SARRP	WWA	Sarp Vishwa	Save wildlife organisation	Individual rescuers	Forest department	
2014	1429	—	—	—	—	—	—	1429
2015	1377	—	—	—	—	—	—	1377
2016	1029	76				—	105	1210
2017	1425	13	1097		2	139	125	2801
2018	388			215	99		414	1116
Total								7933

—, data not obtained.

the higher is the intensity and volume of human–animal conflict, which is supported by several studies. Delon (2021) pointed out that urbanisation curtails animals' freedom. Murray et al. (2019) concluded that there was a 'small but significant negative relationship between urbanisation and wildlife health'. Urbanisation causing altered prey and predator behaviour has been reported by Ellington and Gehrt (2019) and Gallo et al. (2019).

Also, this finding indicates the higher level of awareness about wildlife conservation and illegality of killing or injuring wild animals on the part of the individual calling for support and regarding record-keeping among the NGO/individual in the urbanised areas. This highlights the necessity of more awareness-raising among the MMR smaller municipalities and the requirement of training for the rescue volunteers regarding formal record-keeping and maintenance in these areas.

In Table 13.3, the source-wise and year-wise volume of data has been provided.

13.3.2 Taxon-wise Analysis of Rescue Operations

As can be observed from Figure 13.3, Class Reptilia is seemingly the most threatened, with more than 70% of the rescue operations involving reptiles. Also, more than 95% of all reptiles rescued were various snake species. This finding is in concurrence with other studies

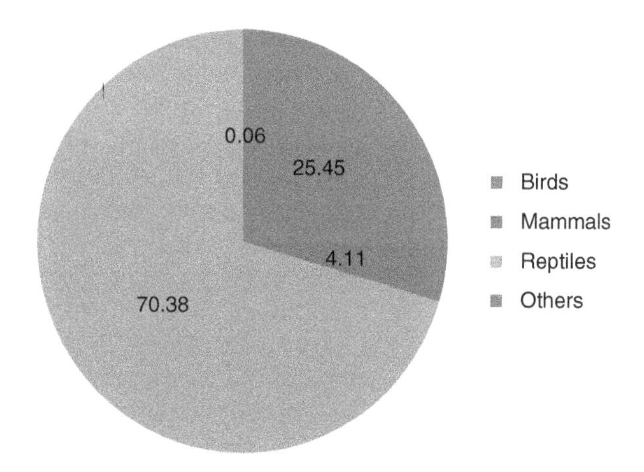

Figure 13.3 Various taxons rescued in the Mumbai Metropolitan Region during 2014–2018.

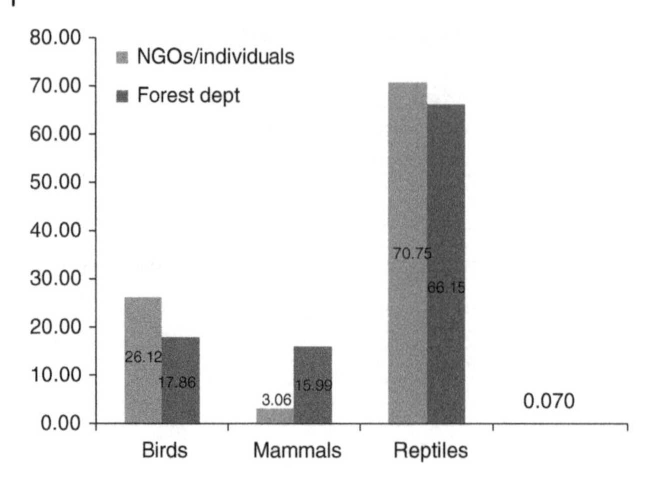

Figure 13.4 Percentage distribution of taxon-based rescue operations – a comparison between NGO/ individuals and forest department in the Mumbai Metropolitan Region during 2014–2018.

conducted by Roshnath and Jayaprasad (2017) and Shine and Koenig (2001). Working with a decade-old series of urban wild animal rescue operations in New South Wales, Australia, Shine and Koenig (2001) observed that more reptiles were being rescued despite not being naturally abundant. They attributed this to the residents' discomfort with snakes and large lizards, wanting these species away from their homes. Roshnath and Jayaprasad (2017) also reported similar findings and attributed a similar reason to the same.

Another observation was the higher proportion of mammals rescued directly by the Forest Department as compared to the NGOs and individuals (Figure 13.4).

Thirty-four percentage of the snakes rescued were Indian Rat Snakes (*Ptyas mucosa*), commonly called *dhamin*. These are very commonly found non-venomous species. However, a substantial number of venomous species were rescued each year, including Spectacled Cobra (*Naja naja*), Russell's Viper (*Daboia russelii*), Bamboo Pit Viper (*Trimeresurus gramineus*), Saw-scaled Viper (*Echis carinatus*), and Common Krait (*Bungarus caeruleus*). More than one-third (31.1%) of the rescued snakes were venomous. This finding raises serious issues pertaining to the safety concerns of local citizens and of the rescue volunteers.

Among other commonly rescued snake species were Indian Rock Python (*Python molurus*), Chequered Keelback (*Fowlea piscator*), and Indian Wolf Snake (*Lycodon aulicus*). Throughout 2014–2018, 28 species of snakes were rescued (Figure 13.5).

Other snake species reported were Striped Keelback, Brahminy Worm Snake, Banded Racer, Earth Boa, Green Keelback, Variegated Kukri, Dumeril's Black-Headed Snake, Gunther's Racer, Barred Wolf Snake, Painted Bronzeback Tree Snake, and Ball Python. Among the unusual reptile species rescued were – two-headed Russell's Viper snake, and Indian Marsh Crocodile.

Among birds, the Rock Dove (*Columba livia*) was the commonest rescue (45.9%), followed by Black Kite (*Milvus migrans*) (14.6%), and House Crow (*Corvus splendens*) (9.2%). Among the rare species rescued were Brown Booby, Indian Scops Owl, Mottled Wood Owl, Pied Cuckoo, Pallas's Gull, Bonelli's Eagle, and Peregrine Falcon. Among birds, too, abandoned pets such as African Love Birds and Cockatiels had to be rescued on a few occasions (Figure 13.6).

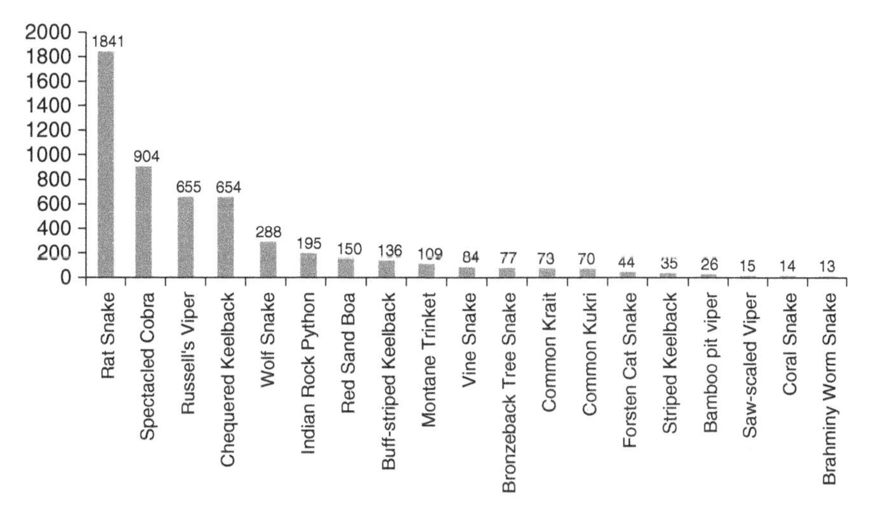

Figure 13.5 Commonly rescued snake species in the Mumbai Metropolitan Region during 2014–2018.

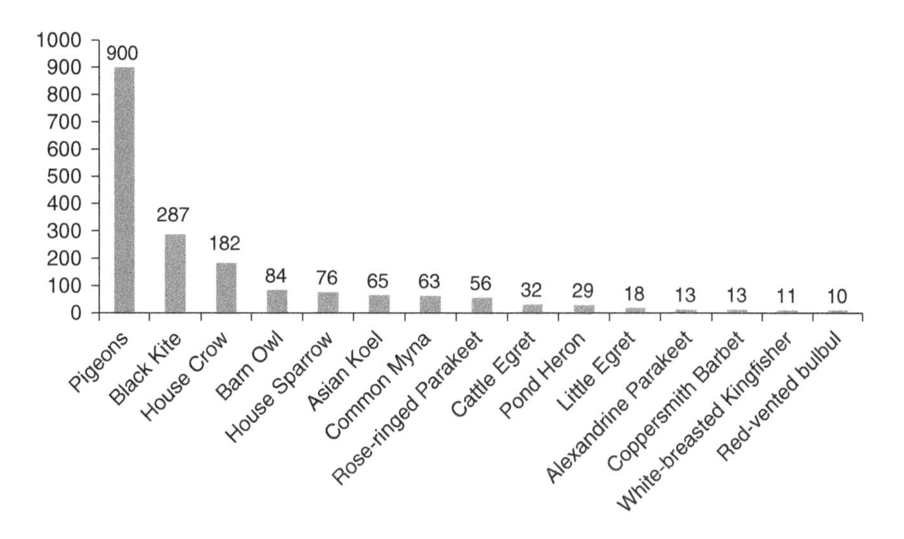

Figure 13.6 Commonly rescued bird species in the Mumbai Metropolitan Region during 2014–2018.

Bonnet Macaque (*Macaca radiata*) and Three-striped Palm Squirrel (*Funambulus palmarum*) were the most common mammal species rescued, apart from two Hanuman Langur (*Semnopithecus entellus*) and four Golden Jackals (*Canis aureus*), one Jungle cat (*Felis chaus*) and a Civet Cat species. Two Golden Jackals were rescued by RAWW volunteers from the Vikhroli (E) mangroves on 24 May 2017 and 14 August 2017. The threat to golden jackals from stray dogs was highlighted. That domestic dogs are drivers of conflict against close wild relatives such as coyotes has been reported by Schell et al. (2021).

RAWW volunteers rescued the civet cat from the residential areas of Bhandup, which is close to the mangroves along Thane creek. On the other hand, the Forest Department staff rescued the jungle cat from the densely populated Sion location of Mumbai.

The only amphibian species to be rescued was one Bombay Caecilian (*Ichthyophis bombayensis*) individual from the IIT-Bombay campus, which has wide green spaces and several old trees, on 10 August 2017 by a volunteer of the Thane-based NGO WWA.

Several rescue operations involved abandoned pets. Fauna such as 51 Indian Star Tortoises (*Geochelone elegans*), 8 Red-eared Slider Turtles (*Trachemys scripta elegans*), 2 individuals of Indian Violet Tarantula (*Chilobrachys fimbriatus*), and 1 Iguana species lizard was rescued.

13.3.3 Impact of Seasons on Rescue Operations

The seasons were differentiated as follows: winter (December, January, and February), summer (March, April, and May), monsoon (rainy) season (June–September), and a post-monsoon period (October–November) (Figure 13.7).

As per Shine and Koenig (2001), capture rates for all reptile species were highest in warmer months, and especially on days with dry, warm weather. Our findings also indicated strong seasonality, but for the months of monsoon (Jun–Sept). The previous findings may perhaps explain this: most of the rescue operations involved reptiles – heavy rains tend to flood reptile habitats, forcing them into conflict situations with the urban life in the surroundings.

13.3.4 Geographical Analysis of Rescue Operations

Based on the data obtained from these various sources, the following map outputs were obtained (Figure 13.8a–g).

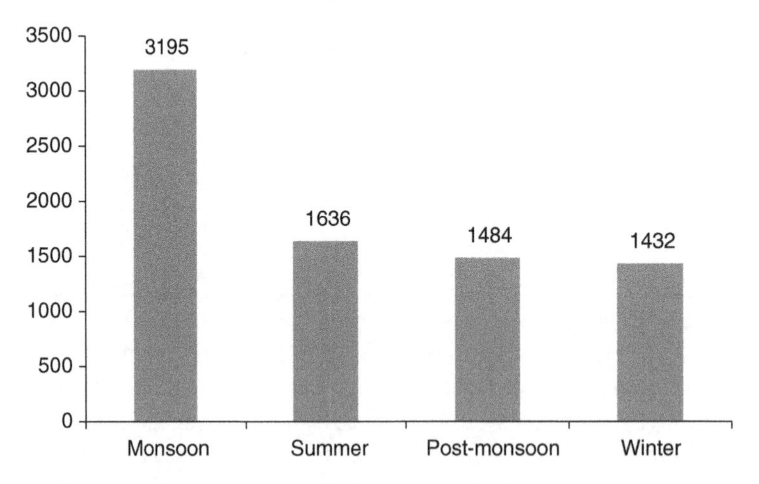

Figure 13.7 Season-wise number of rescue operations in the Mumbai Metropolitan Region during 2014–2018.

(a)

Wildlife rescued from Mumbai–Mumbai Suburban–Thane Tehsils by RAWW NGO (2014).

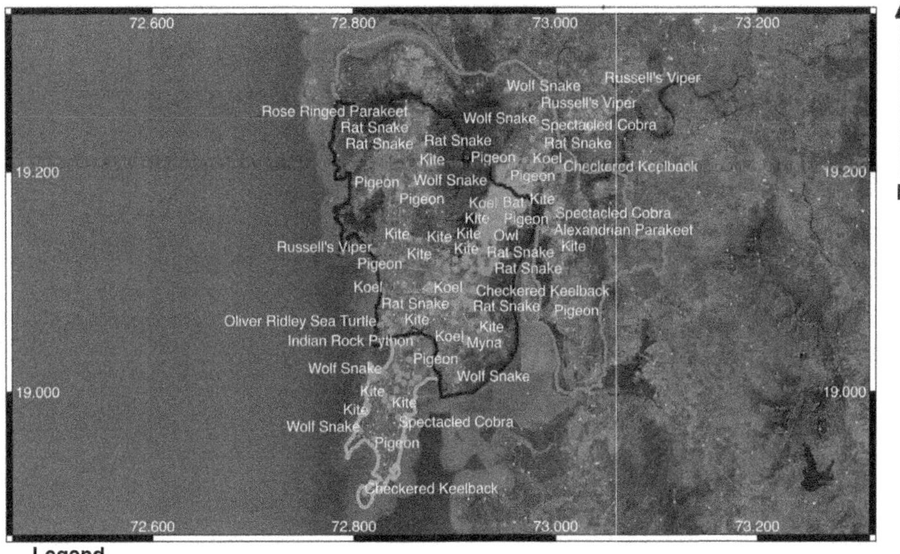

Legend

Mumbai suburban tehsil boundary.
Thane tehsil boundary.
Wildlife rescue locations. google satellite hybrid imagery.
Mumbai tehsil boundary.

(b)

Wildlife rescued from Mumbai–Mumbai Suburban–Thane–Panvel Tehsils by RAWW NGO (2015).

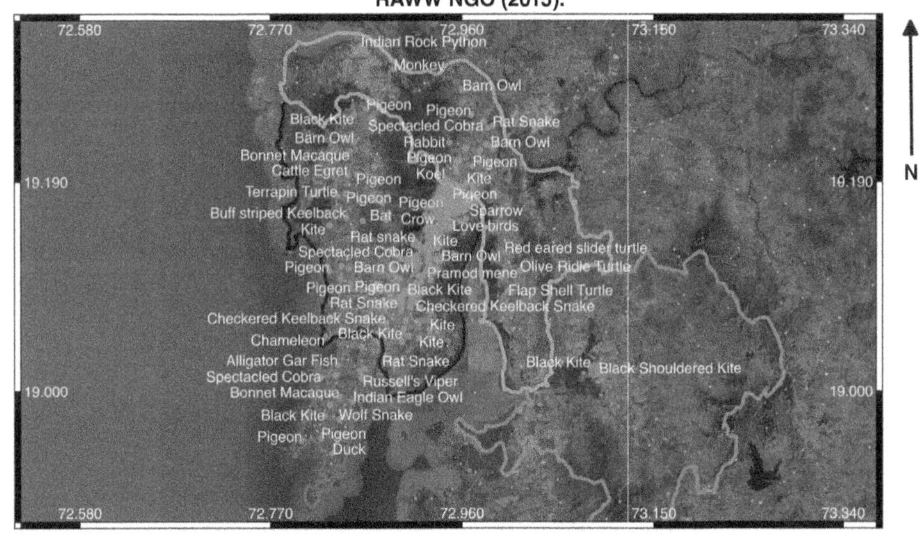

Legend

Thane tehsil boundary.
Mumbai suburban tehsil.
Mumbai tehsil bounsary.
2015 Release data
Panvel tehsil bounsary.
Google satellite hybrid imagery.

Figures 13.8 (a–g) Wildlife rescue locations in the Mumbai Metropolitan Region.

(c)

Wildlife rescued from Mumbai–Mumbai Suburban–Thane–Bhiwandi Tehsils by RAWW NGO (2016).

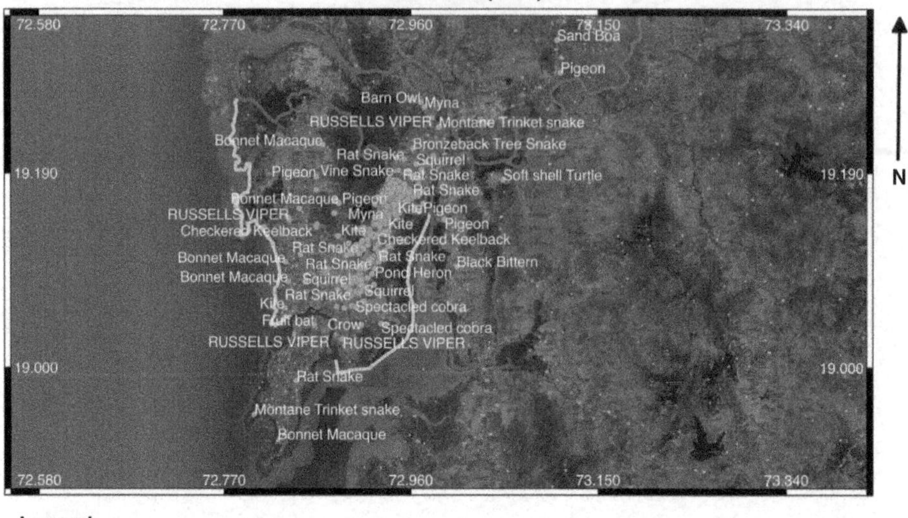

Legend

Thane tehsil boundary. Mumbai tehsil boundary. Mumbai suburban tehsil boundary. Bhiwandi tehsil boundary. Wildlife rescue locations. google satellite hybrid imagery.

(d)

Wildlife rescued from Mumbai Suburban Tehsils by SARRP NGO (2016).

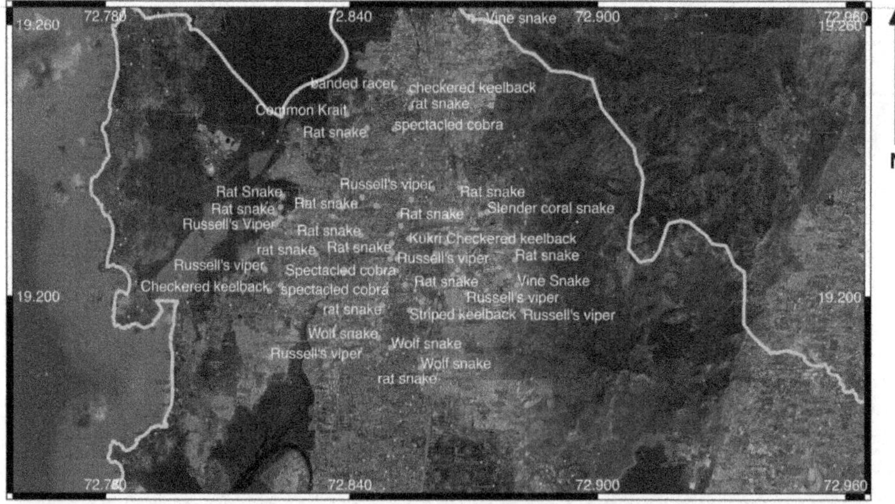

Legend

Mumbai suburban tehsil. Google satellite hybrid imagery. Wildlife rescue locations.

Figures 13.8 (Continued)

(e)

Wildlife rescued from Mumbai–Mumbai Suburban–Thane–Kalyan Tehsils, Mumbai Range Forest office Data (2016–2017).

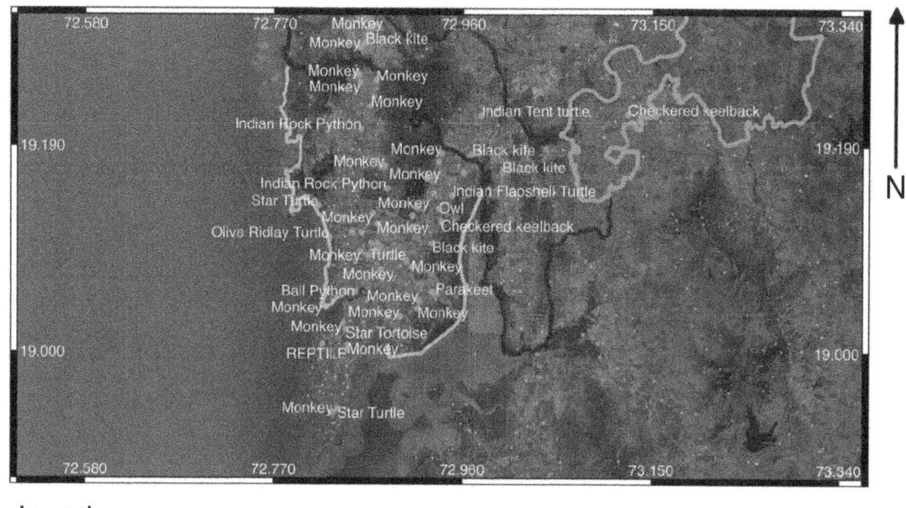

Legend

Kalyan tehsil boundary. Thane tehsil boundary. Mumbai tehsil boundary. Mumbai suburban tehsil boundary. Wildlife rescue locations. google satellite hybrid imagery.

(f)

Wildlife rescued from Kalyan–Bhiwandi–Uihasnagar Tehsils by Datta Bombe.

Legend

Bhiwandi tehsil boundary. Kalyan tehsil boundary. Ulhasnagar tehsil boundary. Google Satellite Hybrid Imagery. Wildlife rescue locations.

Figures 13.8 (Continued)

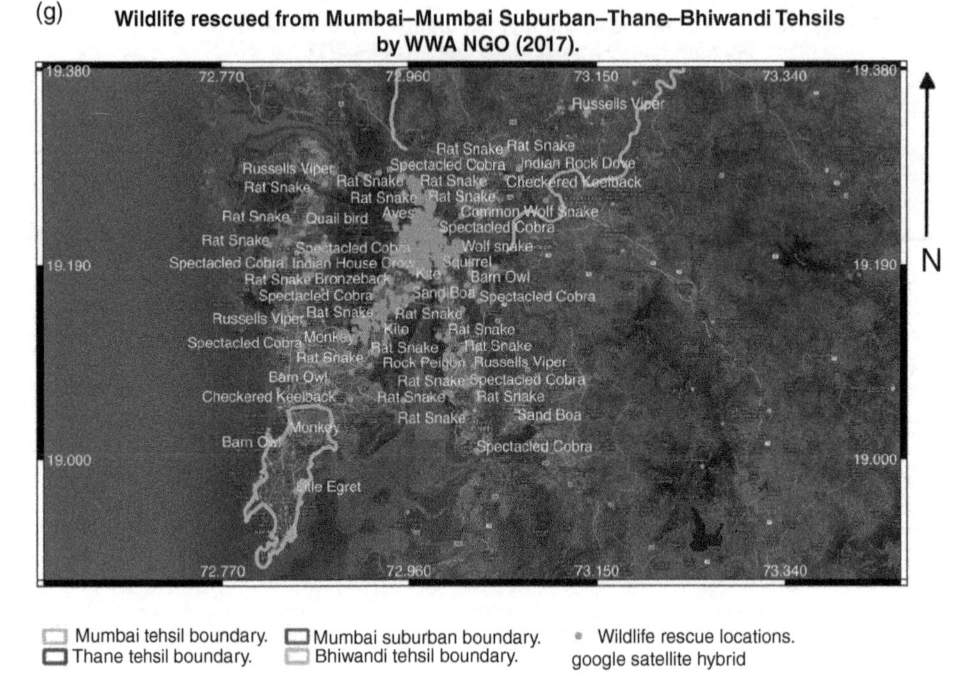

Figures 13.8 (Continued)

With more than 7900 records of human–animal conflicts within four years, from which the fauna species had to be rescued, this study's results highlight how significant the concept of human–animal conflict can be in a highly urbanised city like Mumbai, its suburbs and exurbs.

13.3.5 Limitations of the Study

It was found that several of the volunteers had failed to maintain formal records of their wild animal rescue operations. In the absence of such records, we had to rely upon the data shared orally or on electronic mail by these volunteers, which may not be entirely accurate. To address this situation, a FGD was conducted with the individuals involved in rescue activities to understand as many as possible details about the work done, which yielded qualitative information to support the quantitative data. The FGD also led to a similar conclusion that most individuals were called out to rescue reptiles, mostly snakes. In addition, on several occasions, animals rescued were injured or unhealthy – it is critical to include a more detailed account of the health condition of the rescued species, something which was missing in this study.

Also, since this study did not involve local citizens as a whole, it is possible that we missed out on human–wildlife conflict cases where volunteers or the Forest Department were not involved and local citizens ended up managing the challenge by themselves.

13.3.6 Implications for Researchers, Citizens, and Policymakers

13.3.6.1 For Researchers

This study throws up interesting findings for researchers, providing local faunal biodiversity abundance and distribution trends. For future research, the present study lays a strong foundation by providing the list of species getting trapped in specific locations. Also, this study highlights the lack of scientific data regarding wildlife losses along the Mumbai local train tracks and opens up new avenues for researchers.

One positive step in the direction of addressing human–wildlife conflict is to provide green corridors, overpasses and underpasses to connect fragmented habitats. Wildlife corridor can be defined as a link of wildlife habitat, with native vegetation, which joins two or larger areas of similar wildlife habitat. Corridors are critical for the maintenance of ecological processes. It can comprise of level grounds dedicated to wildlife movement, with overpasses and underpasses to go across manmade barriers like roads and railways or natural ones like rivers or valleys. Wildlife corridors are significant from the point of view of reduced loss of animal life due to road and rail accidents, prevention of gene pool isolation and providing habitat to fauna – underpasses and overpasses are natural habitats by themselves. While larger mammals tend to use it transiently, smaller creatures can complete their life-cycle therein.

However, the concept of ecological corridors in cities and towns and for small mammals or reptiles has been largely understudied in the Indian context. Indeed, ecological corridors have been thought of only in the context of tigers and elephants here and urban biodiversity restricted to avifauna. The importance of ecological corridors for reptiles, small mammals, and birds in the urban context is immense regarding addressing the negative effects of habitat fragmentation and gene pool isolation and reducing road kills. However, town planners have grossly ignored this aspect in the up and coming cities of India (Figures 13.9 and 13.10).

The design and construction material of wildlife corridors depend on several aspects – including the type of species that is likely to use it and the local geographical conditions – which has been discussed in the present study.

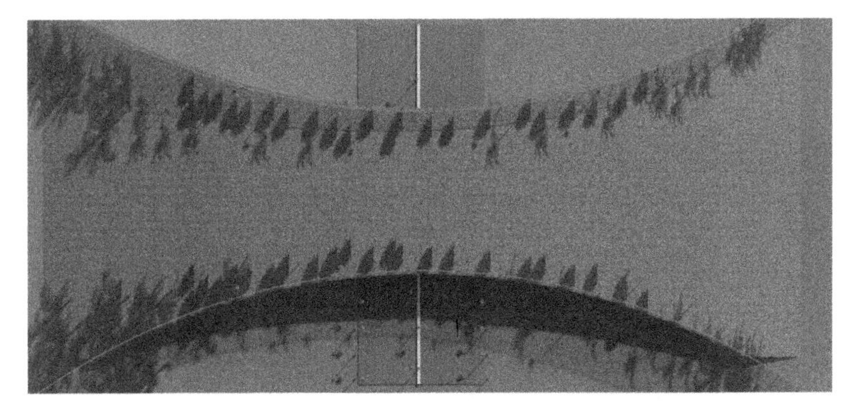

Figure 13.9 Pictorial/diagrammatic representation of overpasses.

Figure 13.10 Pictorial/diagrammatic representation of underpasses and overpasses.

Such information needs to be taken forward by a team of urban planners and engineers who can design suitable underpasses and overpasses at the points where they are most required and most suitable. Planet Earth's future lies in the peaceful coexistence of all species, and the design and construction of such ecological infrastructure will go a long way towards realising species conservation goals.

13.3.6.2 For Citizens

The substantial number of venomous snakes rescued highlights the safety issues of local citizens and necessitates and justifies the expenditure of substantial funds towards creating a win–win situation for human beings and wildlife alike.

It also brings to fore the importance of collating more and more volunteers from among local citizens who are willing to be trained as wildlife rescuers. However, given the significant number of venomous snakes rescued, this study also rings out a warning bell about the degree of training, practice, and precautions that wildlife rescue volunteers need to take. Even among those citizens who are not active or aspiring volunteers, this study helps raise awareness about how to avoid human–animal conflict and whom to approach when caught in such situations.

13.3.6.3 For Policymakers

For policymakers, this study underscores the importance of training provided to volunteers about maintaining formal records of rescue operations and working completely in cohesion with the forest department. For instance, the Thane Division of Maharashtra had come up with a mobile app called 'Praani Mitra' for individual rescuers and NGO volunteers to enter data easily.

This study also highlights the importance of establishing wildlife rescue and rehabilitation centres with dedicated infrastructure, wildlife-specific ambulances, and medical staff capable of handling wild animals.

13.3.7 Recommendations

a) Awareness-raising, through actual and virtual training sessions and with the help of adequate IEC (information, education, communication) material, about how to avoid wildlife–human conflicts, with special emphasis on:
   - Understanding what creates the human–wildlife conflict situation, thereby becoming better equipped at avoiding it
   - Identifying common venomous snakes, so that local citizens would know when to panic and when not to
   - Understanding basic animal behaviour so as to become better able to avoid antagonising the creatures

b) Veterinary doctors experienced in handling wild fauna are fewer, and their number needs to be augmented

c) Similarly, facilities to house injured animals, basic medical equipment to diagnose their ill-health conditions (such as portable X-ray machines, portable ultrasound machines) and medications necessary for their well-being need to be enhanced significantly

d) Data collation and record-keeping training for volunteers must be provided.

13.4 Conclusion

With the aim of collating and analysing wild animal rescue operation-related data for the ultimate goal of avoiding human–animal conflict and ensuring species conservation, 7 forest offices, 13 NGOs, and 5 individuals were contacted. These were active in the MMR for wild animal rescue to collate data from 2014 to 2018. Records of approximately 7933 odd rescue events were obtained. On the basis of this study, reptiles, especially snakes, were found to be the most vulnerable to human–wildlife conflict. The Mulund and Bhandup localities in the MMR emerged as the region reporting the highest incidents of conflict. Hence, awareness-raising among local citizens, training and capacity-building of volunteers, and more funds for trained staff and facilities to treat and rehabilitate the rescued wild animals, apart from more investment in creating overpasses and underpasses for safe transit of wild animals within cities are of high importance.

Acknowledgements

We acknowledge the funding given by the Mumbai Metropolitan Region Environment Improvement Society (MMREIS) to support this study. The guidance provided by Dr. J. Ramgaonkar, DFO (Thane circle), and Mr. N. Vasudevan, APCCF (Mangrove cell) is gratefully acknowledged. We are thankful to Mr Pawan Sharma of RAWW for his immense help in data collation. Adwait Jadhav, Chinmay Joshi, Kalyani Gaikwad, and Priyanka Dharne from TerraNero are duly acknowledged for their valuable role.

References

Bajaru, S., Pal, S., Prabhu, M. et al. (2020). A multi-species occupancy modeling approach to access the impacts of land use and land cover on terrestrial vertebrates in the Mumbai Metropolitan Region (MMR), Western Ghats, India. *PLoS One* 15 (10): e0240989.

Beisner, B.A., Heagerty, A., Seil, S.K. et al. (2015). Human–wildlife conflict: proximate predictors of aggression between humans and rhesus macaques in India. *American Journal of Physical Anthropology* 156 (2): 286–294.

Bhatia, S., Athreya, V., Grenyer, R., and Macdonald, D.W. (2013). Understanding the role of representations of human–leopard conflict in Mumbai through media-content analysis. *Conservation Biology* 27 (3): 588–594.

Chace, J.F. and Walsh, J.J. (2006). Urban effects on native avifauna: a review. *Landscape and Urban Planning* 74 (1): 46–69.

Choudhury, A. (2004). Human–elephant conflicts in Northeast India. *Human Dimensions of Wildlife* 9 (4): 261–270.

Clevenger, A.P., Wierzchowski, J., Chruszcz, B., and Gunson, K. (2002). GIS-generated, expert-based models for identifying wildlife habitat linkages and planning mitigation passages. *Conservation Biology* 16 (2): 503–514.

Cox, D.T. and Gaston, K.J. (2018). Human–nature interactions and the consequences and drivers of provisioning wildlife. *Philosophical Transactions of the Royal Society, B: Biological Sciences* 373 (1745): 20170092.

Das, J.P., Lahkar, B.P., and Talukdar, B.K. (2012). Increasing trend of human-elephant conflict in Golaghat District Assam, India: issues and concerns. *Gajah* 37: 34–37.

Delon, N. (2021). Animal capabilities and freedom in the city. *Journal of Human Development and Capabilities* 22 (1): 131–153.

Donihue, C.M. and Lambert, M.R. (2014). Adaptive evolution in urban ecosystems. *Ambio* 44 (3): 194–203.

Ellington, E.H. and Gehrt, S.D. (2019). Behavioral responses by an apex predator to urbanisation. *Behavioral Ecology* 30 (3): 821–829.

Freitas, S.R., Sousa, C.O., and Bueno, C. (2013). Effects of landscape characteristics on roadkill of mammals, birds and reptiles in a highway crossing the Atlantic Forest in southeastern Brazil. *International Conference on Ecology and Transportation (ICOET 2013)*, Arizona.

Gallo, T., Fidino, M., Lehrer, E.W., and Magle, S. (2019). Urbanisation alters predator-avoidance behaviours. *Journal of Animal Ecology* 88 (5): 793–803.

Hatkar, P., Vinhenkar, D., and Kansara, D. (2019). Rescue and rehabilitation of loggerhead sea turtles Caretta caretta from Dahanu Coast, Maharashtra, India. *Marine Turtle Newsletter* 156: 26–29.

Jadesh, M., Kamble, P., Manjunath, K. et al. (2014). A preliminary survey of amphibians and reptiles in around Gulbarga University Campus, Karnataka, India. *International Letters of Natural Sciences* 27: 67–71.

Johnson, M.T.J. and Munshi-South, J. (2017). Evolution of life in urban environments. *Science* 358 (6363).

Hathaway, R.S., Bryant, A.E.M., Draheim, M.M. et al. (2017). From fear to understanding: changes in media representations of leopard incidences after media awareness workshops in Mumbai, India. *Journal of Urban Ecology* 3 (1): jux009.

Li, F., Wang, R., Paulussen, J., and Liu, X. (2005). Comprehensive concept planning of urban greening based on ecological principles: a case study in Beijing, China. *Landscape and Urban Planning* 72 (4): 325–336.

Longkumer, T., Armstrong, L.J., Santra, V., and Finny, P. (2016). Human, snake, and environmental factors in human-snake conflict in North Bihar-a one-year descriptive study. *Christian Journal for Global Health* 3 (1): 36–45.

Manral, U., Sengupta, S., Hussain, S.A. et al. (2016). Human wildlife conflict in India: a review of economic implication of loss and preventive measures. *Indian Forester* 142 (10): 928–940.

McKinney, M.L. (2002). Urbanisation, biodiversity, and conservation the impacts of urbanisation on native species are poorly studied, but educating a highly urbanised human population about these impacts can greatly improve species conservation in all ecosystems. *Bioscience* 52 (10): 883–890.

Murray, M.H., Sánchez, C.A., Becker, D.J. et al. (2019). City sicker? A meta-analysis of wildlife health and urbanisation. *Frontiers in Ecology and the Environment* 17 (10): 575–583.

Nielsen, C.K., Anderson, R.G., and Grund, M.D. (2003). Landscape influences on deer-vehicle accident areas in an urban environment. *The Journal of Wildlife Management* 67: 46–51.

Nulkar, G. (2016). Silent conflicts–human-wildlife interactions in urban spaces. *Journal of Ecological Society* 29: 34–43.

Odden, M., Athreya, V., Rattan, S., and Linnell, J.D. (2014). Adaptable neighbours: movement patterns of GPS-collared leopards in human dominated landscapes in India. *PLoS One* 9 (11): e112044.

Penteriani, V., del Mar Delgado, M., Pinchera, F. et al. (2016). Human behaviour can trigger large carnivore attacks in developed countries. *Scientific Reports* 6 (1): 1–8.

Priston, N.E. and McLennan, M.R. (2013). Managing humans, managing macaques: human–macaque conflict in Asia and Africa. In: *The Macaque Connection* (eds. S. Radhakrishna, M.A. Huffman and A. Sinha), 225–250. New York, NY: Springer.

Ranade, P., Londhe, S., Mishra, A., and Bhatnagar, P. (2015). Geo-informatics approach to human-leopard conflict in urban forest areas-a case review of Sanjay Gandhi National Park (SGNP), Borivali, Mumbai. *International Journal of IT, Engineering and Applied Sciences Research* 4 (3): 12–15.

Ravenelle, J. and Nyhus, P.J. (2017). Global patterns and trends in human–wildlife conflict compensation. *Conservation Biology* 31 (6): 1247–1256.

Richardson, S., Mill, A.C., Davis, D. et al. (2020). A systematic review of adaptive wildlife management for the control of invasive, non-native mammals, and other human–wildlife conflicts. *Mammal Review* 50 (2): 147–156.

Roshnath, R. and Jayaprasad, D. (2017). A review on wildlife rescue activities in North Kerala, India. *Indian Forester* 143 (10): 1004–1010.

Rudd, H., Vala, J., and Schaefer, V. (2002). Importance of backyard habitat in a comprehensive biodiversity conservation strategy: a connectivity analysis of urban green spaces. *Restoration Ecology* 10 (2): 368–375.

Russo, D. and Ancillotto, L. (2015). Sensitivity of bats to urbanization: a review. *Mammalian Biology-Zeitschrift für Säugetierkunde* 80 (3): 205–212.

Santos, S.M., Carvalho, F., and Mira, A. (2017). Current knowledge on wildlife mortality in railways. In: *Railway Ecology* (eds. L. Borda-de-Água, R. Barrientos, P. Beja and H.M. Pereira), 11–22. Cham: Springer.

Schell, C.J., Stanton, L.A., Young, J.K. et al. (2021). The evolutionary consequences of human–wildlife conflict in cities. *Evolutionary Applications* 14 (1): 178–197.

Seiler, A. and Olsson, M. (2017). Wildlife deterrent methods for railways—an experimental study. In: *Railway Ecology* (eds. L. Borda-de-Água, R. Barrientos, P. Beja and H.M. Pereira), 277–291. Cham: Springer.

Sen, A. and Pattanaik, S. (2015). Alienation, conflict, and conservation in the protected areas of urban metropolis: a case study of Sanjay Gandhi National Park, Mumbai. *Sociological Bulletin* 64 (3): 375–395.

Sen, A. and Nagendra, H. (2019). Mumbai's blinkered vision of development: sacrificing ecology for infrastructure. *Economic and Political Weekly* 54 (9): 20–23.

Sharma, G., Ram, C., and Rajpurohit, L.S. (2011). Study of man-monkey conflict and its management in Jodhpur, Rajasthan (India). *Journal of Evolutionary Biology Research* 3 (1): 1–3.

Shine, R. and Koenig, J. (2001). Snakes in the garden: an analysis of reptiles "rescued" by community-based wildlife carers. *Biological Conservation* 102 (3): 271–283.

Sidhu, S., Raghunathan, G., Mudappa, D., and Raman, T.S. (2017). Conflict to coexistence: human–leopard interactions in a plantation landscape in Anamalai Hills, India. *Conservation and Society* 15 (4): 474.

Trombulak, S.C. and Frissell, C.A. (2000). Review of ecological effects of roads on terrestrial and aquatic communities. *Conservation Biology* 14 (1): 18–30.

Underhill, J.E. and Angold, P.G. (1999). Effects of roads on wildlife in an intensively modified landscape. *Environmental Reviews* 8 (1): 21–39.

Vyas, R. (2013). Snake diversity and voluntary rescue practice in the cities of Gujarat State, India: an evaluation. *Reptile Rap* 15: 27–39.

Wu, J. and Hobbs, R. (2002). Key issues and research priorities in landscape ecology: an idiosyncratic synthesis. *Landscape Ecology* 17 (4): 355–365.

Section 4

Urbanisation, Sustainable Development Goals (SDGs), and Climate Change

14

Building Knowledge on Urban Sustainability in the Czech Republic: A Self-assessment Approach

Svatava Janoušková[1,2] *and Tomáš Hák*[1,3]

[1] *Environment Centre, Charles University, Prague, Czech Republic*
[2] *Faculty of Science, Charles University, Prague, Czech Republic*
[3] *Faculty of Humanities, Charles University, Prague, Czech Republic*

14.1 Introduction

The world has experienced unprecedented urban growth in recent decades. More than half of its population – 3.9 billion people – live in cities and this number is expected to increase to 5 billion by 2030. Rapid urbanisation results in a growing number of environmental, social, and economic impacts, and therefore, there is an urgent call for making cities more sustainable, i.e. resilient, safe, inclusive, green, compact, etc. (UN 2015). However, meeting these challenges requires understanding the urban concept and definition of sustainability (which environmental, social, and economic phenomena we need to consider) and monitoring progress towards predefined sustainability targets (finding relevant indicators to the monitored phenomena) (Hák et al. 2016).

A wide array of sustainable city definitions and countless indicators (assessment tools) have been developed and implemented by various providers (Cohen 2017; EC 2015). Some have intended to assess cities independently based on publicly available data with the general aim to make 'league tables' ranking cities (Phillis et al. 2017). Other providers offer systems to cities for self-assessment based on ready-to-use assessment tools (Pires et al. 2014). The role of municipal governments is relatively easy – they either get an assessment without the need to provide any input or they just choose the most appropriate assessment concept and indicators best capturing the municipal's needs.

However, existing concepts, frameworks, and indicators cannot fit all national, and most of all, local specifics nor all urban types and structures. Therefore, city officials and officers, experts, and/or the public may not always fully accept them (Ameen et al. 2015). We agree that an efficient approach towards the conceptualisation of urban sustainability lies in the definition of relevant themes (governance, environment, health, education, etc.) by a municipality and external experts, and the subsequent development of thematic indicators

Urban Ecology and Global Climate Change, First Edition. Edited by Rahul Bhadouria, Shweta Upadhyay, Sachchidanand Tripathi, and Pardeep Singh.

which the municipality can analyse, i.e. collecting data, calculating indicators, interpreting results, etc. (Valentin and Spangenberg 2000). Well-balanced cooperation between the municipality and external experts enables the development of a theme- and indicator-relevant assessment tool tailored to a city's needs that makes it acceptable and beneficial for users. Such cooperation brings together diverse sources of knowledge, multiple perspectives, quantitative and qualitative measures (indicators), and secures key characteristics of modern policy frameworks (Elgert 2016).

We argue that the self-assessment approach using an appropriate assessment tool(s) enables municipal representatives to understand the strengths and weaknesses of their path towards sustainability and allows them to take effective measures to eliminate weaknesses. In this study, we demonstrate this in the self-assessment method entitled 'Urban Sustainability Audit' applied by Czech municipalities under the Local Agenda 21 (LA21) programme. The study describes the original participatory methodology applied by municipalities of various types and sizes which has been in use for nearly a decade. Key findings and conclusions – i.e. benefits and challenges from different types of LA21 implementations reflected by municipal representatives and external evaluation experts (hereinafter referred to solely as experts) – may serve as inspiration for other cities and countries.

14.2 Sustainable Development at a Local Level

Sustainable development has become a fundamental policy concept in the field of environmental, economic, and social development in recent years (Purvis et al. 2019). However, the term is one of the most difficult to define in practical terms – 'sustainable development' was coined by the International Union for the Conservation of Nature (IUCN) at the 1980 World Conservation Strategy (IUCN, UNEP, and WWF 1980) where it took on the meaning of conserving the earth's natural resources. *Our Common Future* (the Brundtland Report) then gave further direction to comprehensive global solutions and defined the concept of sustainable development as a development 'which meets the needs of the present generation without compromising the ability of future generations to meet their own needs (WCED 1987). Since the early 1990s, the one-word term sustainability has been widely used, and several authors have stated that it has become one of the most overused, abused, and misused terms in society, as well as in development literature (Luke 2005; Károly 2011; James 2014). We hear about sustainable cities, sustainable economy, sustainable growth, or sustainable debt, but we rarely know what these concepts mean.

The specific term 'sustainable urban development' is a concept strongly related to sustainable development. A straightforward way to define a sustainable city follows closely the Brundtland Commission's definition: it can be analogically defined as 'urban development that meets the needs and demands of the present generation without compromising the ability of future generations to meet their needs and demands'. The main idea has been reformulated many times, e.g. Zwart et al. (2012) cite a goal-based definition – that a sustainable city improves its citizens' quality of life, including its ecological, cultural, political, institutional, social, and economic components without leaving a burden on future generations (a burden which is the result of reduced natural capital and excessive local debt). However, such general definitions would face the same critique for their vagueness, ambiguity, and lack of consensus on what to sustain and what to develop (Jabareen 2008; Pesqueux 2009). Due to the complexity of cities and the pragmatics of day-to-day politics and related local measures, the definition of the sustainable urban development concept is often formulated along more practical lines or not embracing all sustainability aspects as, e.g. 'the ability of communities to consistently thrive over time as they make decisions to improve the community today without sacrificing the future' (McGalliard 2012). Robinson (2004) critically examines the theory and practice of sustainable development and argues for an approach to sustainability that is integrative, is action-oriented, goes beyond technical fixes, incorporates recognition of the social construction of sustainable development, and engages local communities in new ways.

Many existing definitions and concepts emphasize a different aspect of urban sustainability; however, they mostly have governance themes. This results from the global inception dating back to 1992 when the United Nations Conference on Environment and Development adopted Agenda 21, a global action plan for sustainable development (UN 1992). Chapter 28 of Agenda 21, entitled 'Local Authorities' Activities in Support of Agenda 21' states that:

> Because so many of the problems and solutions being addressed by Agenda 21 have their roots in local activities, the participation and cooperation of local authorities will be a determining factor in fulfilling its objectives. Local authorities construct, operate, and maintain economic, social, and environmental infrastructure, oversee planning processes, establish local environmental policies and regulations, and assist in implementing national and sub-national environmental policies. As the level of governance closest to the people, they play a vital role in educating, mobilising, and responding to the public to promote sustainable development.

Based on Agenda 21, derived Local Agenda 21 programmes started worldwide as typical instruments for extensive governance (Joas and Grönholm 2004). Local Agenda 21 (hereafter 'LA21') is a local government-led, community-wide, and participatory effort to establish a comprehensive action strategy for environmental protection, economic prosperity, and community well-being in the local jurisdiction or area. This requires the integration of planning and action across economic, social, and environmental spheres with key elements such as community participation (direct and via stakeholder groups), target setting, assessment and monitoring, and reporting. It is primarily a voluntary policy tool with the goal of activating local government and engaging local stakeholder organisations.

14.3 Sustainability Assessment – A General View

According to Kates et al. (2001), the purpose of sustainability assessment is to provide decision-makers with an evaluation of global to local integrated nature-society systems over short- and long-term perspectives to assist them in determining which actions should or should not be taken in an attempt to make society sustainable. We can define sustainability assessment as any process that directs decision-making towards sustainability (Bond et al. 2012). This definition encompasses many potential forms of decision-making from, choices of individuals in everyday life through to projects, plans, programmes, or policies at various levels. The diversity of sustainability assessment practice is reflected in the recent explosion of published works employing the term 'sustainability assessment'. There are many frameworks, tools, and techniques supporting sustainability assessment processes such as the eco-efficiency framework (WBCSD 1999) or capital framework (Smith 2004), multicriteria analysis (Geneletti 2019), cost-benefit analysis (Barbier et al. 1990), life cycle sustainability assessment (Finkbeiner et al. 2010), sustainability indicators (Waas et al. 2014), etc. Other authors propose more inclusive, deliberative techniques (Gasparatos and Scolobig 2012).

There are many characteristics for making a typology of sustainability assessments: subjective and objective, quantitative and qualitative, bottom-up and top-down, monetary and physical, global/national and local, integral/holistic and one dimensional, expert-based and participatory, etc. From an early stage of the sustainable development concept, it was clear that information and namely quantitative indicators would play a key role. Chapter 40 of Agenda 21 called for 'indicators that show us if we are creating a more sustainable world' (UN 1992); since then, many indicators, indicator sets and dashboards, compound (composite and aggregated) indicators, and indices have been introduced (Hák et al. 2012). An indicator-based approach has underpinned all major sustainability assessments. However, despite the efforts of many national and international organisations and governments – including long-term programmes such as the European Commission's 'Beyond GDP' and the OECD's 'Measuring the Progress of Societies' – there has not been any theoretical consensus on how to measure current well-being nor sustainability (e.g. UNECE, OECD, and Eurostat 2008; Stiglitz et al. 2009). Also, despite many national and international efforts on measuring sustainability, only a few of them have an integrated approach considering environmental, economic, and social aspects. In most cases, the focus is on one of the three aspects without supplementing each other.

Another relevant feature of the sustainability assessment approaches is the role of external experts and local capacities. Traditionally, indicator-based sustainability assessment methods have been developed with experts, mostly academics or practitioners, playing a dominant role (Hák et al. 2007). Recent extensive engagement of local knowledge, expertise, and participatory methods have raised the issue of the legitimacy and subsequently credibility of the results and conclusions. Bhagavatula et al. (2013) emphasize the importance of the link between research and practice, and between science and policy. They argue that many tools and instruments with a focus on urban sustainability have been produced in recent decades but the potential of this wealth of knowledge is not being fully utilised. In reality, both research and practice communities, more often than not, operate in isolation or disconnect (Mascarenhas et al. 2010).

In general, any sustainability assessment method may be applied within an organisation (a building, a firm, a city, etc.) as well as externally through governmental, non-governmental, or private bodies. For example, an organisation may develop and apply performance indicators for assessing the sustainability level itself based on its specific conditions. Alternatively, such assessment may be performed using externally developed indicators with organisation-specific customisation (Coyne 2006). Well-known examples of the latter are the frameworks for standardised sustainability reporting such as triple bottom line (TBL) reporting (e.g. Archel et al. 2008), global reporting initiative (GRI) (e.g. Marimon et al. 2012), and corporate social responsibility (CSR) reporting (e.g. Pompper 2015). Such types of sustainability reporting enable the organisation to consider its impacts on a wide range of sustainability issues, thus being more transparent about the risks and opportunities it faces. Sustainability reporting is, therefore, a key platform for communicating sustainability performance and impacts. The unified standards allow reports to be done quickly, judged fairly, and compared simply and thus as useful as possible for managers, executives, analysts, and other stakeholders.

14.4 Sustainability Assessment at the Local Level

While cities are mostly centres of innovation and cross-cultural collaboration, the urban environment is often degraded, and the environmental footprint of the world's cities extends far beyond urban physical boundaries. Moreover, the socioeconomic situation of many urban dwellers is difficult, the environmental quality inadequate, and serious disparities within and between cities exist. Some data and indicators are generally available, in rich countries even abundant, but in such a confusing and/or unsystematic way that city managers do not understand which indicators to use. How to recognise, then, a sustainable city?

A literature review reveals that there is an array of types of potential urban sustainability assessment frameworks (Cohen 2017; EC 2015; Harris and Moore 2015). These frameworks are based on assessment methods (impact assessment, asset-based assessment, eco-efficiency), principles (socioecological system integrity, intra-, and intergenerational equity), or concepts (urban metabolism, urban carrying capacity), etc. Some authors conclude that there is no clear organisational structure for urban sustainability assessment across the literature but a myriad of frameworks, methods, principles, categories, goals, objectives, themes, criteria, indicators, sub-indicators, etc. (EC 2004). Regardless of the conceptual and terminological confusion, it is obvious that the assessment method should rather relate to the urban sustainability concept. For example, if a city follows the concept of urban metabolism, then it can assess its state, progress, or meeting targets by material flow analysis indicators – e.g. consumption of non-renewable and renewable materials, material addition to the city infrastructure, material outputs in the form of various emissions, etc. Similarly, if a city promotes a green city strategy, it can assess the quality and quantity of its green infrastructure, ecosystem services, etc. Regardless of the approach leading to local sustainability, assessment of the state and trend of local conditions, and monitoring set targets are common steps.

In recent decades, promoting a more sustainable urban future has become the focus of urban studies and, as a result, a multitude of city concepts have been introduced to promote urban sustainability in some way. The concept of sustainable urbanisation is not new and recent decades have witnessed a proliferation of innovations by municipalities and city authorities in its promotion worldwide. Despite there being no single or prevailing model of a sustainable city (Hamman et al. 2017), according to the results of extensive bibliometric analysis by Fu and Zhang (2017), the dominating city concepts promoting a more sustainable urban form are the sustainable city and the smart city. However, many other concepts and frameworks use indicator-based assessment of urban sustainability, e.g.

Global City Indicators Programme, Reference Framework for Sustainable Cities, Covenant of Mayors, European Common Indicators, World Bank Eco-City Initiative, OECD Green Cities Programme, etc. (EC 2015). Besides these internationally based or lead initiatives, there are many other initiatives developed and supported by national governments or cities themselves as, e.g. Monet and Circle Indicators in Switzerland (Altwegg et al. 2004; ISDC 2012), Urban Pioneer programme in Manchester, UK (Nature Greater Manchester 2021), Pearl Community Rating System in Abu Dhabi (Abu Dhabi Urban Planning Council 2010), etc.

We may roughly distinguish (i) locally specific approaches and methods developed and applied by a city itself usually featuring the participation of citizens (a set of specific urban sustainability indicators as, e.g. in Seattle in 1986; see Holden 2006), (ii) standardised approaches and methods applied internally by a municipality itself – municipal officers with lesser or greater public involvement (e.g. a set of the European Common Indicators developed by the European Commission and applied by many cities; see Ambiente Italia 2003), and (iii) standardised approaches and methods applied by an external expert (e.g. carbon footprint calculated and interpreted by a hired consultant). The first two groups may be called sustainability self-assessment methods emphasizing the instrumental role of the assessor. Besides these three distinct groups, there are many mixed approaches with lesser or greater engagement of the assessed municipality. Turcu (2012) recognised tensions between expert-led and citizen-led approaches raising much debate in the literature and suggested an integrated approach effectively utilising knowledge of sustainability.

According to the World Bank (Farvacque-Vitkovic and Kopanyi 2019), the urban self-assessment approach is attractive for central governments due to its impact on local government's capacity and performance building, and how it improves the implementation of transformative actions for policy change. City leaders and policymakers find it instructive and informative, with each issue placed in its context, and municipal staff in charge of day-to-day management may benefit from strong connections between data and information and municipal actions towards sustainability. Finally, partners of cities – such as banks and funds, utility companies, civil society, and private operators – find their foundations for more effective collaborative partnerships. Examples of self-assessment approaches on local sustainability supported by governments and participating organisations are the Green City Tool and the Sustainability Tools for Assessing and Rating Communities (STAR Communities). They differ in many aspects (Pace et al. 2016; Lynch and Mosbah 2017) but their common characteristics relevant for this study is the self-assessment approach.

The Green City concept is one of the latest responses to the various efforts and research conducted to address the problems caused by the dispersed model of city development and to help cities to become more sustainable (greener), less dispersed, and more livable (Brilhante and Klaas 2018). The Green City Tool developed by the EU (2019) is a simple self-assessment and benchmarking tool for cities that complement other city governance approaches taken to sustainable urban planning. This tool does not contain specific indicators of sustainability, for example, it does not ask what the ambient air quality is but whether there is real-time air quality information available via the internet (Yes–No question). It covers 12 key environmental topic areas – Air, Mobility, Energy, Climate change adaptation, Nature and Biodiversity, Noise, Governance, Water, Climate change mitigation, Green growth and innovation, Land use, and Waste. This assessment has been designed for towns and cities with over 50 000 inhabitants, and while it can be run for smaller cities, it may prove difficult for them (EU 2020). With a focus on sustainable urban planning, the tool emphasizes city governance and approach, rather than quantitative measures (it does not rank cities). Cities can use the tool anonymously or, if they wish, register themselves officially, and put themselves on the Green City map to show others that they are committed to becoming more sustainable. Registration also allows them to compare their results to those of other cities. Working both as a means of benchmarking and self-assessment, the tool allows cities also to communicate their efforts.

STAR Communities is a toolbox developed for community leaders in the United States to assess the sustainability of their community, set targets for the future, and measure progress along the way (Lynch et al. 2011). STAR was initiated in 2007 and currently, STAR operates as a Washington, DC-based nonprofit organisation (STAR Communities 2016). The rating system is part of a certification programme that bases a final rating of three, four, or five stars on points accumulated through the measurement of sustainability indicators (USGBC 2016). The certification process is managed by local governments: a city is registered for the programme and then proceeds with the data collection and reporting procedure via an online reporting tool. Certification does not require cities to submit data for every one of the seven objectives – built environment, climate and energy, economy and jobs, education, arts and community, equity and empowerment, health and safety, and natural systems. STAR is among the more detailed sustainability rating systems in existence in the United States; in all, the seven goal areas comprise 44 objectives and over 500 indicators. The indicators have been developed over time by several technical advisory committees. Cities compile their data from broad stakeholder-driven processes, giving the system credibility, and submit the data for evaluation. The verification process of submitted data and indicators takes about six months. Cities are required to recertify every three years. The tool is freely accessible or local governments can engage in the rating system via three service subscriptions (with different fees): Participating, Reporting, and Leadership STAR Community (Lynch and Mosbah 2017). Over 90 communities in North America, mostly in the United States, have registered so far. The system is adjusted based on new challenges – recently, STAR has connected to the New Urban Agenda and Sustainable Development Goals.

There is a vast amount of literature covering the topic of sustainability assessment at the local level (e.g. Merino-Saum et al. 2020). This study does not deal with many theoretical issues briefly reviewed above – it focuses on the indicator-based assessment approach applied at the local level in the Czech Republic.

14.5 Local Agenda 21 and Its Assessment in the Czech Republic

In the Czech Republic, LA 21 is a voluntary policy instrument based on the declared endorsement of local governments that enables building more sustainable cities (Kveton et al. 2014). Local governments in many countries have institutionalised LA21 as a means of performance measurement and management (Yetano 2013). In the Czech Republic LA21 is coordinated and supported by the Ministry of Environment and its information agency CENIA (Czech Environmental Information Agency). LA21 is implemented by municipalities according to a criterial framework – municipalities are put into one of five advancement categories (from the lowest to the highest: pre-category 'applicant', D, C, B, and A categories) based on the successful fulfilment of agreed criteria (CENIA 2021a). They fulfil and monitor three general areas – implementation of sustainable development concept, public engagement, and quality of strategic management. The LA21 Working Group at the Ministry of Environment, composed of representatives of relevant sectors, non-governmental organisations, academia, etc., decides on awarding a category. Figure 14.1 shows that there are 81 A–D category municipalities in the Czech Republic (latest data as of 2019).

Pre-category applicants involve municipalities that have started thinking about the LA21 process but do not want or cannot create any formal structures or initiate processes for implementing LA21. This phase serves as a step in familiarising themselves with LA 21 rules and the urban sustainable development concept. Category D requires a municipality to create organisation support within a municipality (e.g. appointment of an official onsite LA21 coordinator). An important part is to launch communication and management procedures, including public engagement in regional planning and decision making, or cooperation with the civil and business sector. Higher categories (C, B, and A) require a gradually more elaborated management structure and higher local political support and civic participation. More advanced municipalities in category C and those in categories B and A perform self-assessment based on the audit following the rules and principles described in the 'Methodology of the sustainable

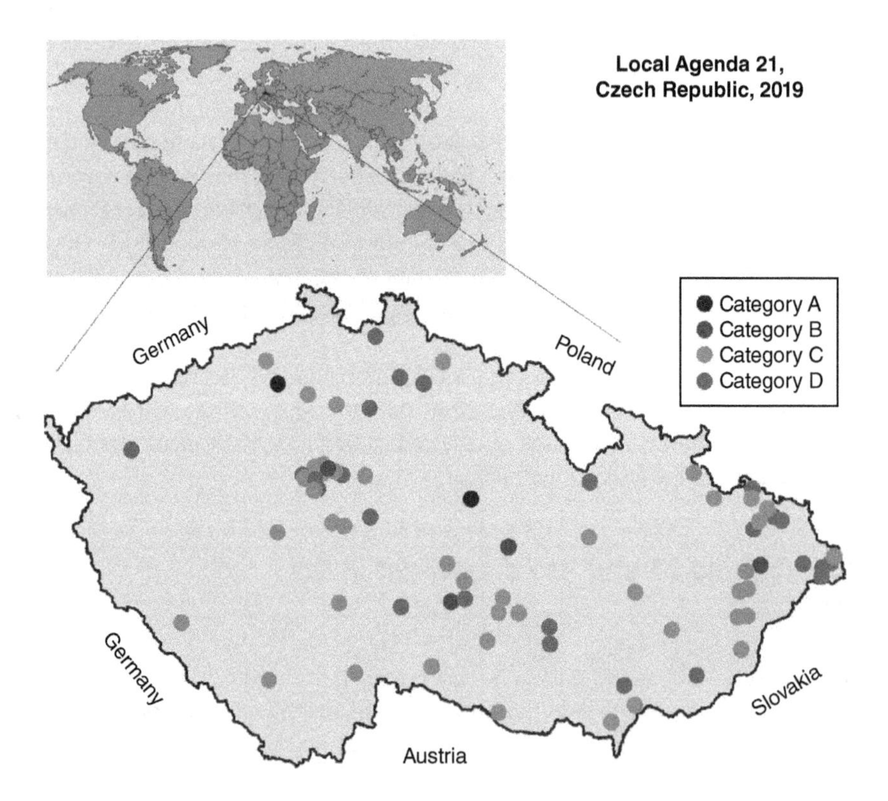

Figure 14.1 The LA21 A–D category municipalities in the Czech Republic, 2019. *Source:* Data from CENIA (2021b).

cities assessment: A sustainability audit for LA21 applicants in the Czech Republic' (Janoušková et al. 2017). The methodology was developed through the cooperation of municipality representatives, NGOs, academia, state administration, and other stakeholders in the field of sustainable development. Development took about a year – initially, there were meetings and discussions about the structure, themes and subthemes, indicators, etc., then pilot testing in four municipalities with previous LA21 experience.

The basic structure of the audit is derived from the Aalborg Commitments (Zimmermann 2007) and contains ten sustainable development themes: governance, environment/natural common goods, responsible consumption and lifestyle choices, transportation and mobility, health, local economy and business, education, culture and leisure, social environment, and global responsibility. Each theme is further divided into 3–4 subthemes and each sub-theme is finally concretised by a set of guiding questions with or without indicators. These are defined in a simple methodological sheet containing recommended methodology, assessment frequency, potential sources of data, and the target value (if available). Several guiding questions are not associated with indicators, especially if a qualitative description of the assessed phenomenon is required (see the example below – Table 14.1). Such guiding questions provide contextual information on the sub-theme.

Table 14.1 An example of guiding questions and related indicators of the environment theme.

Theme: **environment**		
Sub-theme: water quality, water-saving, and efficient water use	*Guiding question:* What is the year-to-year change in water consumption?	*Main indicator*: total consumption of drinking water in households (litres/cap/year) (Ind. = total amount of invoiced or produced water/population). Municipal data on the water production or consumption will be provided by the relevant water utility company
		Target or target value: the long-term goal is the efficient use of water
	Guiding question: Was the size of the built-up area that manages rainwaters changed?	*Main indicator:* the built-up area from which rainwater is drained out to detached sewerage, water flows (rivers), or soaked into the ground (% of the total city built-up area)
		Optional indicator: management of rainwater (description of the measure the city performs for rainwater recycling, retaining in the area, etc.)
		Target or target value: increasing the size of the city area; increasing the number of houses taking care of rainwater
	Guiding question: Is there any information on drinking water saving in municipal buildings? Was an awareness campaign on water saving organised in the city?	*No indicator: verbal description for the guiding question is required*
		Description of measures taken and supported by the municipality on saving drinking water (leaflets, articles in newspapers, signs in toilets in schools, hospitals, offices, awareness campaigns, etc.)

Guiding questions enable municipality representatives to understand what the important urban sustainability themes are, and they keep users on track when looking for data and information about themes. Indicators for each sub-theme may be of two types – obligatory or optional. Obligatory indicators provide important information on a municipality's progress towards sustainability. This kind of information is supposed to be widely available, and a municipality can, with lesser or greater effort, obtain it. Optional indicators also provide important information, but the data may be less available – the production of indicators may require specialist knowledge or technology (geographic information systems, etc.); data from non-municipal sources can be bought. Therefore, these indicators are recommended standards, but, in practice, the municipality may substitute them with suitable alternatives. The alternative indicator, however, must also provide relevant information on the assessed phenomenon. In practice, municipalities mostly apply the proposed optional indicators – looking for alternatives and/or developing their own indicators is exceptional.

14.6 Urban Sustainability Audit – Implementation of the City Self-assessment

The methodology for sustainable city self-assessment brings a step-by-step description of the whole auditing process a city needs to undertake (Janoušková et al. 2017). At the heart of the methodology is the Urban Sustainability Audit, which defines sustainability themes and sub-themes and together with related guiding questions and indicators for making the self-assessment. The auditing process itself is the same for municipalities of different types (village or small town, city, city district, and region). The guiding questions and indicators differ in number (fewer questions for smaller municipalities) and detail to capture the municipal type specifics while still being comprehensive and relevant.

A basic feature of the methodology is the self-assessing principle – the municipality itself answers all the guiding questions and works out all indicators. Each question is assessed by grades (−2, −1, 0, 1, 2) on a five-point scale: −2 denotes a very bad situation while +2 denotes a very good situation from the sustainability point of view. Also, each sub-theme receives a total (average) grade of −2 to +2 resulting from the particular grades of all pertinent guiding questions. Besides the grades, each sub-theme also receives a textual (qualitative) assessment summarising its strengths and weaknesses in interpreting the grades.

'C' category municipalities need only work out partial (incomplete) audits. They may choose just a few themes to be assessed, e.g. environment, education, and transport. After completing three themes the municipality receives a star. This means that 'C' category municipalities are differentiated according to their advancement – a C*municipality has three audited themes whereas a C***municipality has received a full audit (all 10 themes). 'A' and 'B' category cities must fulfil all the themes to be eligible for awarded this status. 'A' cities must also fulfil the prescribed criteria with no exceptions in the Governance theme; at least one theme must receive excellent evaluation and no theme can receive an unsatisfactory evaluation (see further).

Each municipality in categories A–C is assigned to a team of external evaluators: one to three experts (based on the category) for each theme. The higher category, the more experts (one expert in the case of a C audit, three experts for A). If there are any omissions or unclear interpretations of guiding questions and/or indicators, the experts ask the municipality for amendments within a certain timetable (such process may be of an iterative character if necessary). If an audit displays important and major omissions or other shortcomings, it can be rejected, and its evaluation stopped.

An audit of a C category municipality is evaluated by an external expert from the point of view of its completeness, i.e. existence and collection of relevant data needed for fulfilling the guiding question and/or indicators and for summary assessment of all sub-themes. Evaluation of A and B category cities by experts is more complex – besides completeness of the audit, the evaluation is focused on the correctness of the sustainability assessment, i.e. grading each guiding question (on the scale from −2 to +2). Evaluation of an A category city is complemented by a visit where experts may verify some information on the spot and discuss problematic issues or differences between the findings of the audit and those of the experts.

In the A and B category audit, experts summarise their findings, including conditional requirements (qualifications needed to be met by the next audit) and recommendations in the final evaluation report. Each theme is evaluated in terms of the progress towards sustainability – meeting the sustainability criteria, conditional requirements, or failure to meet the criteria. The complete evaluation of the audit for each municipality is submitted to the LA21 Working Group which decides on awarding the municipality a particular category. The audit must be repeated every three years (re-audit) and is extended by assessment of the change since the last audit (both the current state and the change/trend are assessed). The whole process is schematically shown in Figure 14.2.

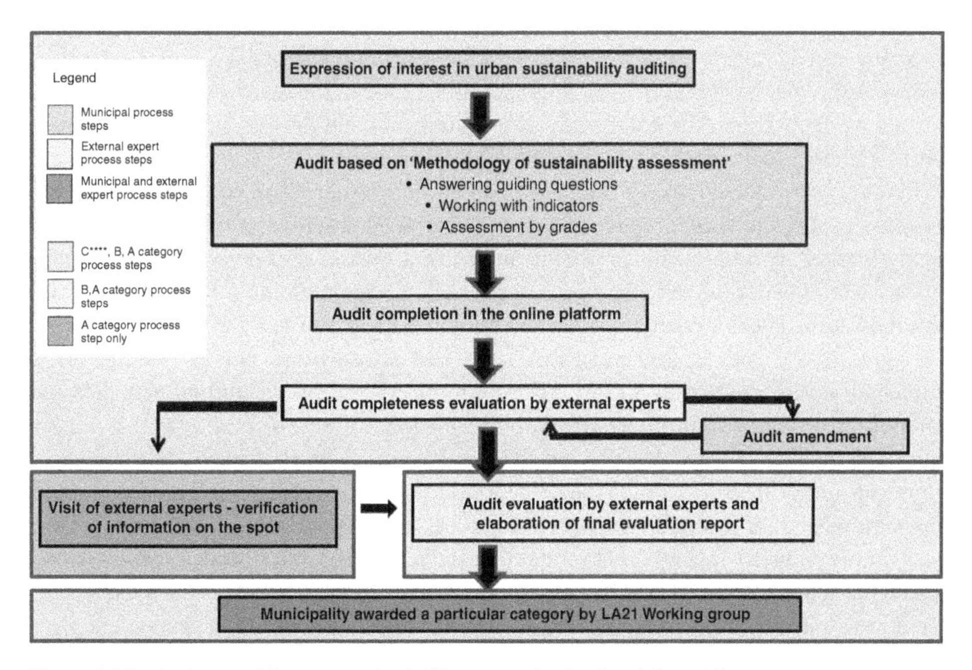

Figure 14.2 A diagram illustrating the LA21 process in the Czech Republic.

14.7 Benefits and Challenges – Municipal Representatives and External Experts' Views

The first study regarding the identification of drivers and barriers for using the guide was done in 2013 after the first audit of the city of Chrudim (Janoušková 2013). Since then, many partial audits, full audits, and re-audits have been performed, not only in cities but also in small municipalities and city districts. As a result of this experience, it was possible to observe and analyse the benefits and challenges raised by the audits over eight years.

Qualitative research methods can be used for this type of research (Elgert 2016). For the benefits and challenges analysis, we used, in particular, interviews and a retrospective think-aloud approach during discussions on the audit outcomes (from the perspective of experts). Another method was a focus group applied by the authors of this study during workshops with experts. The interviews and retrospective thinking aloud methods were applied every year in one or two municipalities implementing complete audits by two experts (the authors of the study). The focus group method was applied twice, in 2018 and 2019. Data were collected utilising extensive notes and analysed by an inductive approach at denoting particular categories of benefits and challenges. Coding was then done independently by both authors of this study to increase research objectivity. Information on C category municipalities was gathered by a discussion with the respective municipal representatives regarding their audits. In our work, we followed the methods of qualitative research as described by Denzin and Lincoln (2011), Leighton (2017), Saldaña (2021).

Both benefits and challenges are summarised in Table 14.2. Challenges primarily viewed as problems or difficulties to be overcome finally turned into benefits for the cities. Both challenges and benefits may be differentiated into two groups: the first group consists of content-specific knowledge and skills – sustainability themes and indicators and related work; the second group comprises process skills including soft skills – communication, teamwork, problem-solving, etc. The boundary line between the two groups is not explicit, but this typology shows the complexity of requirements and necessary know-how of the audit's authors to secure a high quality of the assessment.

This study shows that benefits for C category municipalities differ, to some extent, from the benefits of A and B category municipalities. This is commensurate with the audit complexity. C category municipalities use the audit in the area of content-specific knowledge and skills for a better understanding of sustainability principles and for learning to work with indicators. These aspects are stressed by both municipalities and experts. The experts also agree that collection and processing data and information may be difficult for C municipalities in some cases and proper communication between the municipality and experts is instrumental for successful audit completion. They also emphasize the willingness of municipal officers to learn new things beyond their routine work. Support by municipal authorities primarily to the officers working on data collection is another important requirement. If internal motivation decreases and support fades away, the audit is usually not completed (experts try to encourage officers by specific advice or recommendation on particular steps in the audit). The audit coordinator who manages the overall coordination within the municipal office and communicates with relevant stakeholders has a specific role. Internal coordinated management of audit preparation is stressed as

Table 14.2 The views of sustainability self-assessment's benefits and challenges.

Benefits/challenges	The view of municipality representatives	The view of experts
In-depth understanding of sustainability systemic knowledge and principles necessary for evidence-based policy	Category C	Category C
	• Understanding sustainability concept and principles	• Need to think on sustainability themes in the scope and detail prescribed by the audit methodology for both the audit authors and municipal government
	• Building a system of sequential sustainability assessment for evidence-based policy (municipalities may start the process only with several selected themes, which may be of priority or easier processing, and proceed to more difficult themes)	• Emerging thoughts on the importance of evidence-based policy concerning municipal sustainable development
	Category B, A	Category B, A
	• Discovering emerging or hidden problems due to auditing	• Discovering emerging or hidden problems due to auditing – decision-making based on evidence-based policy
	• A suitable argumentation place for designing measures aimed at sustainability weaknesses	• Drawing attention to important themes that have not previously been a priority
	• Feedback from experts allows other perspectives (more general, broader, external, etc.)	
Knowledge and experience to work on/with sustainability indicators (data collection, analysis, presentation, interpretation)	Category C	Category C

(*Continued*)

Table 14.2 (Continued)

Benefits/challenges	The view of municipality representatives	The view of experts
	• Work on/with sustainability indicators often requires motivation for learning new things (data collection and processing). Such capacity building and new know-how increase the cooperation potential of municipal officers (NGOs, business sector, regional partners)	• Data acquisition is difficult for some municipalities in category C, but most, in cooperation with an expert – external evaluator, will provide the necessary data and accomplish the audit. The help of an expert is often extensive, but it contributes to the municipality's capacity in obtaining important know-how • The audit quality depends on the willingness of the authors to invest time in obtaining data/information and their processing, as well as on their willingness to learn new things that are not part of their daily workload
	Category B, A	Category B, A
	• Data processing, its analysis, and interpretation often requires additional knowledge and know-how (e.g. obtained by consultations with experts)	• Acquisition or work with data is usually good and only in rare cases do problems occur. These are, however, usually solved together with experts • In some cases, help with the interpretation of data and indicators is needed, especially for more complex sustainability phenomena. Cooperation with experts increases the know-how of the audit authors • Audit quality depends on the willingness of the authors to invest time in obtaining data/information and their processing, as well as on their willingness to learn new things that are not part of their daily workload
Cooperation within the municipal office as well as with other organisations	Category C	Category C

Table 14.2 (Continued)

Benefits/challenges	The view of municipality representatives	The view of experts
	• Support from the municipal government	• In some cases, only one municipal officer tries to conduct a partial audit, but some assistance from other people in the office or from outside the office is sometimes necessary. Experts try to explain the need for cooperation leading to quality audit processing • The support of the municipality's government in preparing the audit is important • The role of the LA21 coordinator, who manages the audit, is crucial. This person is often also an intermediary between external experts and audit authors. He communicates intensively with important stakeholders inside and outside the office and has a general overview of the audit process
	Category B, A	Category B, A
	• Support from the municipal government to motivate officers to make a good quality sustainability audit. LA21 coordinator's role for communication, coordination, and keeping momentum	• Data were obtained from many sources. In some municipalities, this led to the strengthening of communication and cooperation with common and new subjects (organisations). In some cases, this required more effort from audit authors but mostly, cooperation on data acquisition and transfer was established
	• Performing the audit improved communication among departments within the municipal office (in particular, during the data collection phase)	• The support of the municipal government in preparing the audit is essential (mayor/mayor and city council, also the role of the municipal manager is important)
	• Data collection outside the municipal office opened up or improved communication and cooperation with other organisations (NGOs, business sector, etc.)	• The role of the LA21 coordinator, who manages the audit, is crucial. This person is often also an intermediary between experts and audit authors. They communicate intensively with important stakeholders inside and outside the office and have a general overview of the audit process

(Continued)

Table 14.2 (Continued)

Benefits/challenges	The view of municipality representatives	The view of experts
Cooperation among the auditing municipalities	Category C	Category C
		• Experts recommend looking at already approved audits of category A and B municipalities. These may inspire and motivate some municipalities in the preparation of their own audits
	Category B, A	Category B, A
	• The LA 21 coordinators have mutual working contacts based on meetings and experience sharing in various municipal and regional sustainability projects. Hence, they may share know-how on sustainability audits in a similar way	• Municipalities share know-how on conducting audits. This usually simplifies the consultancy work of expert-external evaluators
Communication with experts	Category C	Category C
	• Cooperation with experts is important in various stages of the audit – not only during finalisation but typically also at data collection (advice on a data source) or results in interpretation	• Although not the responsibility of expert-external evaluators, most seek to provide advice in completing audits. Most municipalities successfully complete audits and the data in the audit is complete. Occasionally, however, audits are not completed, particularly, if they are processed by only one person or people with limited know-how in a given theme. The help of an expert consists mainly in drawing attention to absent data and referring to their possible sources and advising on their processing
	Category B, A	Category B, A
	• A key benefit from experts lies in the opportunity of viewing strengths and weaknesses from another (external, objective) perspective, including national and international experience with sustainability assessment. Also, discussions on sustainability problems and different opinions of experts bring added value to current audits and local capacity building	• Experts find their use in discussions on interpretations of results. Also, discussions about various approaches applied in sustainability assessment at national and international levels take place

important even for C municipalities, although they may start assessment of just a few selected themes. Similar stress is put on experience sharing using successful freely accessible audits.

The study further shows that A and B municipalities have succeeded in completing all their audits so far. There is quite intensive cooperation between city officers and experts because the experts do not execute just a controlling function, but they work out a detailed evaluation report on the audit. During the audit preparation, both sides exchange views on data, indicators, and interpretations, learning from each other, resulting in well-made assessments. Also, more intensive cooperation and experience sharing (e.g. on the comments and recommendations from experts) about the audit preparation among the municipalities is observed.

All category municipalities (A, B, and C) acknowledge that sustainability audits are beneficial for diagnosing important problems in the municipality and for monitoring these issues regularly. They appreciate the methodological support from the experts, who bring a broader perspective and different experiences. On the other hand, even re-auditing municipalities report that the sustainability self-assessment of the more advanced categories A and B is demanding and requires resources, knowledge, and time capacity. If important success factors – motivation and support from the city government – are lacking or gradually fades, the probability of failure is high. A political cycle, which may elevate political parties that do not support LA 21, may jeopardise the process built up over a long period.

14.8 Key Findings and Conclusions

Setting and organising a sustainability agenda in a municipality requires innovative approaches to effectively address current and emerging challenges. Local sustainability practice takes place through a variety of actors (not only municipal authorities but also citizens and any interested business and/or other stakeholders), approaches, methods, and instruments, within different time, spatial, financial, and organisational frameworks. Despite municipalities being the level of governance closest to the people, sustainability results may be diffuse and difficult to track. The literature review shows that the municipal level is quite underestimated in terms of sustainability research, particularly regarding sustainability assessment. Because the one-size-fits-all approach is not often appropriate at this level, LA21 co-developed by municipal representatives and external experts in the Czech Republic brought a standardised yet flexible assessment system for communities of different sizes and types. It represents an important departure from traditional city management when initial difficulties and challenges often become benefits for both city representatives and citizens.

LA21 requires municipalities to apply the methodology for sustainable city self-assessment. Besides procedural steps, the Urban Sustainability Audit defines sustainability themes and sub-themes and related guiding questions and indicators for making the self-assessment. Thus, it stands for a conceptual framework not using the often three pillars-based sustainability concept (economy, social, environment) but a comprehensible theme-based framework for practical use. The 10 themes and sub-themes are internationally agreed upon and in good compliance with the agenda structure of municipalities. The framework does not explore mutual relationships among the themes and sub-themes or indicators, but it encourages local governments to obtain the right data/information on their progress towards sustainability and to share it with other municipalities as well as to inform central government (via awarding inter-sectoral Working Group, etc.).

The overall methodology and the assessment process are the same for municipalities of different advancement levels (C–A). All these categories of municipalities acknowledge that sustainability audits are beneficial for diagnosing important problems in the municipality and for monitoring these issues regularly. While B and A category municipalities have gradually built capacities for self-assessment and used its results for argumentation for designing sustainability measures, C category municipalities use the audit in the area of content-specific knowledge and skills for better understanding sustainability principles and for learning to work with indicators. A slow start – the partial audit focusing only on a few selected themes – is an important success factor since a full audit is a time-consuming activity.

Co-development of the self-assessment methodology by municipal representatives and external experts seems to be the cornerstone for the successful course of the whole initiative. Cooperation at the early stages of the process, in particular, brings a pragmatic benefit: correct design of the audit (relevant themes and sub-themes, relevant, and adequate number of indicators) ensures that the results will be relevant and accepted by the community. This acceptance is a prerequisite for any further use in analyses and actions.

The support of a municipality's government in the LA21 process and in preparing the audit is instrumental for success. At a practical level, the LA21 coordinator, who manages the audit, serves as an intermediary among the in-house resources (the audit authors from various departments of the municipal office) and between the in-house resources and external experts. If support from the city government is lacking or fades, there is a probability of failure (incomplete audits, discontinuing, or terminating the whole LA21 process). Such changes ensue from the political cycle (elections), which may elevate political parties that do not support LA21.

14.9 Limits of the Study

Our research intends to understand a complex reality and the meaning of processes in the context of sustainability self-assessment provided by LA21 in the Czech Republic. To do this, we applied qualitative research methods to obtain a comprehensive, in-depth picture of the surveyed issue; such methods best suit the situation under analysis. However, qualitative research methods have their limits that need to be taken into account when reading this study. The study findings cannot be generalised to all municipalities seeking self-assessment, not even in the Czech Republic, although data saturation was achieved (no new information appeared after the research was finished). Furthermore, the results are time, people, and situation dependent. This means that new municipal representatives or experts (external evaluators), at other times and in different situations (influenced for example by crises caused by Covid-19) can bring other insights to the theme. Results of qualitative studies may be influenced by the researcher's personal biases and idiosyncrasies. To avoid this, more methods were used and the data were analysed independently by two researchers. Also, different respondents-stakeholders were surveyed. Finally, comparable results were brought by another foreign study applying similar methods exploring similar phenomena (Elgert 2016).

Despite these limits, we believe that the study provides important findings that complement existing studies from abroad. Inspiration can be found in both the self-assessment process itself and at the assessment framework (themes and indicators). Moreover, the study provides an in-depth insight into challenges that need to be considered during sustainability assessment implementation.

References

Abu Dhabi Urban Planning Council (2010). Pearl community rating system: design & construction, version 1.0. http://www.dmt.gov.ae/en/Urban-Planning/Pearl-Community-Rating-System (accessed 1 February 2021).

Altwegg, D., Roth, I., and Scheller, A. (2004). Monitoring sustainable development MONET. Final report - methods and results, Swiss Federal Statistical Office, Neuchâtel.

Ameen, R.F.M., Mourshed, M., and Li, H. (2015). A critical review of environmental assessment tools for sustainable urban design. *Environmental Impact Assessment Review 55*: 110–125.

Archel, P., Fernández, M., and Larrinaga, C. (2008). The organizational and operational boundaries of triple bottom line reporting: a survey. *Environmental Management 41* (1): 106–117.

Barbier, E.B., Markandya, A., and Pearce, D.W. (1990). Environmental sustainability and cost-benefit analysis. *Environment and Planning A 22* (9): 1259–1266.

Bhagavatula, L., Garzillo, C., and Simpson, R. (2013). Bridging the gap between science and practice: an ICLEI perspective. *Journal of Cleaner Production 50*: 205–211.

Bond, A., Morrison-Saunders, A., and Pope, J. (2012). Sustainability assessment: the state of the art. *Impact Assessment and Project Appraisal 30* (1): 53–62.

Brilhante, O. and Klaas, J. (2018). Green city concept and a method to measure green city performance over time applied to fifty cities globally: influence of GDP, population size and energy efficiency. *Sustainability 10* (6): 2031.

CENIA (2021a). LA21 in maps (in Czech). http://ma21.cenia.cz/cs-cz/ma21vdatech/ma21vmap%c3%a1ch.aspx (accessed 1 February 2021).

CENIA (2021b). LA21 criteria. https//ma21.cenia.cz/cs-cz/%C3%BAvod/prorealiz%C3%A1tory/krit%C3%A9riama21kesta%C5%BEen%C3%AD.aspx (accessed 1 February 2021).

Cohen, M. (2017). A systematic review of urban sustainability assessment literature. *Sustainability 9* (11): 2048.

Coyne, K.L. (2006). Sustainability auditing: evaluating organisations' progress towards sustainable development. *Environmental Quality Management 16* (2): 25–41.

Denzin, N.K. and Lincoln, Y.S. (ed.) (2011). *The Sage Handbook of Qualitative Research*. SAGE.

EC (2004). *Study on Indicators of Sustainable Development at the Local Level*. Bruxelles: European Commission.

EC (2015). Indicators for sustainable cities. In-depth report 12 produced for the European Commission DG Environment by the Science Communication Unit, UWE, Bristol. http://ec.europa.eu/science-environment-policy (accessed 1 February 2021).

Elgert, L. (2016). The double edge of cutting edge: explaining adoption and nonadoption of the STAR rating system and insights for sustainability indicators. *Ecological Indicators 67*: 556–564.

EU (2019). *Introducing the EU's Green City Tool – Compendium*. Luxembourg: Directorate-General for Environment.

EU (2020). Green city tool. http://webgate.ec.europa.eu/greencitytool/topic/e (accessed 1 February 2021).

Farvacque-Vitkovic, C. and Kopanyi, M. (2019). *Better Cities, Better World: A Handbook on Local Governments Self-Assessments*. World Bank Publications.

Finkbeiner, M., Schau, E.M., Lehmann, A., and Traverso, M. (2010). Towards life cycle sustainability assessment. *Sustainability 2* (10): 3309–3322.

Fu, Y. and Zhang, X. (2017). Trajectory of urban sustainability concepts: a 35-year bibliometric analysis. *Cities 60*: 113–123.

Gasparatos, A. and Scolobig, A. (2012). Choosing the most appropriate sustainability assessment tool. *Ecological Economics 80* (0): 1–7.

Geneletti, D. (2019). Principles of multicriteria analysis. In: *Multicriteria Analysis for Environmental Decision-Making*, 5–16. Anthem Press.

Hák, T., Moldan, B., and Dahl, A.L. (ed.) (2007). *Sustainability Indicators: A Scientific Assessment (SCOPE Vol. 67)*. Washington, DC: Island Press.

Hák, T., Moldan, B., and Dahl, A.L. (2012). Editorial. *Ecological Indicators 17*: 1–3.

Hák, T., Janoušková, S., and Moldan, B. (2016). Sustainable development goals: a need for relevant indicators. *Ecological Indicators 60*: 565–573.

Hamman, P., Anquetin, V., and Monicolle, C. (2017). Contemporary meanings of the 'sustainable city': a comparative review of the French-and English-language literature. *Sustainable Development 25* (4): 336–355.

Harris, A. and Moore, S. (2015). Convergence and divergence in conceptualising and planning the sustainable city: an introduction. *Area 47* (2): 106–109.

Holden, M. (2006). Sustainable Seattle: the case of the prototype sustainability indicators project. In: *Community Quality-of-Life Indicators*, 177–201. Dordrecht: Springer.

ISDC (2012). *Sustainable Development in Switzerland – A Guide*. Interdepartmental Sustainable Development Committee: Berne http://www.are.admin.ch/are/en/home/media/publications/sustainable-development/nachhaltige-entwicklung-in-der-schweiz-ein-wegweiser.html (accessed 1 February 2021).

Italia, A. (2003). *European Common Indicators. Towards a Local Sustainability Profile*. Milano: Ambiente Italia Research Institute.

IUCN, UNEP, and WWF (1980). *World Conservation Strategy*. Gland: International Union for the Conservation of Nature.

Jabareen, Y. (2008). A new conceptual framework for sustainable development. *Environment, Development and Sustainability 10* (2): 179–192.

James, P. (2014). *Urban Sustainability in Theory and Practice: Circles of Sustainability*. Hoboken: Routledge.

Janoušková, S. (2013). Implementation of an evaluation system – an indicator set – in the Healthy City of "Chrudim", Czech Republic. WP3 report for the BRAINPOOL project. http//neweconomics.org/uploads/images/2018/01/WP3-case-study-Chrudim.pdf (accessed 21 February 2021).

Janoušková, S., Hák, T., and Švec, P. (ed.) (2017). *Methodology of the Sustainable Cities Assessment: A Sustainability Audit for LA21 Applicants in the Czech Republic*, 3e (in Czech). Praha: HCCR and MoE.

Joas, M. and Grönholm, B. (2004). A comparative perspective on self-assessment of Local Agenda 21 in European cities. *Boreal Environment Research 9* (6): 499–507.

Károly, K. (2011). Rise and fall of the concept sustainability. *Journal of Environmental Sustainability 1* (1): 1.

Kates, R.W., Clark, W.C., Corell, R. et al. (2001). Sustainability science. *Science 292* (5517): 641–642.

Kveton, V., Louda, J., Slavik, J., and Pelucha, M. (2014). Contribution of Local Agenda 21 to practical implementation of sustainable development: the case of the Czech Republic. *European Planning Studies 22* (3): 515–536.

Leighton, J.P. (2017). *Using Think-aloud Interviews and Cognitive Labs in Educational Research*. Oxford University Press.

Luke, T.W. (2005). Neither sustainable nor development: reconsidering sustainability in development. *Sustainable Development 13* (4): 228–238.

Lynch, A., Andreason, S., Eisenman, T., et al. (2011). Sustainable urban development indicators for the United States. U.S. Department of Housing and Urban Development. https://www.researchgate.net/publication/281241372_Sustainable_Urban_Development_Indicators_for_the_United_States (accessed 1 February 2021).

Lynch, A.J. and Mosbah, S.M. (2017). Improving local measures of sustainability: a study of built-environment indicators in the United States. *Cities 60*: 301–313.

Marimon, F., del Mar Alonso-Almeida, M., del Pilar Rodríguez, M., and Alejandro, K.A.C. (2012). The worldwide diffusion of the global reporting initiative: what is the point? *Journal of Cleaner Production 33*: 132–144.

Mascarenhas, A., Coelho, P., Subtil, E., and Ramos, T.B. (2010). The role of common local indicators in regional sustainability assessment. *Ecological Indicators 10* (3): 646–656.

McGalliard, T. (2012). Reframing the sustainability conversation from what to how. *Public Management* 94: 2.

Merino-Saum, A., Halla, P., Superti, V. et al. (2020). Indicators for urban sustainability: key lessons from a systematic analysis of 67 measurement initiatives. *Ecological Indicators 119*: 106879.

Nature Greater Manchester (2021). Urban pioneer. http://naturegreatermanchester.co.uk/project/urban-pioneer/ (accessed 1 February 2021).

Pace, R., Churkina, G., and Rivera, M. (2016). How green is a "Green City"? A review of existing indicators and approaches. IASS Working paper. Institute for Advanced Sustainability Studies, Postdam, Germany. http://publications.iass-potsdam.de/pubman/item/item_1910926_3/component/file_1910931/IASS_Working_Paper_1910926.pdf (accessed 1 February 2021).

Pesqueux, Y. (2009). Sustainable development: a vague and ambiguous "theory". *Society and Business Review* 4 (3): 231–245.

Phillis, Y.A., Kouikoglou, V.S., and Verdugo, C. (2017). Urban sustainability assessment and ranking of cities. *Computers, Environment and Urban Systems 64*: 254–265.

Pires, S.M., Fidélis, T., and Ramos, T.B. (2014). Measuring and comparing local sustainable development through common indicators: constraints and achievements in practice. *Cities 39*: 1–9.

Pompper, D. (2015). *Corporate Social Responsibility, Sustainability and Public Relations: Negotiating Multiple Complex Challenges*. Routledge.

Purvis, B., Mao, Y., and Robinson, D. (2019). Three pillars of sustainability: in search of conceptual origins. *Sustainability Science 14* (3): 681–695.

Robinson, J. (2004). Squaring the circle? Some thoughts on the idea of sustainable development. *Ecological Economics* 48: 369–384.

Saldaña, J. (2021). *The Coding Manual for Qualitative Researchers*. SAGE.

Smith, R. (2004). A Capital-based Sustainability Accounting Framework for Canada. In: *Measuring Sustainable Development: Integrated Economic, Environmental and Social Frameworks* (ed. OECD). Paris: OECD Publishing.

STAR Communities (2016). STAR Community rating system. STAR Communities, Washington, DC. http://www.cedar-rapids.org/STAR%20Manual.pdf (accessed 1 February 2021).

Stiglitz, J.E., Sen, A., and Fitoussi, J.-P. (2009). *Report by the Commission on the Measurement of Economic Performance and Social Progress.*

Turcu, C. (2012). Re-thinking sustainability indicators: local perspectives of urban sustainability. *Journal of Environmental Planning and Management* 56: 1–25.

UN (1992). *Earth Summit Agenda 21. The United Nations Programme of Action from Rio.* New York: United Nations Department of Public Information.

UN (2015). Transforming our world: the 2030 agenda for sustainable development. RES/70/1, New York.

UNECE, OECD, and Eurostat (2008). *Measuring Sustainable Development: Report of the Joint Working Party on Statistics for Sustainable Development.* New York and Geneva: United Nations.

USGBC (2016). STAR Community rating system, version 2.0. http://usgbc-rec.org/star-community-rating-system/ (accessed 21 February 2021).

Valentin, A. and Spangenberg, J.H. (2000). A guide to community sustainability indicators. *Environmental Impact Assessment Review 20* (3): 381–392.

Waas, T., Hugé, J., Block, T. et al. (2014). Sustainability assessment and indicators: tools in a decision-making strategy for sustainable development. *Sustainability 6* (9): 5512–5534.

WBCSD (1999). *Eco-efficiency Indicators and Reporting: Report on the Status of the Project's Work in Progress and Guidelines for Pilot Application*. Geneva: World Business Council for Sustainable Development.

WCED (1987). *Our Common Future. World Commission on Environment and Development.* Oxford: Oxford University Press.

Yetano, A. (2013). What drives the institutionalization of performance measurement and management in local government? *Public Performance & Management Review 37* (1): 59–86.

Zimmermann, M. (2007). Local governments and sustainable development. *Environmental Policy and Law 37* (6): 504.

Zwart, R., Kamphof, R., Hollander, K., and Iwaarden, A. (2012). *Activities of the European Union on Sustainable Urban Development. A Brief Overview*. European Metropolitan Network Institute: The Hague.

15

A Sustainable Approach to Combat Climate Change: Case Studies from Some Urban Systems

Meenakshi Chaurasia[1], *Kajal Patel*[1], *Ranjana Singh*[2], *and Kottapalli S. Rao*[1]

[1] *Department of Botany, University of Delhi, Delhi, India*
[2] *Government Model Degree College, Bulandshahar, Uttar Pradesh, India*

Abbreviations

BREEAM	Building Research Establishment Environment Assessment Method
CIUP	Community Infrastructure Upgrading Programme
CO_2	carbon dioxide
CSR	corporate social responsibility
GDP	gross domestic product
GHGs	greenhouse gases
ICT	information and communication technologies
IPCC	Intergovernmental Panel on Climate Change
IT	information technology
JNNURM	Jawaharlal Nehru National Urban Renewal Mission
MCGM	Municipal Corporation of Greater Mumbai
MDGs	millennium development goals
NAPCC	National Action Plan on Climate Change
ROAP	Regional Office for Asia and Pacific
SDGs	sustainable development goals
TFM	technology facilitation mechanism
UN	United Nations
UN DESA	United Nations Department of Economic and Social Affairs
UNCED	UN Conference on Environment and Development
UNEP	UN Environment Programme
UNFCCC	UN Framework Convention on Climate
UNRA	UN Refugee Agency

Urban Ecology and Global Climate Change, First Edition. Edited by Rahul Bhadouria, Shweta Upadhyay, Sachchidanand Tripathi, and Pardeep Singh.

15.1 Introduction

Today, the world is facing great challenge for sustainable development, threatened by recent trends in growing proportion of global urban population and associated activity and consequent global climate change. Currently, 55.7% of the total world population is urbanised, and it is expected to increase to 68% by 2050 at the current growth rate (UN DESA 2019). Urbanisation presents huge opportunity for socioeconomic progress and improves the living standard, and at the same time, increased resource consumption and waste generation accompanied by process of rapid urbanisation generates myriad of severe environmental issues (Seto and Satterthwaite 2010). Recent studies have confirmed that impromptu and large urban growth is intensifying the challenge of climate change and global warming (Min et al. 2011; Pall et al. 2011). Cities have been accused of being the chief contributor of inclusive greenhouse gases (GHGs) emission (Carter et al. 2015). About 90% of anthropogenic carbon emission is attributed to urban activities (Svirejeva-Hopkins et al. 2004). Immoderate demand for food and energy brought about by improper and disproportionate development in and around cities has transformed the land use pattern, leading to desertification and biodiversity loss. Infrastructure development profoundly affects the environment functioning of the region by altering surface heat fluxes, water cycle, biogeochemistry of soil, etc. All these elements directly or indirectly affect the sources and sinks of earth warming GHGs and thus imbalance their concentration in atmosphere, causing complex shifts in weather and climate pattern of the planet (Grimmond 2007).

Concurrently, the changing climate is increasing the vulnerability and risk for the environmental, social, and financial aspects of human well-being. Urban settlements and their residents are at the forefront of the consequences of extreme weather events such as storms, heat wave, heavy precipitation, rise in sea level, and other climate-related natural disasters (Rosenzweig et al. 2015). Nearly, all the urban centres around the world, ranging from metropolitan cities to urban clusters of developing countries, are dealing with or endangered by the pressing effects of extreme climatic variability. Much of the vulnerability to climate forcing is concentrated in the urban settlement with limited or almost no provision for basic amenities such as health, education, employment, and accommodation (Mitlin and Satterthwaite 2013). Disproportionate impact of climate processes such as rising sea level, water inaccessibility, and desertification and climatic events such as drought, flooding, hurricane, and heatwave is inducing the human migration across the globe. Estimation by Warner (2010) suggests that annually 50 million peoples facing the risk of climate change migrate to urban area. This estimate of migrants, driven by environmental change, is predicted to reach 200 million, which may consequently result in increasing social and environmental vulnerability of the destination centres (Biermann and Boas 2010).

Interaction of urbanisation and impact of climate change is complex. Incessant urban expansion escalates the perils associated with climate extremes, whilst climate change further propels the trend for urbanisation, forming a feedback loop between these two presages (Jaramillo et al. 2013). Impacts of these two critical processes are converging in pernicious way, jeopardising the three basic criteria of sustainability, i.e. economic, social, and ecological stability. Understanding the interplay between cities and climate forces is a pivotal step to instigate the strategies to combat the impact of climate change threats, increase the urban resilience, and cater sustainable environment for development (Revi et al. 2014).

Urban centres are centres to major demographic, social, and economic activities, and therefore, cities are integral to sustainable development. Under current prospects for urbanisation and climate change, cities will play decisive role in achieving the sustainable development goals (SDGs), i.e. equity, health, water, energy for all, and climate change action. For this reason, SDGs 11 with the goal of developing safe, inclusive, resilient, and sustainable cities was adopted by United Nation General Assembly in 2015 (UN 2018). The trajectory for sustainable urbanisation can be realised by focussing on developing resilience against disturbance, enhancing resource efficiency, and improving the quality of life (UNEP 2012). This helps in reducing the ecological footprint combined with sustaining socioeconomic growth of urban settings in the current scenario. Cities represent dynamic comprehensive system that holds immense potential to combat climate change and to consummate the goal of sustainable development. Urban adaptation and mitigation approach is increasingly recognised by local, national, and international governments and organisations (UN HABITAT 2011; IPCC 2018). Several schemes and actions have been implemented by high-, middle-, and low-income countries in different contexts of environment (Stafford-Smith et al. 2017). However, in the absence of proper institutional, technical, and financial assistance, these opportunities remain untapped. UN Environment Programme (UNEP) is promoting environment-friendly energy consumption and production practices in Asia and the Pacific to address inefficient resource-use patterns (UNEP 2012). Initiatives such as 'Cities and Climate Change Initiative' by UN Habitat promote collaborations between different national and local government bodies in developing countries to address the issue of climate change by following sustainable urbanisation (de Coninck et al. 2018). Undoubtedly, these are important steps in the right direction, but only a collective effort on a global scale will be commensurate to the task at hand. To maintain the life quality on the face of climate change challenge, multidisciplinary approaches at varying scales are needed. We need to embrace sustainable transformative pathways to build resilience, to minimise the risk posed by extreme climatic events, and to encourage preparedness to cope with the negative impacts of climate change. More resource-efficient systems to provide the essential services to their residents, adopting greener technology to reduce pollution and minimising the disparity in society, are urgently needed.

With the aim to comprehend the challenges faced by urban centres and the opportunities it offers to address the issue of global climate change and sustainable environment, the present chapter highlights the (i) relation between the two most concerned issues of the twenty-first century, i.e. urbanisation and climate change, (ii) SDGs in relation to urbanisation and climate change mitigation and adaptation, (iii) the potential approaches for achieving the goal of sustainable urban development, with focus on the role of green infrastructure to combat climate change, and (iv) case studies that will advance the systematic learning on urban mitigation and adaptation options.

15.2 Urbanisation and Climate Change

Over the past few decades, demographic and economic growth has impelled rural to urban transformation, leading to rapid urbanisation across the globe with unequivocal consequences on different dimensions of the environment. Today, around approximately

4.2 billion urban dwellers live in urban centres. Between 2018 and 2050, this proportion is projected to grow by 2.5 billion, with 90% of growth concentrated in India and Africa (UN DESA 2019). This spatial transformation process is accompanied by finite energy consumption, resource depletion, and increased GHG emissions, which can have adverse impact on the environment. Cities consume 60–80% of world energy and accounts for 75% of anthropogenic GHG emission, with transportation and built being the major contributors (UN HABITAT 2011). Excessive urban demand for energy, food, and housing has prompted changes in land use and land cover, aggravating problems such as desertification and habitat loss (Foley et al. 2005). One of the notable results of anthropogenic alterations is urban head island effect (Kalnay and Cai 2003). Urbanisation can also have significant influence on local winds and precipitation patterns through the modification of natural surfaces and atmospheric conditions and further changes the local weather and climate system (Li et al. 2016). All these activities directly or indirectly contribute to global issue of earth warming and climate change.

Urban centres not only concentrate on large and growing agglomeration and associated activities that drive climate change but also amass resources and infrastructure and, hence, are particularly vulnerable to the impact of changing climate. Scientific evidence suggest that climate change is expected to affect more than hundred million of urban dwellers in the coming decades (Revi et al. 2014). Even lower levels of anticipated climate change are expected to produce a cascade of intricate direct and indirect impacts on natural environments, economies, and societies, varying from local to global scale (IPCC 2007; USGCRP 2009). However, vulnerability to climate change impact and natural hazard varies, depending on their geographical location, economic structure, and resource availability (Filho et al. 2019). As projected, much of the urbanisation process will unfold in low- and middle-lower income countries where poverty and poor health is already exacerbated and lack proper infrastructure and framework to confront and manage climate extremities, further intensifying the disproportionate social, economic, and environmental transformations in these regions (DePaul 2012). Similarly, countries located on riverine and coastal regions are especially at high risk to the impact of changing climate. Frequent flood and cyclone in these areas causing soil erosion, salinisation, and flooding damages the infrastructure and affect water supply, health, and other social services, ultimately leading to natural ecosystem degradation and mortality (Aboulnaga et al. 2019).

The probable impact of climate change on cities has been categorised by a world bank study into three dimensions: (i) social – increased risk to health, loss of employment opportunities, and spatial shift of population; (ii) economic – impact on activities contributing to gross domestic product (GDP) such as agricultural and fisheries industries, tourism, and reduced water and energy security, all of which could have negative impacts on humankind; and (iii) environmental – change in land use and land cover, leading to loss of biodiversity, shrinking of forest cover, and threats to coastal and marine systems of high ecological and economic value (Singh and Singh 2012). Increased global warming along with increase in energy consumption and production creates urban heat island effect in urban areas and increases pollutant concentration and heat wave, which eventually cause surge in GHG emission (Akbari 2005; Aboulnaga et al. 2019). Based on their assessment, several studies have concluded climate change as a significant factor affecting the societal health.

Impact of climate change will be further complicated by migration around the world, forced by extreme climatic conditions such as rise in sea level, flooding, drought, storms, and conversion of arable land (Aboulnaga et al. 2019). As per UN Refugee Agency (UNRA), climate forcing has compelled approximately 24 million to migrate between 2002 and 2012, and it is expected to dislodge 200 million people by 2050 (Barnett and Webber 2010). Mass population displacement will have serious consequences on the destination centres, i.e. increased pressure to supply food, infrastructure, health, and other social services and subsequent waste generation. Climatic disaster episodes such as rainfall downpour, drought, and hurricane may deteriorate urban infrastructure, worsening the situation further.

In addition to these direct impacts of climate change, cities are likely to experience other indirect influence of climate change. Climate change-induced migration will lead to shift in land use pattern, having implication on urban ecosystem. Altered climatic patterns and periodic events of drought and floods may amplify the global challenge of food security for all (Shankar 2018). Global warming together with expanding population and rising food and energy cost is expected to add millions of more people at risk of hunger by 2030 (Godfray et al. 2010). Decrease in crop yield and productivity will further affect the economic growth of urban centres. With rapid expansion in urban population, cities are in frontline to bear the impacts of climate change and, thus, play crucial role in combating and minimising the impact of climate change.

Alteration in hydrological cycle and precipitation pattern triggered by global warming is probable to intensify the water-scarcity problems such as shortage of portable water and poor water quality (Aboulnaga et al. 2019). Currently, approximately 768 million of people do not have access to safe water and sanitisation. Studies have projected that 43–50% of the total world population will be residing in water-deficit regions by 2080 (UN DESA 2015). Climate change risk to water is more pronounced in poor countries, where poor sanitation and water quality might increase the chances of water-borne diseases. Furthermore, the loss of agricultural sector and fishery industries, water security issues, disturbed tourism, and other factors altogether result in economic loss (World Bank 2019).

Environmental changes faced by humankind are deeply intertwined with complex urbanisation processes and happen at a previously unseen rate and magnitude. Climate change presents unique sets of challenges in urban regions. The net result is a general scientific acceptance that some climate change effects are now inevitable and unavoidable (IPCC 2007). However, research on the intersection of climate change and urbanisation are still at initial stage. Studies on climate change require coherent analysis of projections on global urbanisation to explore emissions and mitigation prospects (Krey et al. 2012; O'Neill et al. 2010, 2012) and to assess the risk and vulnerability related to climate change impacts (McDonald et al. 2011; McGranahan et al. 2007; Zhou et al. 2004).

Urbanisation is an inevitable process for the socioeconomic development of a nation, at the same time unplanned and unsustainable urbanisation will induce and exacerbate the climate change and its impact on the urban systems (Chan 2017). Growing urban cluster has been blamed for deteriorating urban climate. Properly managed urban settlement can provide opportunity to mitigate climate change, one of the key goals of sustainable development. Integrating climate actions into development agendas will benefit the cities to maintain equilibrium between economic growth and liveable environment (Figure 15.1).

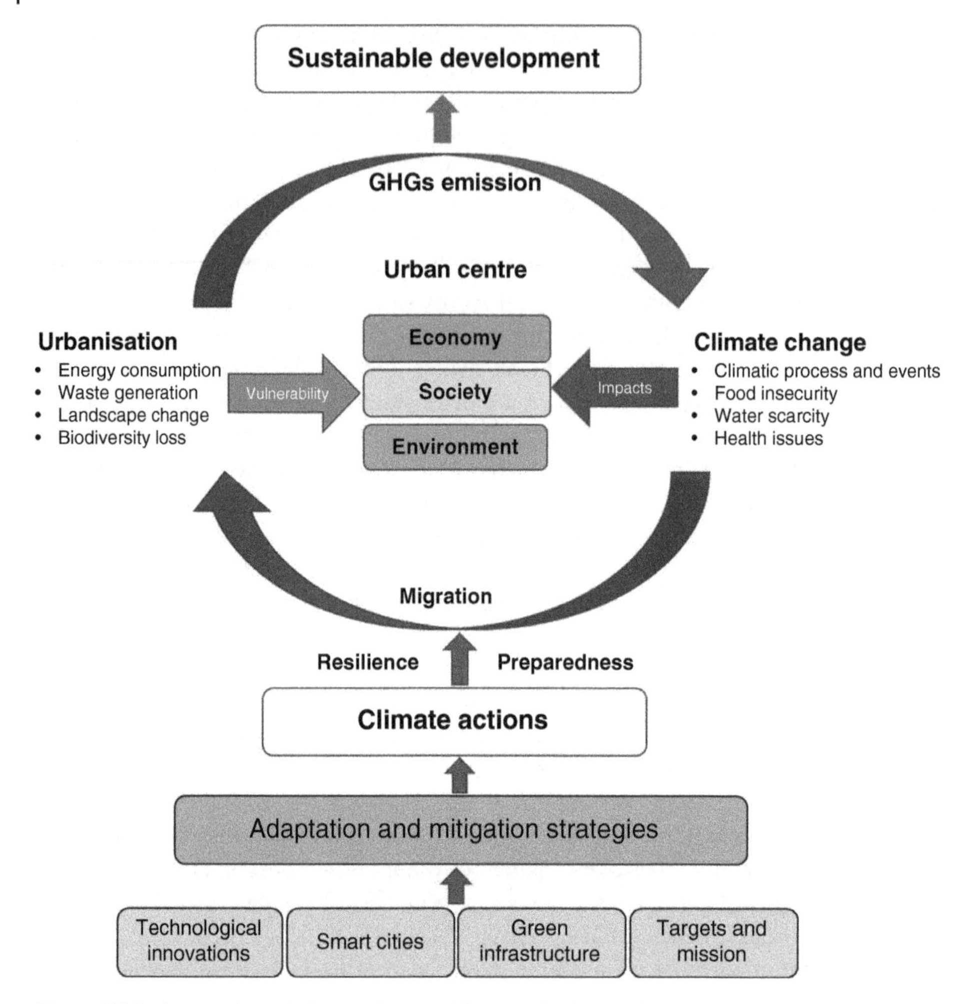

Figure 15.1 Interaction of climate change drivers and urbanisation process and its impact on urban centres.

15.3 Sustainable Development Goals (SDGs)

In order to develop sustainably, which focuses on meeting the current generation needs without compromising the ability of future generations to meet their needs, the United Nations (UN) adopted a set of guidelines comprising 17 goals with 169 targets known as SDGs. After the 2015 deadline of earlier guidelines which were millennium development goals (MDGs), the UN committed governments to create a set of SDGs to follow-up MDGs in Rio+20 summit in which protection and conservation of earth system, which comprises atmosphere, land, ocean, life, etc., and curtailment of poverty were the essence of the whole talk (Griggs et al. 2013).. The term sustainable development has come into consideration in World Charter for Nature (UN 1982) for the first time, and in 2012 summit document named 'The Future We Want' it was accepted completely. The SDGs definition may vary between and within societies, but ultimately it embraces the aim for economic development, environmental sustainability, and social inclusion; thus, this approach is named as triple bottom line (Sachs 2012). The changing present world, which is entering into a totally new geological epoch, is a consequence of fluctuating earth dynamics for which rise in various anthropogenic activities is the responsible factor. The global economic growth led by emerging economies with high rate of human activities occurs in order to fulfil the needs of the exploding population that is expected to reach eight million by 2024; therefore, both the factors are causing an unprecedented stress on the earth's ecosystem. In the present time, humanity is facing many overlapping crises that comprise climate change due to GHG emissions, massive environmental pollution, water pollution, unsustainable forests, changes in land use patterns, and depletion of fossil fuels resources at both global and local levels. Thus, to protect humanity, there is a need for an action plan which comes in the framework of 17 SDGs and is a prerequisite for the present era (Stafford-Smith et al. 2017).

Climate change is arising as one the most critical concerns because it causes serious threat to humankind, their survival, and sustainable growth by affecting various natural resources such as water, air, and soil. According to the projections of Intergovernmental Panel on Climate Change (IPCC), it is estimated that the global average temperature of the earth is rising at a pace which will lead to around 1.4–5.8 °C increase by 2100. Subsequently, due to this rise, the earth will get hotter and thus have severe impacts on water cycle, ecosystem diversity, oceanic biodiversity, food production, ecological balance, and other related processes. The main cause for the rise in temperature which ultimately results in global warming is considered to be the escalated GHG emissions which are the repercussions of human activities. Growth and development lead to exponential promotion of urbanisation, which is critical for sustainability as the processes related to urbanisation cause an exponential rise in CO_2 and other GHG levels. Hence, all of 17 SDGs play an interlinked, dynamic, and paramount role to make the process of urbanisation sustainable and beneficial for all components of an ecosystem, so as to maintain an ecological balance.

15.4 Sustainable Approaches to Combat Climate Change

The global warming and climate change is one of the most major challenges for the sustainable development, and thus, they act as biggest obstacle for SDGs' 2030 Agenda. Many serious attempts to fight the challenge of climate change have been taken, both at global and regional levels. Among them, some of the most pronounced actions are listed below (Sathaye et al. 2006; Nerini et al. 2019):

1) 1992 UN Conference on environment and development (UNCED) at Rio de Janeiro resulted in FCCC, i.e. Framework Convention on Climate Change.
2) 1997 Kyoto Protocol, which guided for stabilisation of GHGs concentration by reducing the emission of carbon dioxide, methane, nitrous oxide, chlorofluorocarbon, hydrocarbons, and perfluorocarbons.
3) 2015 Paris Agreement on Climate Change.
4) Recently, SDGs (2030 Agenda) adopted in 2015 under which SDGs 13 specifically framed to fight the climate change, thus named as Climate Action.

Nerini et al. (2019) found that the climate change could weaken the 72 targets of other 16 SDGs as it will influence the aim of attainment of prosperity and welfare, poverty eradication and employment, food, energy and water availability, and health. According to them, climate action can play a remarkable role to build prosperous, equal, and peaceful societies, both at global and local levels. Climate Action has interlinkages with achievement of SDGs. Due to synergistic existence, it contributes towards sustainable growth and development in a consolidated manner through improving integrated resource management as well as maintaining balance between various ecosystems. The gradual progress and population growth demonstrate a proportionate relationship with urbanisation and creation of many cities or urban colonies of varied sizes. Hence, it is evident that sustainable cities, human settlements, infrastructure, industrialisation, etc. will be the key players for climate change mitigation as well as for adaptation strategies. Besides, Sachs (2012) described the indispensable role of technological, innovative, and social ideas in achieving the targets set under the SDGs. However, every factor has its own role to play, but public participation and private–public partnerships are essential keys of the framework, which could be more impactful with various awareness campaigns, setting of short-term goals and targets, incentivised completions, and progresses (Stafford-Smith et al. 2017).

15.4.1 Sustainable Cities

It has been concluded from various studies and surveys that urbanisation is a major aspect for sustainable development and that the process of urbanisation could be a blessing if it is controlled, regulated, and planned, but in its contrast, it could prove to be curse for human society if it would become uncontrolled, unplanned, and unregulated (Mohan et al. 2011; Liu et al. 2011; Shen et al. 2012; Roders 2013). The UN 2010 report defined urbanisation as the mobilisation of people from rural to urban areas termed as urban migration is found to be the most predominant trend of the twenty-first century. The rapid increase in urbanisation occurs due to many attractive benefits such as jobs, education, market efficiency, improved health, better lifestyle, and raised living standards. But all these benefits come at the cost of air, food and water pollution, depletion of cultivated land, global warming, climate change, and others (Dye 2008; Liu et al. 2011; United Nations 2010).

The UN report (2010) estimated the rise in urban populations from 13% in 1900 to 29% in 1950 to 50% in 2009 and projected to be 69% in 2050. Therefore, making the cities or the urban centres more sustainable could be the best remedy in order to achieve SDGs as well as to fight against prime issues such as socioeconomic inequality, climate change, pollution, ecological imbalance, and degradation of nature and natural resources. The whole discussion will give rise to another term, i.e. sustainable urbanisation, which needs a holistic approach that includes broad perspectives of economic, social, political, demographic, and environmental performances. This brings another set of challenges, which includes disproportionate economic growth and nature management, inequity due to absence of political and social inclusion, flawed governance system, and lack of coordination for the development among various factors (Drakakis-Smith and Dixon 1997; Habitat/DFID 2002; Liu et al. 2011). There are six main principles for sustainable urban planning listed by Pivo (1996), i.e. compactness, completeness, conservation, comfort, coordination, and collaboration. Northam (1975) studies depicted urbanisation as a dynamic process with various stages expressed in the form of attenuated 'S' curve with three stages, i.e. *initial* stage (slow pace till the population reaches about 30%), *acceleration* stage (begins with conspicuous pace), and when the population is over 70%, it is the *terminal* stage (Shen et al. 2012). Based on the urbanisation rate and their occurrence stages, it could be categorised in four classes (Matthiessen 1980; Enyedi and Hungary 1990; Liu et al. 2011):

1) Concentrated urbanisation: The growth rate of urban population is very high, and the industrial development pacc is similar to that of the initial stage.
2) Suburbanisation: The growth rate occurs due to urban expansion and is found to be similar to the rate found in the outskirts of cities. This class is observed in the intermediate stage.
3) Counter urbanisation: It is the reverse mode of urbanisation where people start to emigrate from urban centres due to severe problems such as intolerance of urban climate, polluted food, and water.
4) Re-urbanisation: In contrast to counter-urbanisation, it takes place due to immigration of people from rural and subrural areas to urban areas due to special benefits of government policies. Both the counter-urbanisation and re-urbanisation take place in terminal stage.

There are diverse forms of methods used to measure urban sustainability such as using specific indicators, flow analysis model, and frameworks. Some of the common methods are Organisation for Economic Co-operation and Development (OECD) Pressure State Response Indicator Model, Pentagon Model, and Quantifiable City Model to determine urban sustainability using the society's responses to environmental changes, factors for implementing renewable energy technologies, and various perspectives, respectively (Oswald and McNeil 2010; Waheed et al. 2009; Nijkamp and Pepping 1998; May et al. 1997; Liu et al. 2011). Hence, based on various perspectives and contribution of various factors such as economic (S_E), social (S_S), and ecological (S_{EN}) towards urban sustainability, Liu et al. (2011) derived a modified formula to quantify the urban sustainability, which is

$$\mu_S = S_E(\text{urbanisation})U\ S_S(\text{urbanisation})U\ S_{EN}(\text{urbanisation})$$

It is concluded that urbanisation shows close relation with economic development of a city and thus with gross national product (GNP) per capita (Berry 1973; Kasarda and Crenshaw 1991; Scott and Storper 2003).

Sathaye et al. (2006) recommended some of the methods that can be implemented to carry urban development with sustainability, which may help to cut down the pollution extent with the reduction in greenhouses gases emissions, the leading cause of global warming and climate change. These recommendations are:

1) Innovation and implementation of green energy which involves the use of renewable energy sources such as wind and water for power production which will be cost effective and eco-friendly too.
2) Utilisation of economically and energy-efficient techniques from the production to distribution to end users for electricity generation in order to cut down losses also.

3) Conservation and protection of biodiversity by conserving their habitat through forest conservation, afforestation, reforestation, and sustainable forest management practices.
4) Implementation of fuel-efficient transportation medium or use of electric vehicles.
5) Building of efficient infrastructures in terms of energy production, power use, waste disposal, and emission reduction.
6) Proper waste management by utilising wastes such as biodegradables used for manure formation, and non-biodegradables recycled or reused.
7) Planned, regulated, and regularly scrutinised industries establishments.
8) Participation of people from ground to top levels by incentivising the progressive contribution at individual to societal and communal levels through various schemes.
9) Promoting the public–private partnerships for investments, infrastructure building, scheme promotions, and to spread awareness for innovative ideas, their uses, and merits.
10) Collaborations at various levels, i.e. regional, local, national, and global, to achieve the above targets in an integrated manner.

15.4.2 Urban Green Infrastructure

The concept of urban green infrastructure recently gained high momentum in the past two decades, being in the spotlight of the policymakers and researchers to achieve sustainability and SDGs. According to Mell (2008), it is found as connective matrices of green spaces established in and around urban and urban fringe landscapes and also be entitled as urban and peri-urban green space systems by Tzoulas et al. (2007). Some of the most common examples of urban green infrastructures are the parks, forests, roadside trees, green roofs, gardens, and cemeteries, which are critical factors of urban centres in the form of filters which purify the air, water, and soil by filtering out pollutants and their other ecosystem services. By their unique capabilities they regulate the microclimate in urban centres in a positive manner. Besides this, they occupy many values at aesthetic, social, cultural, ecological, educational, and economical levels (Breuste et al. 2015). Over the times, many scientists and policymakers proved their important contribution towards the improvement of atmosphere through various studies. Among them, European Commission (1996) described the green infrastructure as important as the other infrastructures for urbanisation, similar to as explained by Mazza and Rydin (1997) in the case studies of Great Britain and Italy. Further, some studies focused on green cover sizes and their importance in order to regulate microclimate and improve the air quality for betterment of urban living (Botkin and Beveridge 1997; Gómez et al. 2001; Sandström 2002).

To evaluate the multifunctional role of urban green spaces established according to 'green plans' in urban centres and for that analysis to be done on the basis of their social, ecological, cultural, and structural values, six criteria are proposed (Boverket 1992):

1) Recreation: related to availability, accessibility, and quality of green matrices.
2) Maintenance of biodiversity: depicts the protection and conservation of species.
3) City structure: based on urban lifestyle and various structures in the region.
4) Cultural identity: relates to promote the awareness towards history and traditions behind the city and its various components.

5) Environmental quality: indicates the improvements observed in climatic and atmospheric conditions of particular region after introducing the green infrastructures.
6) Biological solutions to technical problems: helps to analyse the interlink between technical and green infrastructure.

Researchers and policymakers can use these criteria to develop a variety of indicators to evaluate the importance and benefits of establishments of urban green infrastructures according to the plan proposed by them, as done by Sandström (2002) to evaluate green plans perceived in seven Sweden cities. Some policymakers evaluate the green plans and value of green structures on the basis of economic values; the approach is known as cost-benefit approach. It includes the costs of various inputs in their establishments and the economic benefits obtained from these establishments in terms of monitoring values. Thus, we enlisted some of the important economic benefits of green infrastructure from various studies done by individuals and agencies, which include climate change adaptation and mitigation, flood alleviation and water management, quality of place, health and well-being, land and property values, economic growth and investment, labour productivity, tourism, recreation and leisure, products from the lands, and land and biodiversity (Horwood 2011).

15.4.3 Technology and Innovations

To accomplish the targets and goals set under Agenda 2030, which focuses on achievement of SDGs action on a number of fronts, harnessing and maximising the potential of technological innovations are prerequisite. The technological innovations also need a proper route where various policymakers should have to set milestones for every field at varying levels, i.e. from local to global. The mid-twentieth century (also known as information age, computer age, digital age, or new media age) is characterised by a rapid epochal shift from the traditional industry established by the industrial revolution to an economy primarily based on information technology (IT). Therefore, the upsurge of the broad knowledge-based collaborative networking via information and communication technologies (ICTs) is an integrative part of various planning and policies (Pearce et al. 2012). Information sharing related to weather, climate, land uses, resources and minerals distribution, crops patterns, and valuation would also be beneficial. Financial tracking, healthcare, transportation services, disasters, and others are some of the most common examples of knowledge-based sharing network. Thus, in order to address the challenges of sustainable development, involvement of numerous agencies and organisations and their collaboration is imperative. But there are many obstacles which hinder the path of achievement of SDGs using ICTs, one of the most prominent obstacles being global inequalities in terms of socioeconomic, demography. However, knowledge and technology transfer from developed countries to developing countries is pivotal for accomplishments of various targets set under SDGs (Schwachula et al. 2014; Managi et al. 2021). In order to eliminate this hindrance, a technology facilitation mechanism (TFM) and a technology bank for least developed countries were the two mechanisms adopted in the 2030 Agenda of SDGs and Addis Ababa Action Agenda in the year 2015 in order to promote and facilitate the transfer of budget and long-lasting technological solutions to maintain a balance between development and ecological balance (Managi et al. 2021).

There are many studies which investigated the significant endowment of various technological methods to protect the environment and develop sustainably, which incorporates low or zero carbon and GHG emission, pollution control, and waste disposal and management (Brown and Southworth 2008). According to Shen et al. (2012), there are some of the advanced technologies (such as formulation of models) which also emerge as supplements to SDGs such as remote sensing, the cellular automata model, the SLEUTH model, and the system dynamics model. These models help to monitor the impact of urbanisation and land use changes. Thinkers around the world developed several types of urban planning models on the diverse grounds such as stages of urbanisation, sustainability of urban plans, and socio-economic benefits. 'Free market', 'redesigning', 'self-reliant', and 'fair shares' are the four types of city plans recommended by Haughton (1999), while Holden (2004) suggested 'urban sprawl', 'the green city', 'large (monolithic) compact city', and 'decentralised concentration' models. Furthermore, 'compact city', 'eco-city', 'urban containment', and 'neo-traditional development' are four more model concepts given by Jabareen (2006). Shen et al. (2012) also enlisted some of the important management strategies prosecuted by various representatives to achieve sustainable development such as Building Research Establishment Environment Assessment Method (BREEAM) established in the United Kingdom in 1990 and the Leadership in Energy and Environmental Design developed by the US Green Building Council in 1993. In addition to it, Green Star is a voluntary sustainability rating system for buildings launched in Australia in 2003 (Green star 2003). There is one more concept of CSR, i.e. corporate social responsibility given by H.R. Brown in 1953 in his famous work on social responsibility and defined it as the 'Social Responsibilities of Businessman' (Bowen 1953). Later, CSR emerged as an international self-regulating business model that helps a company to be socially accountable to itself, its stakeholders, and the public (IPCC 2018).

15.4.4 Targets, Campaigns, and Missions

The SDGs 13 calls for an urgent action to combat change and its impacts. It is found that 'Climate Action' is intrinsically linked to all 16 of the other Goals of the 2030 Agenda for Sustainable Development (Hák et al. 2016). According to the United Nations, the year 2019 was the second warmest year on record and the end of the warmest decade (2010–2019) ever recorded. It also coincided with high carbon dioxide and other GHG levels in the atmosphere. Climate change is disrupting national economies and affecting lives through influencing the weather patterns, sea level rise, and other extreme weather events. The UN and IPCC have given some facts and figures which can be used by various agencies, organisations, and governance system in order to evaluate the measures and also to formulate new plans, missions, and policies. Some are listed below:

1) Rise in average global temperature by 0.85 °C between 1880–2012 and 2017 till it became around 1 °C above than the pre-industrial era.
2) Global warming causes rise in the temperature of oceans, and the amounts of snow and ice melting cause the rise in sea levels of about 20 cm (8 inches) since 1880 and are projected to rise further to 30–122 cm (1–4 feet) by 2100.
3) Global emissions of carbon dioxide (CO_2) have increased by almost 50% since 1990. In order to limit the warming to 1.5 °C, global net CO_2 emissions must drop by 45% between 2010 and 2030 and reach net zero emissions around 2050.

4) Climate pledges under The Paris Agreement cover only one-third of the emission reductions needed to keep the world below 2 °C.
5) In terms of monitory values also, the bold actions against climate change could help to save approximately US$26 trillion by 2030.
6) The achievement of sustainable energy is the integrative part of SDG accomplishments which can create almost 18 million more jobs by 2030.

The climate action SDGs comprise five targets, the first three are the 'output targets', and remaining two are 'means of achieving targets' (https://www.undp.org/content/undp/en/home/sustainable-development-goals/goal-13-climate-action.html; https://www.un.org/sustainabledevelopment/climate-change/):

13.1 Strengthen resilience and adaptive capability to climate-related disasters
13.2 Integrate climate change measures into policies and planning
13.3 Build knowledge and capacity to meet climate change by awareness raising
  13.A. Implement the UN Framework Convention on Climate Change (UNFCCC)
  13.B. Promote mechanisms to raise capacity for effective planning and management.

Globally, to fight climate change several governments and organisations pledged to work collaboratively for which innumerable accords and agreements got signed as per the time needed. Some of the most admired accords include Montreal Protocol, Kigali Agreement, UN Climate Change Conference, Paris Agreement, and many more described in Table 15.1. India itself launched a well-pronounced national-level action plan to combat and mitigate climate change which is known as 'National Action Plan on Climate Change (NAPCC)'. It was launched by the Government of India on 30 June 2008 to have a combined policy to tackle climate issue. In order to meet the targets of NAPCC, eight missions (Table 15.2) were initiated by the Government with the collaborative support of respective ministries, inter-sectoral groups, commissions, experts from various sectors, academia, and civil society.

15.5 Case Studies

The expansion of urban population and their ecological footprint can greatly contribute to global warming and are most vulnerable to forcing of climate change. Nevertheless, cities possess potential to mitigate and adapt to various threats posed by climate change. Over the past few years, the concept of sustainable urban development to build resilience to climate change is globally recognised by governments, policymakers, and scientific communities. Many large to small cities are already commencing the strategies to develop capacity to combat the environmental disturbances and to adopt more sustainable alternative urbanisation pathways. Curitiba, Brazil, resolved the challenge of uncontrolled urban sprawl and associated problems such as traffic congestion, unemployment, and lack of basic services through transit-oriented development. The City improved the quality of life of the residence, while reducing its carbon footprint by integrating its land-use and transportation planning. Additionally, emphasis was given on developing waste management system, greener landscape, and raising environment consciousness among citizens (Soltani and Sharifi 2012).

Table 15.1 Global initiatives set to fight climate change.

S.no.	Accord/Protocol	Year	Place	Number of parties	Target
1	Montreal protocol	1987	Montreal, Canada	197	To eliminate major ozone-depleting substances
2	Kyoto protocol	1997	Kyoto, Japan	192	It is legally binding to reduce GHG emissions
3	National Action Plan on Climate Change (NAPCC)	2008	India	Central and state governments of India	To combat and mitigate climate change
4	United Nations Climate Change Conference (UNFCCC)	2009	Copenhagen, Denmark	183	To cap the global temperature rise below 2 °C
5	Nagoya protocol	2010	Nagoya, Japan	128	Legal framework for the implementation of convention on biological diversity
6	Cancun agreements	2010	Cancun, Mexico	197	To reduce GHG emissions and protect against climate change
7	Durban climate change conference	2011	Durban, South Africa	197	Second largest meeting for review earlier plans
8	Rio+20 earth summit	2012	Rio de Janeiro, Brazil	172	To develop ‘Rio Declaration on Environment and Development’, ‘Agenda 21’, and ‘Forest Principles’
9	Paris agreement	2015	Paris, France	195	To avoid dangerous climate change by limiting global warming to $<2\,^{\circ}C$
10	Kigali amendment	2016	Kigali, Rwanda	197	To reduce the manufacture and use of hydrofluorocarbons (HFCs)

Source: Data from Climate Change, Vikaspedia. Centre for Development of Advanced Computing. https://vikaspedia.in/energy/policy-support/environment-1/climate-change. Accessed 30 August 2021.

Table 15.2 Various missions set under the umbrella of National Action Plan on Climate Change (NAPCC) by India.

S.no.	Mission	Launched	Target
(i)	National Solar Mission (NSM):	2010	Set a target of 20 000 MW of grid-connected solar power by 2022 but revised in June 2015 as 1 00 000 MW by 2022
(ii)	National Mission on Sustainable Habitat (NMSH):	2010	Aim to make habitats more sustainable improvements in energy efficiency of buildings
			Management of municipal solid waste and to promote urban public transport
(iii)	National Mission for Enhanced Energy Efficiency (NMEEE):	2011	To strengthen the market for energy efficiency by creating favourable policies and regulations
			To promote the innovations to make affordable and energy-efficient appliances/products in certain sectors
(iv)	National Water Mission (NWM)	2011	Aims at conserving waste and ensuring more equitable distribution
			To develop a framework which helps to increase the water use efficiency by 20%
(v)	National Mission for Sustainable Agriculture (NMSA)	2012	To make agriculture more productive by promoting integrated or composite farming systems
			To spread awareness regarding soil and moisture conservation measures
(vi)	Green India Mission (GIM)	2014	To protect, preserve, and enhance India's diminishing forest cover and ecosystem services
			Target to increase the vegetation cover from 23% to 33% by afforestation
(vii)	National Mission for Sustaining Himalayan Ecosystem (NMSHE)	2014	To plan calls for empowering local communities
			To develop a sustainable national capacity to assess the health status of Himalayan ecosystem
(viii)	National Mission for Strategic Knowledge on Climate Change (NMSKCC)	2014	To work in collaborations with global communities in research and technology development
			To encourage private sector initiatives for developing innovative technologies for adaptation and mitigation

Source: Data from Climate Change, Vikaspedia. Centre for Development of Advanced Computing. https://vikaspedia.in/energy/policy-support/environment-1/climate-change. Accessed 30 August 2021.

Fredrikstad, Norway municipality, mapped the city vulnerability to future climate change with the technical assistance of the Community Adaptation and Vulnerability (NORADAPT) project to develop climate adaptation strategy. Vulnerability assessment reviewed the local susceptibility under present climate conditions, societal characteristics of vulnerability, and possible solutions to future challenges (Urban Nexus 2012).

City of Toronto has undertaken several climate actions towards building sustainable city. Incentives such as Toronto atmospheric fund, transform TO, and waste management strategies aim to cut off emission and reduce waste, while programmes such as eco-green roof are being implemented to develop city resilience to climate forcings. To adapt hot climate, many community agencies in Toronto developed heat alert and response system by integrating health data with social and geographic data to provide health-related services (https://www.c40.org/cities/toronto/case_studies).

The Government of Tokyo focussed on city planning measures such as increased urban plantation and heat absorbing pavement to cope with the effect of urban heat island (Rosenzweig et al. 2011). Faenza, Italy, included incentive programmes such as bio-neighbourhood in their town planning, which permits extension of buildings only if it meets certain environment standards. Likewise, in Berlin, several urban planning regulations such as addition of some green urban space have been incorporated. Cities such as Basel and Chicago are encouraging the inclusion of environment-friendly structures such as green roof in their regulations for building construction. These actions will not only be economic but also help in conserving biodiversity. Under climate change-adapting strategy, measures for wetland restoration and conservation were taken by the State of Louisiana and the City of New Orleans to protect from flood condition due to rise in sea level. Toronto, Canada, implemented several long- and short-term actions since 2008 to reduce the flooding risk (Carbonell and Meffert 2009). Fort Collin city of Colorado developed a resilient strategy to manage the 500-year floodplain area without damaging the neighbouring infrastructure of the city (https://toolkit.climate.gov/case-studies/building-smart-floodplain).

The Ministry of Urban Development in India initiated national scheme in 2005, i.e. Jawaharlal Nehru National Urban Renewal Mission (JNNURM) covering 65 important cities, to monitor climate mitigation risk assessment and management. It involves actions such as infrastructure improvement and poverty eradication (Revi 2008). The Municipal Corporation of Greater Mumbai (MCGM) in association with IIT-Mumbai built up systems for collecting climate-related data and awareness among municipality officials to combat flood-related risk faced by the city (Sharma and Tomar 2010).

The Regional Office for Asia and Pacific (ROAP) of UN Habitat in Fukuoka, Japan, in collaboration with the local institutes is providing low-cost, environmental technologies to the developing countries. One such solution is the Fukuoka method for managing urban solid waste which is being used by various cities such as Kiambu, Kenya and Yangon, Myanmar. Another example is the underground tank for rainwater harvesting, named Tametotto. Several countries such as Vietnam and Laos have deployed this sustainable solution to the issue of water crisis (UN Habitat 2021).

The Blue Lake Rancheria tribe in Humboldt County, California, has reduced its energy consumption and carbon foot print by employing several measures such as utilisation of renewable energy sources, installing microgrid, and switching to green transportation system. Since 2014, the tribe has cut off energy consumption by 40% (https://toolkit.climate.gov/case-studies/blue-lake-rancheria-tribe-undertakes-innovative-action-reduce-causes-climate-change). In Dar se Salaam, Tanzania, the governing authorities implemented several programmes to improve the existing city infrastructure and provide better public services, i.e. the UN–Habitat Citywide Action Plan, the Community Infrastructure Upgrading Programme (CIUP), the Water and Sanitation Improvement Programme, and the Citywide Action Plan (Kiunsi 2013).

15.6 Conclusion and Future Perspectives

Urban population and urban areas are growing at unprecedented rate, and thereby, the underlying pressure on ecosystem to support its inhabitant. Rigorous anthropogenic activities such as landscape transformation, energy consumption, and waste generation have increased the GHG emission immensely, leading to aggravation of environmental crisis such as extreme climatic process and events such as surface warming, altered precipitation pattern, rise in sea level, storms, and flooding. Subsequently, changing climate has significant influence on urban infrastructure, community, and human well-being. Nearly all corners of world are facing the adverse impact of climate change. However, the developing countries of Africa, Asia, and South America are particularly at risk, as they have limited resources to deal with these issues. Disproportionate distribution of risk and threats of climate change is driving urban rural migration at mass scale. Convening of urbanisation process and climate change impact implies dual stress on urban centres. Thus, it is becoming increasingly important to comprehend the real time situation and respond accordingly. Assessing and projecting these processes will facilitate the preparedness and resilience of cities towards climate change in long term. Notwithstanding the impact of urban sprawl on environment and their vulnerability to challenges posed by climatic forces, cities are central to climate adaptation and mitigation actions and undoubtedly key to sustainable development. As targeted by SDGs 11 resilient, safe, inclusive, and sustainable cities present apparent opportunities to alleviate the ecological, social, and economic crisis as well as mitigate the climate change the world is likely to face in the recent future. For this, several multispectral approaches have been adopted worldwide at national, regional, and local levels. International organisations such as UNEP, United Nations Department of Economic and Social Affairs (UN DESA), and IPCC have formulated numerous programmes to aid nations in their climate actions. Large numbers of cities are adopting sustainable incentives to develop resilience and resource utilisation efficiency, while improving the quality of life. Along with city management and planning and raising environment consciousness, technological innovations are being explored to reduce GHG emission, analyse the climate change risk and vulnerability, and to develop measures to alleviate it. However, these efforts are limited by lack of proper infrastructure and administrative forces. We need more inclusive and dynamic approach involving participation of local bodies, stakeholder, and institutions to develop more flexible and empirical mitigation measures. On the basis of our knowledge, the following recommendations are proposed:

1) Advancement in research to better understand the interaction of urbanisation and environmental issues to take apparent actions towards sustainable urbanisation.
2) Continuous monitoring and assessment of actions and policies implemented to the goal of sustainable development.
3) To promote collaborations, participation of peoples from different sectors and disciplines to have inclusive solution for environmental issues that are socioeconomically sound as well.
4) Innovations and technology advancements should be encouraged to develop new measures that are more feasible and have wide implication.
5) Review of successful case studies to set examples and to learn lessons.

Given that urban centres concentrate affluence and population and are integrally linked to environmental changes. Conception of greener and sustainable cities will not only reduce the threat of climate change but can as well promote environmental sustainability.

References

Aboulnaga, M.M., Elwan, A.F., and Elsharouny, M.R. (2019). *Urban Climate Change Adaptation in Developing Countries*. Springer https://doi.org/10.1007/978-3-030-05405-2.

Akbari, H. (2005). Energy saving potentials and air quality benefits of urban heat island mitigation. United States. https://www.osti.gov/servlets/purl/860475 (accessed 20 February 2021).

Barnett, J. and Webber, M. (2010). Accommodating migration to promote adaptation to climate change. *World Bank Policy Research Working Paper*. (5270). https://doi.org/10.1596/1813-9450-5270.

Berry, B.J.L. (1973). *The Human Consequences of Urbanization: Divergent Paths in the Urban Experience of the Twentieth Century*. New York: St. Martin's Press https://doi.org/10.1007/978-1-349-86193-4_5.

Biermann, F. and Boas, I. (2010). Preparing for a warmer world: towards a global governance system to protect climate refugees. *Global Environmental Politics* 10 (1): 60–88. https://doi.org/10.1162/glep.2010.10.1.60.

Botkin, D.B. and Beveridge, C.E. (1997). Cities as environment. *Urban Ecosystem* 1: 3–19. https://doi.org/10.1023/A:1014354923367.

Boverket, S. (1992). *Storstadsuppdraget: En Forstudie om Storstadernas Miljo (A Preliminary Study of the Environment in Big Cities)*. Karlskrona: The National Board of Housing, Building and Planning.

Bowen, H.R. (1953). *Social Responsibilities of Businessman*. New York: Harper & Row https://doi.org/10.2307/j.ctt20q1w8f.

Breuste, J., Artmann, M., Li, J., and Xie, M. (2015). Special issue on green infrastructure for urban sustainability. *Journal of Urban Planning and Development* 141 (3): A2015001. https://doi.org/10.1080/02697450216356.

Brown, M.A. and Southworth, F. (2008). Mitigating climate change through green buildings and smart growth. *Environment and Planning A* 40 (3): 653–675. https://doi.org/10.1068/a38419.

Carbonell, A. and Meffert, D.J. (2009). Climate change and the resilience of New Orleans: the adaptation of deltaic urban form. Cambridge, MA. http://deltacityofthefuture.com/documents/Neworleans_Session3_Meffert.pdf (accessed 20 February 2021).

Carter, J.G., Cavan, G., Connelly, A. et al. (2015). Climate change and the city: building capacity for urban adaptation. *Progress in Planning* 95: 1–66. https://doi.org/10.1016/j.progress.2013.08.001.

Chan, N.W. (2017). Urbanization, climate change and cities: challenges and opportunities for sustainable development. *Asia-Pacific Chemical, Biological & Environmental Engineering Society (APCBEES) International Conference*, Universiti Sains Malaysia, Penang, Malaysia (9 January). https://www.researchgate.net/

publication/312281988_urbanization_climate_change_and_cities_challenges_and_opportunities_for_sustainable_development (accessed 20 February 2021).
de Coninck, H., Revi, A., Babiker, M. et al. (2018). Chapter 4: Strengthening and implementing the global response. In: *Global Warming of 1.5°C: Summary for Policy Makers*, 313–443. IPCC - The Intergovernmental Panel on Climate Change https://research.rug.nl/en/publications/strengthening-and-implementing-the-global-response.
DePaul, M. (2012). Climate change, migration, and megacities: addressing the dual stresses of mass urbanization and climate vulnerability. *Paterson Review of International Affairs* 12: 145–162. http://diplomatonline.com/mag/pdf/PatersonReview_Vol12_Full.pdf#page=153.
Drakakis-Smith, D. and Dixon, C. (1997). Sustainable urbanization in Vietnam. *Geoforum* 28 (1): 21–38. https://doi.org/10.1016/S0016-7185(97)85525-X.
Dye, C. (2008). Health and urban living. *Science* 319: 766–769. https://doi.org/10.1126/science.1150198.
Enyedi, G. and Hungary, B. (1990). Specific urbanization in East-Central Europe. *Geoforum* 21 (2): 163–172. https://doi.org/10.1016/0016-7185(90)90035-5.
European Commission (1996). *European Sustainable Cities*. Luxembourg: Directorate General XI https://edz.bib.uni-mannheim.de/www-edz/pdf/sonstige/sustcities.pdf.
Filho, W.L., Balogun, A.L., Olayide, O.E. et al. (2019). Assessing the impacts of climate change in cities and their adaptive capacity: towards transformative approaches to climate change adaptation and poverty reduction in urban areas in a set of developing countries. *Science of the Total Environment* 692: 1175–1190. https://doi.org/10.1016/j.scitotenv.2019.07.227.
Foley, J.A., DeFries, R., Asner, G.P. et al. (2005). Global consequences of land use. *Science* 309 (5734): 570–574. https://doi.org/10.1126/science.1111772.
Godfray, H., Charles, J., John, R. et al. (2010). Food security: the challenge of feeding 9 billion people. *Science* 327 (5967): 812–818. https://doi.org/10.1126/science.1185383.
Gómez, F., Tamarit, N., and Jabaloyes, J. (2001). Green zones, bioclimatic studies and human comfort in the future development of urban planning. *Landscape and Urban Planning* 55: 151–161. https://doi.org/10.1016/S0169-2046(01)00150-5.
Green Star (2003). Green star setting the standard. www.greenstar.ie (accessed 19 February 2021).
Griggs, D., Stafford-Smith, M., Gaffney, O. et al. (2013). Sustainable development goals for people and planet. *Nature* 495 (7441): 305–307. https://doi.org/10.1038/495305a.
Grimmond, S.U. (2007). Urbanization and global environmental change: local effects of urban warming. *Geographical Journal* 173 (1): 83–88. https://doi.org/10.1111/j.1475-4959.2007.232_3.x.
Hák, T., Janoušková, S., and Moldan, B. (2016). Sustainable development goals: a need for relevant indicators. *Ecological Indicators* 60: 565–573. https://doi.org/10.1016/j.ecolind.2015.08.003.
Haughton, G. (1999). Searching for the sustainable city. *Urban Studies* 36 (11): 1891–1906. https://doi.org/10.1080/0042098992665.
Holden, E. (2004). Ecological footprints and sustainable form. *Journal of Housing and the Built Environment* 19 (1): 91–209. https://doi.org/10.1023/B:JOHO.0000017708.98013.cb.
Horwood, K. (2011). Green infrastructure: reconciling urban green space and regional economic development: lessons learnt from experience in England's north-west region. *Local Environment* 16 (10): 963–975. https://doi.org/10.1080/13549839.2011.607157.

IPCC (Intergovernmental Panel on Climate Change) (2007). Summary for policymakers. In: *Climate Change 2007: Impacts, Adaptation and Vulnerability. Contribution of Working Group II to the Fourth Assessment Report of the Intergovernmental Panel on Climate Change* (eds. M.L. Parry, O.F. Canziani, J.P. Palutikof, et al.), 7–22. Cambridge, UK: Cambridge University Press https://www.ipcc.ch/site/assets/uploads/2018/03/ar4_wg2_full_report.pdf.

IPCC (Intergovernmental Panel on Climate Change) (2018). *Global Warming of 1.5°C. An IPCC Special Report on the Impacts of Global Warming of 1.5°C Above Pre-industrial Levels and Related Global Greenhouse Gas Emission Pathways, in the Context of Strengthening the Global Response to the Threat of Climate Change, Sustainable Development, and Efforts to Eradicate Poverty* (eds. V.P. Masson-Delmotte, P. Zhai, H.-O. Pörtner, et al.) In Press. (2005). Corporate social responsibility and social sustainability: a role for local government. Master thesis. Simon Fraser University https://www.ipcc.ch/site/assets/uploads/sites/2/2019/06/SR15_Full_Report_High_Res.pdf.

Jabareen, Y.R. (2006). Sustainable urban forms: their typologies, models, and concepts. *Journal of Planning Education and Research* 26 (1): 38–52. https://doi.org/10.1177/0739456X05285119.

Jaramillo, J., Setamou, M., Muchugu, E. et al. (2013). Climate change or urbanization? Impacts on a traditional coffee production system in East Africa over the last 80 years. *PLoS One* 8 (1): 51815. https://doi.org/10.1371/journal.pone.0051815.

Kalnay, E. and Cai, M. (2003). Impact of urbanization and land-use change on climate. *Nature* 423 (6939): 528–531. https://doi.org/10.1038/nature01675.

Kasarda, J.D. and Crenshaw, E.M. (1991). Third world urbanization: dimensions, theories, and determinants. *Annual Review of Sociology* 17: 467–501. https://doi.org/10.1146/annurev.so.17.080191.002343.

Kiunsi, R. (2013). The constraints on climate change adaptation in a city with a large development deficit: the case of Dar es Salaam. *Environment and Urbanization* 25 (2): 321–337. https://doi.org/10.1177/0956247813489617.

Krey, V., O'Neill, B.C., van Ruijven, B. et al. (2012). Urban and rural energy use and carbon dioxide emissions in Asia. *Energy Economics* 34: S272–S283. https://doi.org/10.1016/j.eneco.2012.04.013.

Li, X.X., Koh, T.Y., Panda, J. et al. (2016). Impact of urbanization patterns on the local climate of a tropical city, Singapore: an ensemble study. *Journal of Geophysical Research – Atmospheres* 121 (9): 4386–4403. https://doi.org/10.1002/2015JD024452.

Liu, Y., Yao, C., Wang, G., and Bao, S. (2011). An integrated sustainable development approach to modeling the eco-environmental effects from urbanization. *Ecological Indicators* 11 (6): 1599–1608. https://doi.org/10.1016/j.ecolind.2011.04.004.

Managi, S., Lindner, R., and Stevens, C.C. (2021). Technology policy for the sustainable development goals: from the global to the local level. *Technological Forecasting and Social Change* 162: 120410. https://doi.org/10.1016/j.techfore.2020.120410.

Matthiessen, C.W. (1980). Trends in the urbanization process: the Copenhagen case. *Geografisk Tidsskrift-Danish Journal of Geography* 80 (1): 98–101. https://doi.org/10.1080/00167223.1980.10649123.

May, A.D., Mitchell, G., and Kupiszewska, D. (1997). The development of the leeds quantifiable city model. In: *Evaluation of the Built Environment for Sustainability* (eds. P.S. Brandon,

P.L. Lombardi and V. Bentivegna), 39–52. London: E & FN Spon https://doi.org/10.4324/9780203362426.
Mazza, L. and Rydin, Y. (1997). Urban sustainability: discourses, networks and policy tools. *Progress in Planning* 47: 1–74. https://doi.org/10.1016/S0305-9006(96)00006-2.
McDonald, R.I., Green, P., Balk, D. et al. (2011). Urban growth, climate change, and freshwater availability. *Proceedings of the National Academy of Sciences* 108 (15): 6312–6317. https://doi.org/10.1073/pnas.1011615108.
McGranahan, G., Balk, D., and Anderson, B. (2007). The rising tide: assessing the risks of climate change and human settlements in low elevation coastal zones. *Environment and Urbanization* 19 (1): 17–37. https://doi.org/10.1177/0956247807076960.
Mell, I.C. (2008). Green infrastructure: concepts and planning. *FORUM International Journal of Postgraduate Studies in Architecture, Planning and Landscape* 8 (1): 69–80. https://doi.org/10.4337/9781783474004.00013.
Min, S.K., Zhang, X., Zwiers, F.W. et al. (2011). Human contribution to more-intense precipitation extremes. *Nature* 470 (7334): 378–381. https://doi.org/10.1038/nature09763.
Mitlin, D. and Satterthwaite, D. (2013). *Urban Poverty in the Global South: Scale and Nature.* New York and London: Routledge https://www.routledge.com/Urban-Poverty-in-the-Global-South-Scale-and-Nature/Mitlin-Satterthwaite/p/book/9780415624671.
Mohan, M., Pathan, S.K., Narendrareddy, K. et al. (2011). Dynamics of urbanization and its impact on land-use/land-cover: a case study of megacity Delhi. *Journal of Environmental Protection* 2 (09): 1274. https://doi.org/10.1038/s41893-019-0334-y.
Nerini, F.F., Sovacool, B., Hughes, N. et al. (2019). Connecting climate action with other sustainable development goals. *Nature Sustainability* 2 (8): 674–680. https://doi.org/10.1038/s41893-019-0334-y.
Nijkamp, P. and Pepping, G. (1998). A meta-analytical evaluation of sustainable city initiatives. *Urban Studies* 35 (9): 1481–1500. https://doi.org/10.1080/0042098984240.
Northam, R.M. (1975). *Urban geography*. New York: Wiley.
O'Neill, B.C., Dalton, M., Fuchs, R. et al. (2010). Global demographic trends and future carbon emissions. *Proceedings of the National Academy of Sciences* 107 (41): 17521–17526. https://doi.org/10.1073/pnas.1004581107.
O'Neill, B.C., Liddle, B., Jiang, L. et al. (2012). Demographic change and carbon dioxide emissions. *The Lancet* 380 (9837): 157–164. https://doi.org/10.1016/S0140-6736(12)60958-1.
Oswald, M.R. and McNeil, S. (2010). Rating sustainability: transportation investments in urban corridors as a case study. *Journal of Urban Planning and Development* 136 (3): 177–185. https://doi.org/10.1061/(ASCE)UP.1943-5444.0000016.
Pall, P., Aina, T., Stone, D.A. et al. (2011). Anthropogenic greenhouse gas contribution to flood risk in England and Wales in autumn 2000. *Nature* 470 (7334): 382–385. https://doi.org/10.1038/nature09762.
Pearce, J., Albritton, S., Grant, G. et al. (2012). A new model for enabling innovation in appropriate technology for sustainable development. *Sustainability: Science, Practice and Policy* 8 (2): 42–53. https://doi.org/10.1080/15487733.2012.11908095.
Pivo, G. (1996). Toward sustainable urbanization on Mainstreet Cascadia. *Cities* 13 (5): 339–354. https://doi.org/10.1016/0264-2751(96)00021-2.
Revi, A. (2008). Climate change risk: an adaptation and mitigation agenda for Indian cities. *Environment and Urbanization* 20 (1): 207–229. https://doi.org/10.1177/0956247808089157.

Revi, A., Satterthwaite, D., Aragón-Durand, F. et al. (2014). Towards transformative adaptation in cities: the IPCC's fifth assessment. *Environment and Urbanization* 26 (1): 11–28. https://doi.org/10.1177/0956247814523539.

Roders, A.P. (2013). How can urbanization be sustainable? A reflection on the role of city resources in global sustainable development. *BDC. Bollettino Del Centro Calza Bini* 13 (1): 79–90. https://doi.org/10.6092/2284-4732/2452.

Rosenzweig, C., Solecki, W.D., Hammer, S.A., and Mehrotra, S. (2011). *Climate Change and Cities: First Assessment Report of the Urban Climate Change Research Network*. Cambridge University Press https://www.researchgate.net/profile/Roberto-Sanchez-Rodriguez/publication/285117257_Climate_change_water_and_wastewater_in_cities/links/585caf3408ae8fce48fad58d/Climate-change-water-and-wastewater-in-cities.pdf.

Rosenzweig, C., Solecki, W., and Romero-Lankao, P.(ed. et al.) (2015). *ARC3.2 Summary for City Leaders. Urban Climate Change Research Network*. New York: Columbia University https://pubs.giss.nasa.gov/docs/2015/2015_Rosenzweig_ro02510w.pdf.

Sachs, J.D. (2012). From millennium development goals to sustainable development goals. *The Lancet* 379 (9832): 2206–2211. https://doi.org/10.1016/S0140-6736(12)60685-0.

Sandström, U.G. (2002). Green infrastructure planning in urban Sweden. *Planning Practice and Research* 17 (4): 373–385. https://doi.org/10.1080/02697450216356.

Sathaye, J., Shukla, P.R., and Ravindranath, N.H. (2006). Climate change, sustainable development and India: global and national concerns. *Current Science* 90 (3): 314–325. http://www.jstor.org/stable/24091865.

Schwachula, A., Vila Seoane, M., and Hornidge, A.K. (2014). Science, technology and innovation in the context of development: an overview of concepts and corresponding policies recommended by International Organisations (ZEF Working Paper Series No. 132 Center for Development Research (ZEF), University of Bonn. https://www.zef.de/fileadmin/user_upload/zef_wp_132.pdf (accessed 20 February 2021).

Scott, A.J. and Storper, M. (2003). Regions, globalization, development. *Regional Studies* 37 (6&7): 579–593. https://doi.org/10.3166/ges.8.169-192.

Seto, K.C. and Satterthwaite, D. (2010). Interactions between urbanization and global environmental change. *Current Opinion in Environmental Sustainability* 2: 127–128. https://doi.org/10.1016/j.cosust.2010.07.003.

Shankar, S. (2018). Impacts of climate change on agriculture and food security. In: *Biotechnology for Sustainable Agriculture* (eds. R.L. Singh and S. Mondal), 207–234. Woodhead Publishing https://doi.org/10.1016/B978-0-12-812160-3.00007-6.

Sharma, D. and Tomar, S. (2010). Mainstreaming climate change adaptation in Indian cities. *Environment and Urbanization* 22 (2): 451–465. https://doi.org/10.1177/0956247810377390.

Shen, L., Peng, Y., Zhang, X. et al. (2012). An alternative model for evaluating sustainable urbanization. *Cities* 29 (1): 32–39. https://doi.org/10.1016/j.cities.2011.06.008.

Singh, B.R. and Singh, O. (2012). Study of impacts of global warming on climate change: rise in sea level and disaster frequency. In: *Global Warming—Impacts and Future Perspective* (ed. B.R. Singh). IntechOpen http://10.5772/2599.

Soltani, A. and Sharifi, E. (2012). A case study of sustainable urban planning principles in Curitiba (Brazil) and their applicability in Shiraz (Iran). *International Journal of Development and Sustainability* 1 (2): 120–134. http://hdl.handle.net/2440/114648.

Stafford-Smith, M., Griggs, D., Gaffney, O. et al. (2017). Integration: the key to implementing the sustainable development goals. *Sustainability Science* 12 (6): 911–919. https://doi.org/10.1007/s11625-016-0383-3.

Svirejeva-Hopkins, A., Schellnhuber, H.J., and Pomaz, V.L. (2004). Urbanised territories as a specific component of the global carbon cycle. *Ecological Modelling* 173 (2–3): 295–312. https://doi.org/10.1016/j.ecolmodel.2003.09.022.

Tzoulas, K., Korpela, K., Venn, S. et al. (2007). Promoting ecosystem and human health in urban areas using green infrastructure: a literature review. *Landscape and Urban Planning* 81 (3): 167–178. https://doi.org/10.1016/j.landurbplan.2007.02.001.

UN (United Nation) (1982). A World Charter for Nature . United Nations, New York.

UN (United Nations) (2010). World urbanization prospects, the 2009 version, United Nations, New York. (United Nations Population Fund, 2009. Counting the people: advocacy and resource mobilization for successful implementation of the 2010 round of population and housing censuses. United Nations Publications. https://www.ipcc.ch › njlite › njlite_download2 (accessed 20 February 2021).

UN (United Nations) (2018). SDG 11 synthesis report 2018: tracking progress towards inclusive, safe, resilient and sustainable cities and human settlements, Nairobi, Kenya. http://uis.unesco.org/sites/default/files/documents/sdg11-synthesis-report-2018-en.pdf (accessed 20 February 2021).

UN DESA (United Nations, Department of Economic and Social Affairs), Population Division (2015). World urbanization prospects: the 2014 revision. (ST/ESA/SER.A/366). https://population.un.org/wup/publications/files/wup2014-report.pdf (accessed 20 February 2021).

UN DESA (United Nations, Department of Economic and Social Affairs), Population Division (2019). World urbanization prospects 2018: highlights (ST/ESA/SER.A/421). https://population.un.org/wup/Publications/Files/WUP2018-Highlights.pdf (accessed 20 February 2021).

UN HABITAT (2011). Cities and climate change: global report on human settlements 2011. United Nations Human Settlements Programme. https://unhabitat.org/global-report-on-human-settlements-2011-cities-and-climate-change (accessed 20 February 2021).

UN HABITAT (2021). Contributing to SDGs through Japanese low-cost and sustainable environmental technologies. https://unhabitat.org/contributing-to-sdgs-through-japanese-low-cost-and-sustainable-environmental-technologies (accessed 18 February 2021).

UNEP (United Nation Environment Programme) (2012). Sustainable, resource efficient cities – making it happen! Nairobi, Kenya. https://sustainabledevelopment.un.org/content/documents/1124SustainableResourceEfficientCities.pdf (accessed 20 February 2021).

United Nations Human Settlement Programme/Department for International Development (UN-Habitat/DFID) (2002). Sustainable urbanisation: achieving agenda 21. UN-Habitat/DFID, Nairobi, Kenya. https://www.alnap.org/help-library/sustainable-urbanisation-achieving-agenda-21 (accessed 20 February 2021).

Urban Nexus (2012). Synthesis report: annex of case studies, urban climate resilience. https://climate-adapt.eea.europa.eu/metadata/publications/synthesis-report-annex-of-case-studies-urban-climate-resilience-1/11240570 (accessed 13 February 2021).

USGCRP (U.S. Global Change Research Program) (2009). *Global Climate Change Impacts in the United States* (eds. T.R. Karl, J.M. Melillo and T.C. Peterson). New York: Cambridge University Press https://www.nrc.gov/docs/ML1006/ML100601201.pdf.

Waheed, B., Khan, F., and Veitch, B. (2009). Linkage-based frameworks for sustainability assessment: making a case for driving force-pressure-stateexposure-effect-action (DPSEEA) frameworks. *Sustainability* 1 (3): 441–463. https://doi.org/10.3390/su1030441.

Warner, K. (2010). Global environmental change and migration: governance challenges. *Global Environmental Change* 20 (3): 402–413. https://doi.org/10.1016/j.gloenvcha.2009.12.001.

World Bank (2019). Resilient cities. https://www.worldbank.org/en/topic/urbandevelopment/brief/resilient-cities-program (accessed 18 February 2021).

Zhou, L., Dickinson, R.E., Tian, Y. et al. (2004). Evidence for a significant urbanization effect on climate in China. *Proceedings of the National Academy of Sciences* 101 (26): 9540–9544. https://doi.org/10.1073/pnas.0400357101.

Section 5

Climate Change and Threats to Ecological Conservation

16

Threats from Sea Level Rise and Erosion: A Case Study of An Estuarine Inhabited Island Ghoramara, Hooghly Estuary

Niloy Pramanick[1], Eyadul Islam[1], Subhasree Banerjee[1], Rohit Mukherjee[1], Arunashish Maity[1], Rituparna Acharyya[3], Abhra Chanda[1], Indrajit Pal[2], and Anirban Mukhopadhyay[2]

[1] *School of Oceanographic Studies, Jadavpur University, Kolkata, West Bengal, India*
[2] *Disaster Preparedness, Mitigation, and Management (DPMM), Asian Institute of Technology, Pathumthani, Bangkok, Thailand*
[3] *Department of Geography, School of Earth Science, Central University of Karnataka, Kalaburagi, Karnataka, India*

16.1 Introduction

Coastal degradation is the removal of materials from the coastal terrain. An imbalance between material supply and outsourcing in a specific coastline segment often leads to scouring at the base of the foreshore, which results in the loss of soil mass from the coastal fringe (Marchand 2010). Coastal erosion is among those physical processes, which strip away and redistribute solid elements of shoreline and sediment, typically through waves, tidal and coastal currents, and deflation (Central Water Commission 2016; Prasad and Kumar 2014). The erosion rates are expressed correctly in the dimensions of volume/length/time, e.g. in $m^3\ m^{-1}\ year^{-1}$; however, the erosion rate is also expressed in m/year as coastal damage (Mangor et al. 2017). Erosion occurs when the rate of sediment loss exceeds the supply rate, resulting in a landward displacement of the shoreline (Central Water Commission 2016).

Coastal erosion refers to the wearing of rock along the shore (National Geographic Society 2020; BBC 2020), primarily due to destructive waves. These waves tend to wash off the rock along the coast (Geological Survey Ireland 2020). Waves are the foremost determinant of the geometry and structure of the beaches (Central Water Commission River Management Wing 2003; Nehra 2016). Sea waves move towards shore after forming in the middle of the ocean (Lumen 2020). Waves carry enormous energy concentrations to the beach, which is dissipated by wave breaking, creating tides, changing water temperatures, and flowing sediment, turbulence, and heat (Mohr 2001). Erosion is attributed mainly to the following processes: hydraulic action, abrasion, attrition, and corrosion. The vigour of these processes depends on the condition of the coast, the shape of rocks, the occurrence

Urban Ecology and Global Climate Change, First Edition. Edited by Rahul Bhadouria, Shweta Upadhyay, Sachchidanand Tripathi, and Pardeep Singh.

of joints and cracks in the rock, marine chemical reaction, and wave strength (Reddy 2010; Vargas-T et al. 2016). There are different spatial and temporal patterns of coastal degradation or erosion (Cai et al. 2009; Farquharson et al. 2018). In spatial forms, the weaker coast is mainly affected by coastal retreats caused by sea-level (SL) rise (Committee on Climate Change 2018). The landward displacement of the 0-m deep contour is caused by the incision of the beach floor on the shore with a seawall (Cai et al. 2009). The tidal current then causes lower beach erosion in the semi-tidal zone, while the top flat retains its original structure (Cai et al. 2009). Erosion could also be partitioned temporarily, i.e. prolonged erosion of the shoreline due to rising SLs, diversion of waterways, or reduced sediment discharges (Cai et al. 2009; FitzGerald et al. 2008). Without permanent coastal improvements, but with severe disruption, short-term erosion may occur due to stormy tides and floods (Cai et al. 2009). It results from human activity and natural climate change, causing the loss of balance in coastal systems by dynamic actions (wave, sea, wind). The long-term degradation of the sediment contributes to beach erosion and retreat of the shore (Zhenye 2017). The extraction of sand from the sand distribution system results in irreversible changes in the shape and structure of the beach (Prasad and Kumar 2014).

The term 'coastal sediment processes' corresponds to the forces along shorelines that erode, move, and deposit sediment (Asif 2010). The coastal climate consists of continuously changing conditions caused by wind forces, waves, currents, and tides (Smithsonian 2020). Wave action regulates the removal and deposition of materials/sediments onshore, and coastal sand frequently migrates by accretion to the accessible section of the shoreline. Thus, the dual accretion and erosion processes play a significant role in determining coastal morphology (Kaliraj et al. 2015). Coastal erosion and accretion-induced shoreline changes are natural processes that occur on various time scales (Prasetya 2001). These processes are temporarily carried out in response to minor events such as hurricanes, waves, tides, winds, and large-scale events such as glaciation, orogenic cycles, or tectonic activity, which cause coastal lands to subside or arise. Beaches consist of sediments of different sizes, from large rocks to fine sand or mud, leading to the refraction of the waves, which causing them to bend (Columbia University in the city of New York 2020). However, when waves first strike at an angle at the shoreline, their energy causes a longshore current. When these waves break down and recede along the coast, they erode and accumulate sediment in a pattern known as longshore transport (National Ocean Service 2020).

The boundary between land and sea begins to change its form and position based on dynamic environmental factors (Muthukumarasamy et al. 2013). Shoreline refers to the dynamic boundary or line of interface between a landmass and a water body (Pajak and Leatherman 2002; Nandi et al. 2016). The dynamic nature of the shoreline makes it challenging to demarcate how much region is eroded or accredited (Dolan and Morton 2001). The transformation of the shores is attributed to numerous natural and anthropogenic processes (Kundu et al. 2014). The shoreline is subject to short and long-term recurrent changes due to hydrodynamic changes (for instance, river fluctuations, the rise of SL), geomorphological changes (for example, the formation of barrier islands and split), and other factors (for example, seismic shifts, and storm) (Scott 2005). Diverse development projects are carried out in shoreline areas that exert a high volume of strain, causing various coastal threats such as soil erosion, intruding seawater, coral bleaching, coastal changes, and so forth (Prasad and Kumar 2014; Senevirathna et al. 2018). When carbon dioxide and other greenhouse gasses are emitted increasingly into the atmosphere, global warming is expected to rise by around 3 °C by 2030 (Mcsweeny 2020). This rise will elevate the global SL by up to 5 m over the next several centuries, which is a short duration for human habitation on the shore (Oliver-Smith 2009).

Sustainable coastal engineering and management planning contains information regarding the current and past position of the coastline and where it is expected to be in the long term (Coastal Engineering Research Center 1984). In the layout of coastal protection, this knowledge is important for the validation and calibration of numerical models, the estimation of SL rise, and the establishment of danger zones (Oppenheimer and Glavovic 2019). Remote Sensing aims to supplement conservative survey results with less cost-effective and rhythmic data. Several studies employing satellite data have demonstrated its utility in investigating various coastal dynamics (Caballero and Stumpf 2020). Remote Sensing and geographic information system (GIS) have significantly contributed to the mapping and monitoring of shoreline transition (Bertacchini and Capra 2010; Abd El-Kawy et al. 2011; Kundu et al. 2014). Comprehensive and precise information on past and current shoreline positions is required for future shoreline projections (Nandi et al. 2016). Future prediction of shorelines is particularly significant as it demonstrates the transition rate in the area and is necessary for planning purposes (Kundu et al. 2014). In the GIS environment, shoreline prediction models can be run. Several techniques for shoreline projection have been evaluated using previous data, such as the end point rate (EPR) model (Fenster et al. 1993), Jackknife model (JK), linear regression (LR), and average of rates (AOR) (Dolan et al. 1991). Using the historical rate of change statistics, the EPR model primarily forecasts the future position of the shoreline. In contrast, the LR model, on the other hand, is primarily based on a stringent linear projection technique for short-term alteration while employing long-term shoreline statistics (Nandi et al. 2016). Several studies have utilised the EPR model to forecast changes in shoreline (Adarsa et al. 2012; Mukhopadhyay et al. 2012).

Recent developments in Remotely Sensed data and GIS techniques have led to improved coastal studies, including multi-temporal mapping, semi-automatic shoreline projections, relative coastal change detection, shoreline forecasts, topographical and bathymetric data processing, etc. Earlier studies indicated that the Sagar Island, located to the South of Ghoramara, had been subjected to erosion due to various mechanisms (Paul and Bandopadhyay 1987; Bandyopadhyay 2000). Ghosh et al. (2001) analysed the pattern of Sagar Island's geomorphological changes using remote satellite sensing technologies due to environmental and anthropogenic processes. Jayappa et al. (2006) demarcated numerous coastal geomorphological landforms related to coastal changes and recommended remedial steps to manage erosion in Sagar Island. Kundu et al. (2014) carried out a coastal mapping of the Sagar Island by geospatial methods during 1951–2011. Long-term (1975–2002) and short-term (2002–2011) erosion and development rates at Sagar Island were estimated for the study region by Nandi et al. (2016). Both of them have used the digital shoreline analysis system (DSAS) application of the United States Geological Survey (USGS) (Himmelstoss et al. 2018) for casting transects and calculating the change by using the net shoreline movement (NSM) method. The current study is being carried out in the Ghoramara Island of Indian Sundarbans, where different researchers have previously carried out several multidisciplinary studies. A study has been conducted by Ghosh et al.

(2003) to monitor the erosion and accretion processes in Ghoramara by time series analysis using GIS, along with the review of the remedial measures required to protect the island by using the 'bio-engineering technique'. The rate of change in the position of shoreline in Ghoramara Island has been calculated by Adarsa et al. (2012), using the statistical LR, EPR, and NSM method and a cross-validated regression coefficient (R2) method.

The current study focuses on shoreline change in Ghoramara Island (Figure 16.1) between 1990 and 2017, based on multi-temporal Landsat satellite data (thematic mapper (TM) and operational land imager (OLI)) using GIS and Remote Sensing. The aim is to

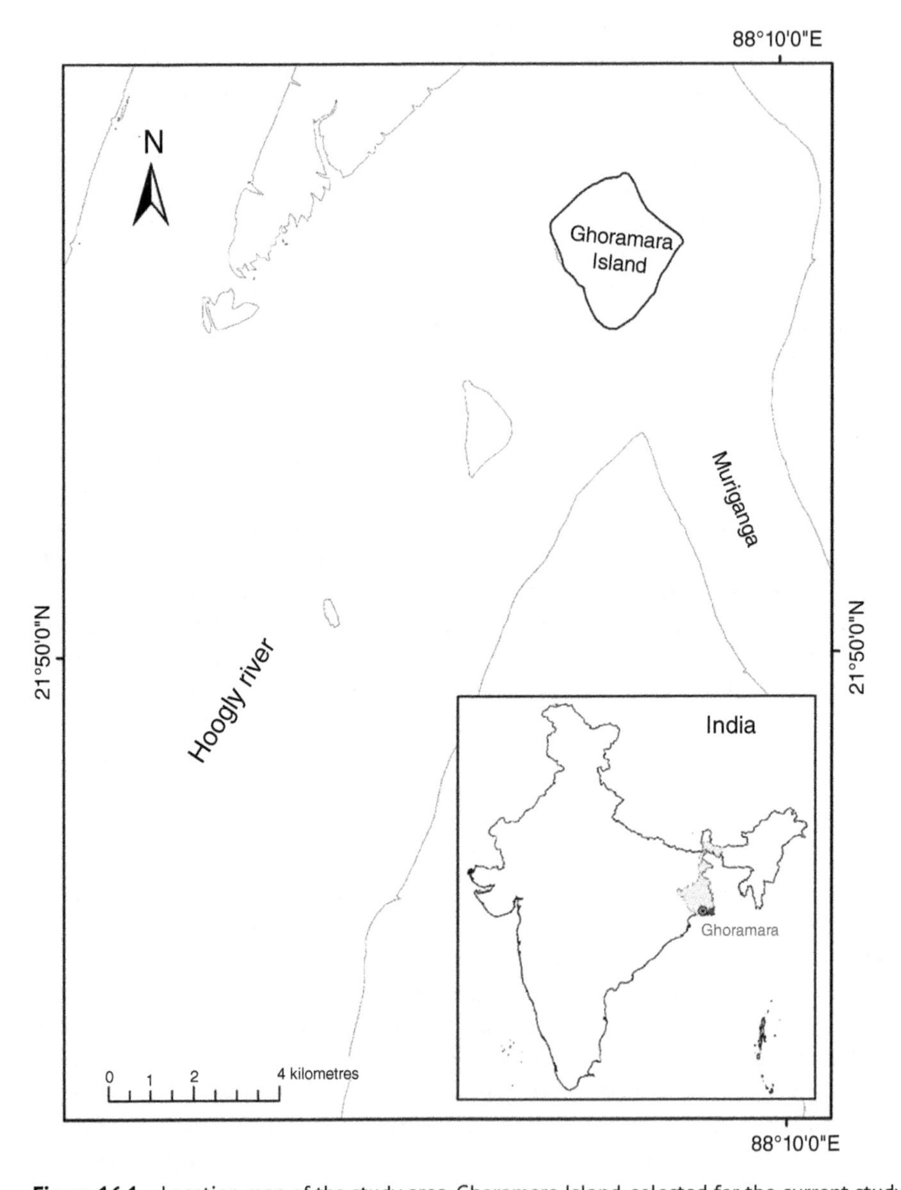

Figure 16.1 Location map of the study area, Ghoramara Island, selected for the current study.

estimate the rate of shoreline change (1988–2017) in the Ghoramara Island by using the EPR and shoreline change envelope (SCE) models by using the tool of DSAS-5.0 (Himmelstoss et al., 2018) as an add-in to ArcGIS 10.5.

16.2 Materials and Methods

16.2.1 Study Area

The island of Ghoramara (Figure 16.1) is situated in the North of Sagar Island. The island is separated from the mainland by rivers. Once, it was part of Sagar Island (Mondal 2015). It extends from 21°53′56″N to 21°55′37″N and 88°06′59″E to 88°08′35″E, covering an area of around 4.8 km^2. Sagar Island lies on the southeastern edge of Ghoramara. The major villages on this island are Khasimara, Hathkola, Raipara, Chunpuri Lakshmi Narayanpur, Mandirtala, Baghpara, and Khasimarachar. Kashimarachar, Lakshmi Narayanpur, and Kashimara villages have already submerged (Adarsa et al. 2012). The northern portion of Ghoramara has drastically degraded over the last few years (Department of Environment, Government of West Bengal 2010).

16.2.2 Data Used in This Study

The multi-temporal satellite data of Landsat TM and Landsat OLI/TIRS (thermal infrared sensor) of varied resolution have been used in this analysis. Landsat data are generally used for coastal countries due to its multi-temporal and multi-spectral data capabilities. Multi-resolution satellite images were obtained at a periodic 5-year interval from 1990 to 2015, except for the last study year of 2017 of the chosen period (1990–2017). The details of the satellite data are provided in Table 16.1. The main discussion in this chapter is regarding the steps used in data processing for the study. Figure 16.2 shows the flow chart comprising an outline of all the steps followed in this study. The steps have been performed in the application of ERDAS IMAGINE 2014. ArcGIS 10.3 has also been used to operate with maps in the GIS environment. The DSAS 4.5 of USGS is the main application used to execute the current study assessments.

Table 16.1 Details of the satellite datasets used for the current study.

Satellite/sensor	Path/row	Date of acquisition
LANDSAT TM	138/45	19 January 1990
LANDSAT TM	138/45	24 January 1995
LANDSAT TM	138/45	22 December 2000
LANDSAT TM	138/45	12 April 2005
LANDSAT TM	138/45	1 February 2010
LANDSAT OLI/TIRS	138/45	2 January 2015
LANDSAT OLI/TIRS	138/45	26 February 2017

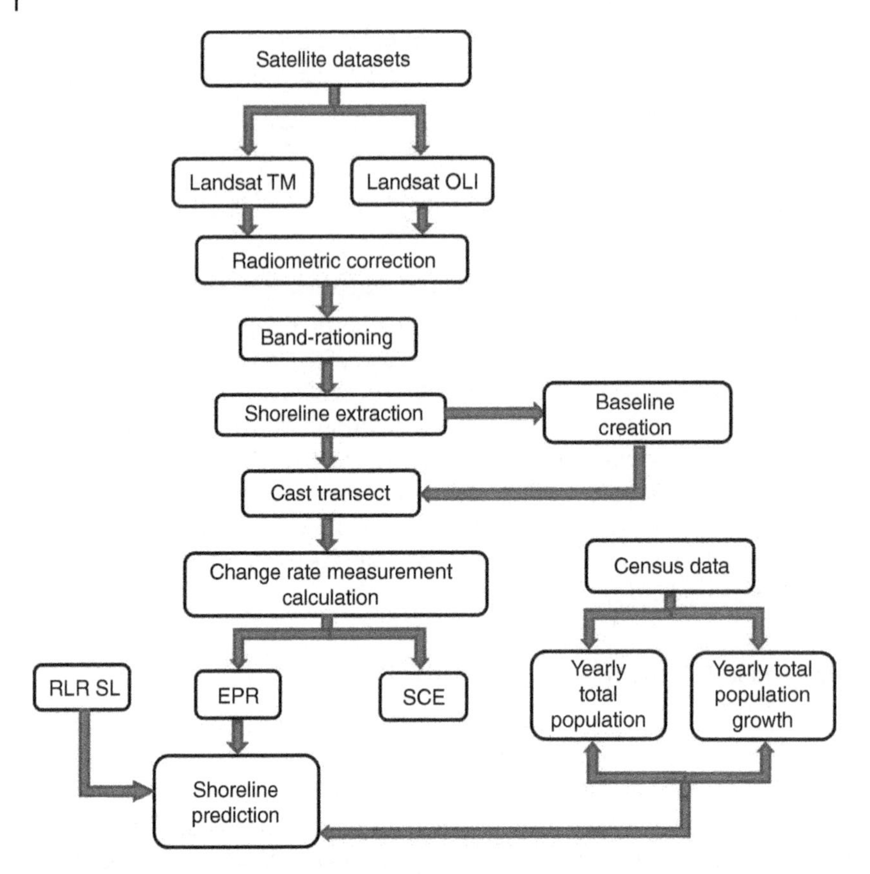

Figure 16.2 Flow chart showing methodology of the current study.

16.2.3 Software

The image processing software, ERDAS IMAGINE 2014 of Leica Geosystems, has been used for optical image processing in the current study. ArcGIS 10.3 has been used to operate with maps in the GIS environment to develop and compile geographic data, vector generation, map composition, and analysis of mapped information.

16.2.3.1 DSAS

The DSAS 4.5 application is a core component of the U.S. Geological Survey's Coastal Change Hazards project and it offers a comprehensive collection of regression rates in a standardised and readily repeatable approach that can be extended to vast amounts of data on a large scale (Himmelstoss et al. 2018). DSAS is a GIS tool for Historical Trends Analysis to assess the position or geometry of former or existing shores. The ability to measure rate change statistics on a time series of shoreline locations is one of the key advantages of using the DSAS in coastal change analysis. The statistics allow measurement and study of the essence of shoreline dynamics and evolving patterns (Oyedotun 2014). It performs five statistical procedures, including SCE, NSM, the EPR, linear regression rate (LRR), and least median of squares (LMS), to measure the rate of change in the shoreline over the long and short term (Oyedotun 2014). For the current study, the DSAS 4.5 application of the USGS as an extension of Arc GIS has been used to study the shoreline changes and all the related calculations linked to the models of EPR and the SCE. The inputs needed for this tool are the vector format of the shoreline, the date of each shoreline, and the transect distance.

16.2.4 Methods

16.2.4.1 Data Processing

As per the methodology of the current study, the multi-resolution and multi-temporal satellite images of the Landsat system over the area of study under Path/Row-138/45 for Landsat TM and Landsat OLI/TIRS were used. At first, geometric corrections were made to all six images. The radiometric correction was done for each band of satellite images. It requires two steps: first, converting the DN values to the radiance values, and then converting the radiance values to the reflectance values.

16.2.4.2 Shoreline Delineation

The automated shoreline delineation is a complex mechanism due to the abundance of water-saturated areas at the land-water boundary (Ryu et al. 2002; Maiti and Bhattacharya 2009). For the shoreline delineation, Huang et al. (2002) have employed the Tasselled Cap Transformation technique for shoreline extraction and derived the coefficients for the transformation of Tasselled Cap Landsat data from the Earth Capital Observation and Analysis (EROS) data centre. The shorelines were drawn using the Landsat TM, and Landsat 8 OLI/TIRS image for Tasselled Cap Transformation techniques (Nandi et al. 2016). Shoreline can also be detected and delineated from a single band image using the Landsat MSS, and TM processed near-infrared (NIR) bands, as the water reflectance is almost negligible in the reflective infrared band. The reflection of the bulk of the ground cover is higher than the water (Adarsa et al. 2012). The present study has been based on band rationing, using a ratio between bands 2 and 4 and between bands 2 and 5. The rationing between bands 5 and 2 takes place on the coastal areas' shoreline to remove the vegetation. Band 2/band, five averages for water, are more than one and ground are less than 1. Disingenuously, due to combined reflectivity, some of the vegetative lands should be assigned water. Both ratios are combined to solve this difficulty. The final images reflecting the shoreline are depicted in Figure 16.3. Thus, in seven various years (1990, 1995, 2000, 2005, 2010, 2015, and 2017), the persistent shoreline locations were mentioned in Table 16.1. Consequently, the binary raster image has been converted into a vector dataset, and the boundary of the shorelines has been delineated, illustrated in Figure 16.3.

16.2.4.3 Method of Casting Transect from Baseline

This approach is focused largely on statistics. A single baseline has been created from the shoreline before drawing the transect. There are mainly two baseline demarcation methods, i.e. firstly, baseline creation from a specific shoreline distance, and secondly, the buffer method (Nandi et al. 2016). The buffer method is the most accurate and precise method to restrict baselines as it assumes the same sinuosity shape as the adjacent shoreline (Nassar et al. 2019).

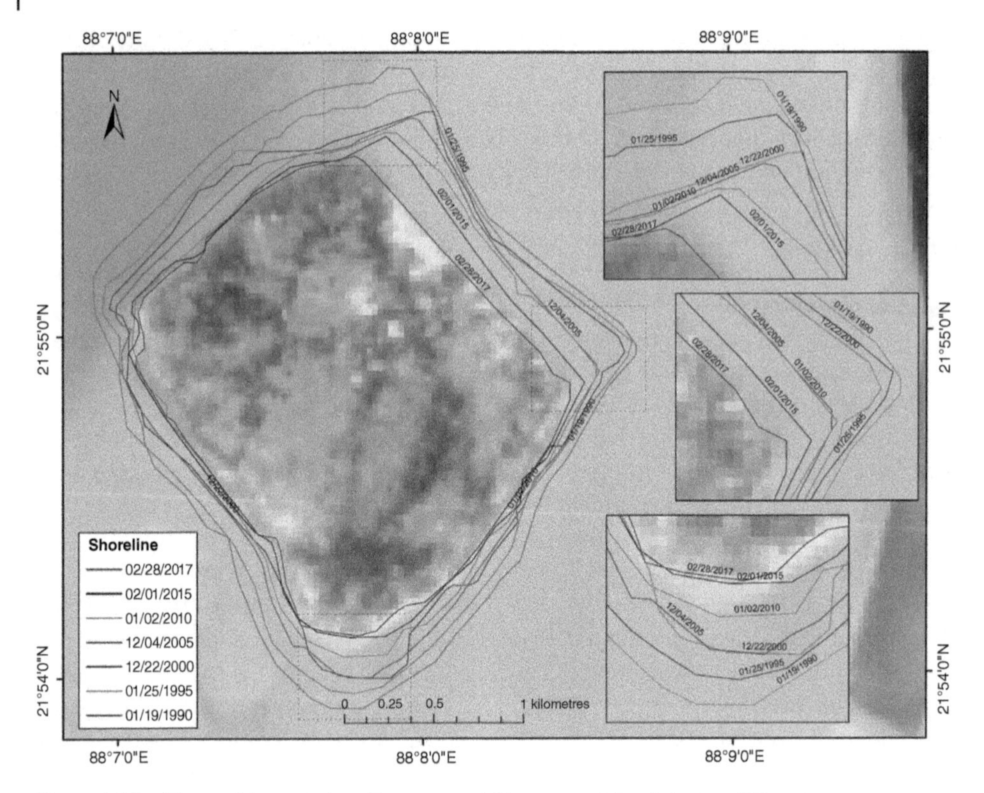

Figure 16.3 The positions of shorelines around Ghoramara Island during different years (1990–2017).

The attribute fields, i.e. OBJECTID, SHAPE, SHAPE Length, ID, Group, OFFshore, and CastDir are required to initiate the baseline calculation in the DSAS programme. These attributes provide the DSAS with required details regarding the order of the transects and the position of baseline concerning the shoreline (onshore or offshore). Thus, the benchmark is set at a buffering distance of 1000 m offshore from the nearest shoreline for the current analysis. The transect lines have been orthogonally casting from the baseline around the shoreline for various years, as shown in Figure 16.4.

16.2.4.4 Shoreline Change Rate Assessment methods

Several statistical methods are being used to measure the rate of change in the shoreline; the most widely used is the EPR, LR and NSM, and SCE (Bheeroo et al. 2016). In this analysis, two major mathematical models, i.e. the EPR and the SCE, have been employed to measure the rate of change in the shoreline of Ghoramara Island. Firstly, the measurement of the EPR is estimated by dividing the distance of shoreline movement by the time difference between the oldest and the youngest shoreline in the data set (Fenster et al. 1993). The biggest drawback of this approach is the simplicity of calculating, and the measurement can only be performed on a minimum of two shorelines. The EPR approach is only reliable for a short-term study of shoreline change as it only considers the current and oldest shoreline location and

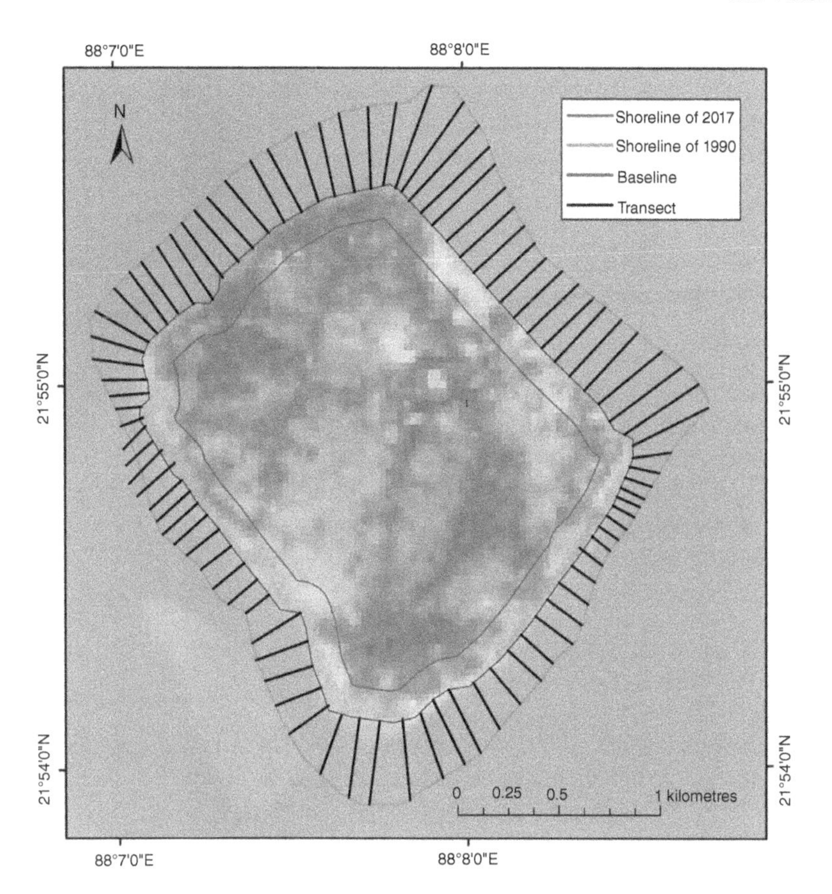

Figure 16.4 The cast transects perpendicular to the baseline for the Ghoramara Island.

suppresses any such long-term analysis. Secondly, the statistical method of SCE is a measurement of the overall change in shoreline movement, taking into account all available shoreline positions and documenting their lengths, without reference to their particular dates (Oyedotun 2014). However, the method of SCE being employed in the current study upon Ghoramara, due to its capacity to include an envelope of uncertainty, which is the only measure to consider the shorelines. In this study, the SCE is calculated as the distance between the furthest and nearest to the baseline of the shorelines. For each transect, the total changes in shoreline movement are being represented for all available shoreline locations. In terms of baseline locations, the rate of accretion and erosion of the shoreline has been considered.

16.2.4.5 Calculation of Net Areal Change

In this method, the areas of erosion and accretion have been demarcated, and the areal changes of the island have been calculated. The spatial analysis is performed in the GIS environment for determining the area of the island during a given year; a polygon is constructed along the boundary of the GIS platform, which yields the area of the island during that year. A polygon is constructed along the shoreline on the GIS platform for the identification of the area of the island within a given year. The polygons for two successive years in five-year intervals (1990–1995, 1995–2000, 2000–2005, 2005–2010, 2010–2015, 2015–2017) are united by Overlay in deciphering net aerial transition quantification of erosion in the island. Two shorelines can be joined in the overlay operation of 'union'. The length and erosion/accretion spectrum of the segments between the polygon was estimated. The vector data are transformed to raster data with a grid scale of 10 m for measuring the aerial changes.

16.3 Results

16.3.1 End Point Rates

The EPR of the study area, Ghoramara Island, estimated using DSAS is shown in Figure 16.5. EPR is calculated by dividing the distance of shoreline movement by the time elapsed between the two shorelines. The transect-wise shoreline transition rate was measured and plotted in Figure 16.5. The plotted graph is highlighting the EPR only in the negative values that suggest that the study area of Ghoramara Island has been exposed to a more or less rapid rate of erosion from the year 1990 to 2017, considered for the study.

The rate of change in shoreline per year to measure the EPR is calculated by using Eq. (16.1), shown as,

$$S_r = f_o - f_y / n \tag{16.1}$$

The rate of change of shoreline per year (metre year^{-1}) is S_r, the distance between the mean and the shoreline at the oldest point at a particular transect (xn); f_y is the distance between the

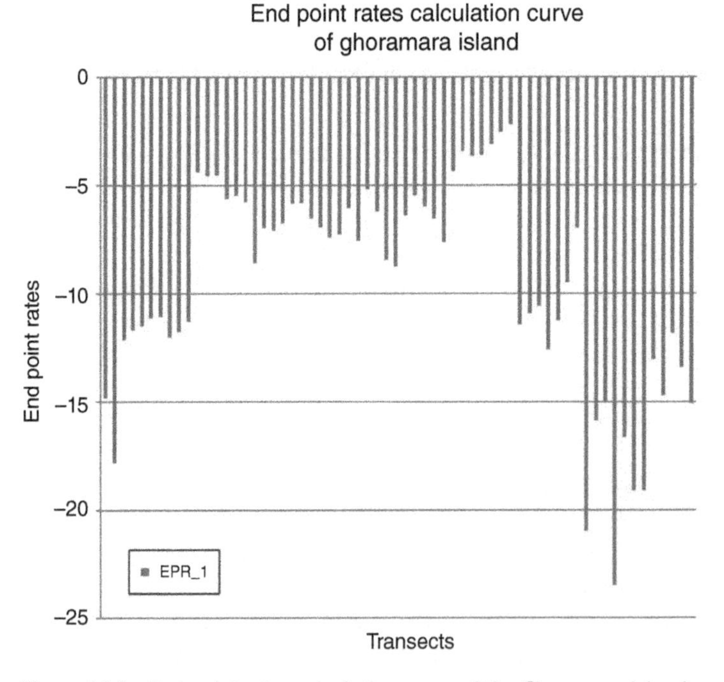

Figure 16.5 End point rates calculation curve of the Ghoramara Island.

Table 16.2 Statistics of changes in EPR of Ghoramara Island.

Maximum	Minimum	Mean
−23.5	−2.16	−9.413

median and the median the baseline. Recently, the shoreline at the same transect above (xn); n is the total number of years from the earliest date to the current date of the study.

The results in the plot shown in Figure 16.5 and Table 16.2 depict that the rate of erosion has been high, reaching up to −20 in the initial section of the calculated plot. Then in the mid-section of the calculated plot, it can be noticed that there has been a fluctuation in the rate of erosions, and compared to the first section, the erosion rate is much lower in the mid-section of the plot. It can be noticed in the right end section of the plot that the rate of erosion is very high in this particular section of the entire calculation curve of EPR.

Table 16.2 depicts the statistics from this calculation curve, where the mean value of EPR is −9.413, then the maximum and minimum value of EPR is −23.5 and −2.16.

The EPR outcomes, which are graphically displayed above in the plot of Figure 16.5, have been highlighted spatially on the map of the study area of the Ghoramara Island in Figure 16.6. It highlights the shoreline portions and direction of the Ghoramara Island,

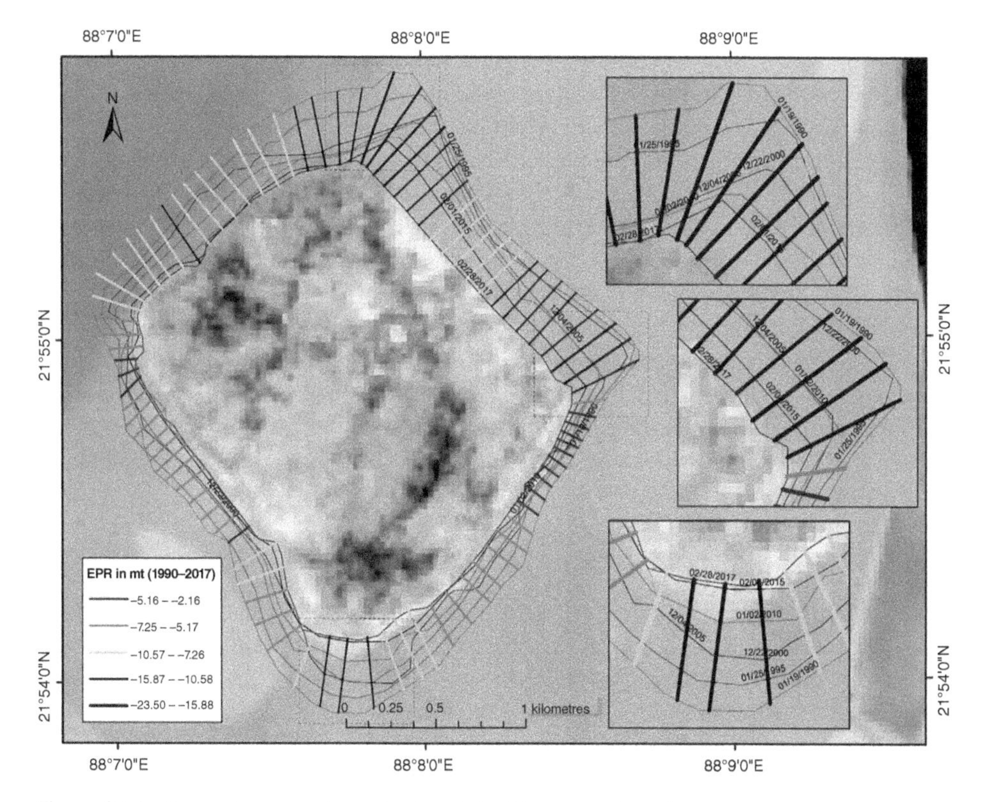

Figure 16.6 Transect-wise end point rates of the Ghoramara Island.

where the erosion of the coast has occurred at maximum, moderate, and minimum scales. Figure 16.6 shows the scenario erosion took place spatially as well as temporally in Ghoramara Island from 1990 to 2017. The maximum level of erosion has occurred on the Northern, Northeastern, and Northwestern sides of the island from 2000 to 2017, denoted by the colour of red, orange, and yellow transect lines in Figure 16.6. A moderate amount of erosion has occurred along the Southern and South-western Shores, denoted by light green coloured transect lines in Figure 16.6. Then the minimum amount of erosion also took place in few areas of the southern sector of the island, denoted by yellow transect lines. Thus the shift in Ghoramara Island has been observed since 1979 (Kundu et al. 2014). The apparent shift in the island of Ghoaramara is mainly due to rapid erosion in its Northwestern and marginal erosion Southeasters side of the shores.

16.3.2 Shoreline Change Envelope

In this study, the SCE statistical method, defined as the distance between the shorelines nearest to the baseline at each transect. The total change in coastal movements has been documented for all available shoreline locations. The rate of change of shore has been considered in respect of the baseline position (Bheeroo et al. 2016).

The function of the SCE is depicted using Eq. (16.2), which is represented as

$$S_d = d_f - d_c \tag{16.2}$$

Where S_d is the change in shoreline distance (m); d_f is the change of distance between the baseline and the farthest shoreline (m) at a given location. The Transect (xn); d_c is the distance between the baseline and the nearest shoreline (m) at the same transect above (xn).

The SCE for the study area is shown in Figure 16.7, which illustrates the distance between the two furthest shorelines and the nearest shoreline to the baseline at each of the transects. This reflects the overall change in the shoreline Movement for all available shoreline locations in the study area of Ghoramara Island from 1990 to 2017. Figure 16.7 shows that the change in the shoreline has taken all around the study area. As the change in uniform all around, the change inshore has been enormous in the Northern portion of the Ghoramara Island, ranging from 429 to 485 m, denoted by deep blue coloured transact lines. The change has been moderate to severe in the Northwestern and Southern portions of the shoreline of Ghoramara Island, with a range of 157–349-m change, denoted by light blue, light green, and yellow-coloured transect lines. The change has been least in the Western and Southeastern shores of Ghoramara Island, with 110–156 m of change, denoted by red-coloured transect lines in Figure 16.7.

16.3.3 Relative Sea-level Change

For the current study, datasets of the SL were obtained from the 'Permanent Service of the Mean Sea Level' (PSMSL), a world repository of the tide-gauge statistics built-in 1933, headquartered in Liverpool at 'National Sea Level Oceanography Centre', UK (National Oceanography Centre 2018). The SL data of PSMSL is provided in two formats: 'metric' and 'revised local reference' (RLR) and is publicly available (National Geographic Society 2020). The datasets of RLR are published after year-to-year analyses are carried out based on a standard date and are ideal for assessing long-term changes in the SL (Nandy and

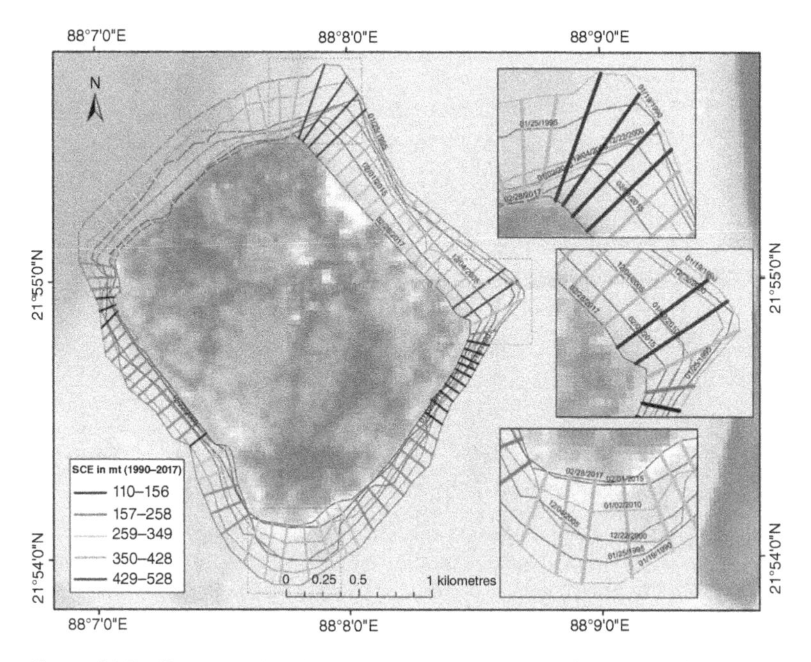

Figure 16.7 Transect-wise shoreline change envelope (SCE) around Ghoramara Island.

Bandyopadhyay 2011). The RLR dataset includes a consistent database of the benchmark observations for each gauge, which implies that the SL is assessed against the recorded land-based data. They indicate variations in the relative SL and depict potential regional and local factors at a gauge station that could cause vertical land adjustments, in conjunction with the continuous surge in global SL (Kumar Mandal et al. 2018). The RLR records of the Indian station Haldia have been considered for the current study since it is located near the island of Ghoramara. The Haldia Gauge Station has more than 50 years of available RLR SL data, and the same period (1948–2013) of SL data is used to assess the annual rate of changes in the SL around the Ghoramara Island. The line graph in Figure 16.8 shows the annual change in seal level, calculated from the RLR time series (1948–2013) recorded at Haldia. Time series data show that there has been an intensity of negative change in SL, but also through fluctuations over many years, there has been a successive increase in SL. As per Haldia's RLR records, the SL was lowest in 1972 (6918 mm) and highest in 2013 (7150 mm), with a mean SL of 7055.69 mm. The change in SL from 1948 to 2013 had a significant impact on the areal extent of the study area of Ghoramra Island. Ghoramara Island has undergone the phase of progressive erosion between 1967 and 2008 as it lost an area of approximately 4 km^2 (Nandy and Bandyopadhyay 2011).

16.3.4 Areal Extent of Erosion

The spatial analysis is performed in the GIS environment to determine the areal extent of the island during the different years of the period (1990–2017). As mentioned in the above section, a polygon is constructed along the land-water boundary of Ghoramara Island in the GIS platform for a given year. It yields the area as well as the perimeter of the island

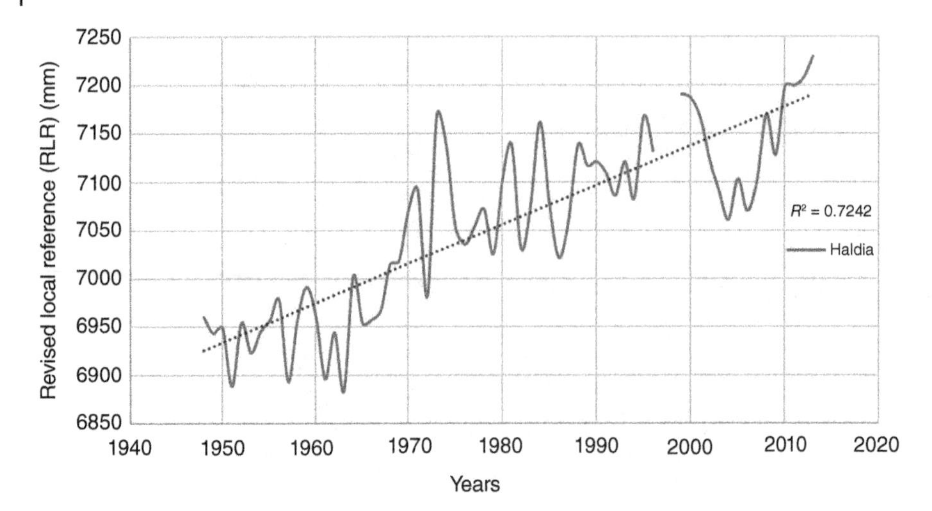

Figure 16.8 Temporal changes in annual sea level (1948–2019) at the Haldia station. *Source:* Based on Revised Local Reference (RLR) Diagram for HALDIA, Haldia Tidal Gauge Station, Station ID:1270, Long: 88.07°E Lat:21.95°N. https://www.psmsl.org/data/obtaining/rlr.diagrams/1270.php

Table 16.3 Year-wise difference in perimeter and area of the Ghoramara Island.

Year	Perimeter (Km)	Area (Km^2)	Changes In Area ($A_{n+1}-A_n$)	Details
1990	9.73151	6.22375		
1995	9.36807	5.68656	−0.53719	EROSION
2000	9.1065	5.06592	−0.62064	EROSION
2005	8.56371	4.89745	−0.16847	EROSION
2010	8.16012	4.487	−0.41045	EROSION
2015	7.87919	4.16835	−0.31865	EROSION
2017	7.57685	3.82449	−0.34386	EROSION

during that particular year. Thus, after the execution of the method mentioned above, the year-wise areal change in the study area has been derived. Table 16.3 shows the extent of area and perimeter of Ghoramara during the successive year of the period (1990–2017) considered for the study. The extent of perimeter and area during the successive years shows that the change in the areal extent has been dynamic as well as negative. The change in the areal extent has been negative because of constant erosion that occurred along the shorelines of Ghoramara Island from 1990 to 2017. The change in the area of the Ghoramara Island has been calculated for the successive years considered for the study, and it shows that the change has been mostly negative (−0.62064 km^2) in 2000, which indicates the erosion has been highest in that year. On the other hand, the area change has been the least (−0.31865) during 2015, indicating a lesser rate of erosion than the previous years.

Figure 16.9 is the graphical representation of the changes in the area of the Ghoramara Island from 1990 to 2017, based on the estimated data mentioned in Table 16.3, using the

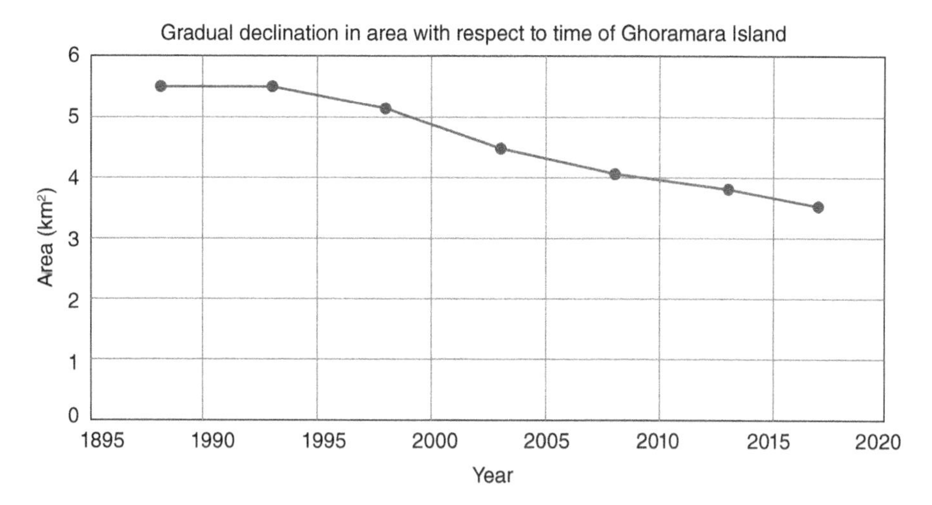

Figure 16.9 Line graph representing the change of the area from 1990 to 2017.

application of DSAS. In Figure 16.6, we can see Ghoramara Island is subjected to considerable changes in the area from 1990 to 2017, subjected to shoreline erosion. The maximum area of Ghoramara Island during 1988 was 5.4838 km^2. The maximum change in the area took place from the year 1990 to 2000, i.e. 0.621 km^2. Between the years 2010 and 2015, there has been a minimum change in the area due to a lower level of erosion than the previous years, considered for the current study. The total change in the area of Ghoramara Island due to coastal erosion from 1990 to 2017 is 2.4 km^2.

Figure 16.10 shows the year-wise shoreline analysis of Ghoramara Island in a merged format. It is the pictorial representation of the change that took place along the shoreline

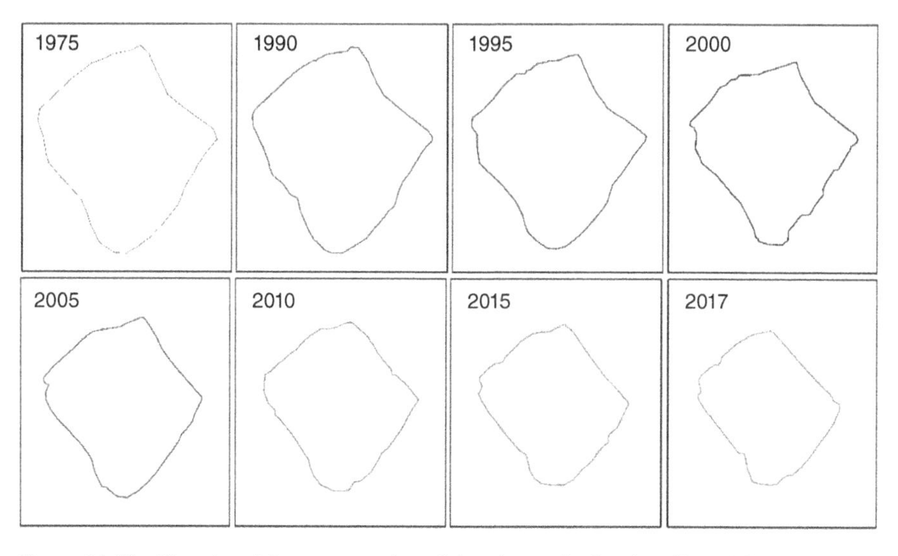

Figure 16.10 The pictorial representation of the change in the shoreline of Ghoramara Island of 1990, 1995, 2000, 2005, 2010, 2015, 2017.

of Ghoramara Island from 1975 to 2017. This figure shows that erosion took place in successive years considered for this study, which led to a shift or change in the shoreline of Ghoramara Island. The shape of the island has gradually changed within this period of 1990–2017. It can be inferred from this composite Figure 16.8 of maps that the change in the island's shape has occurred rapidly between 1995 and 2010. Figure 16.8 shows that there has been a rapid shift of the shoreline due to erosion, mainly in the Northwestern and Southeastern sides of the Ghoramara Island.

16.3.5 Shoreline Change Rate Prediction

The EPR model is an important method used in the current study to predict changes on the coast of Ghoramara Island, depending on the rate of change over the years. In this study, the EPR model has been used only on the shorelines of the year 1990 and 2017 to predict the 2027 shoreline.

$$\text{Shoreline position} = \text{Slope} * \text{Date interval} + \text{Intercept} \tag{16.3}$$

As mentioned above, the total change in Ghoramara Island shoreline erosion area from 1990 to 2017 has been 2.4 km^2. Thus, Ghoramara Island's shoreline of 2027 has been forecasted with the shorelines of the years 1990 and 2017. The map in Figure 16.10 illustrates the predicted shoreline of the year 2027. The predicted map (Figure 16.10) shows that a further change in the shoreline position may occur in the immediate future by the year 2027. It can be inferred from the projected shoreline of 2027 on the island of Ghoramara that further retreats have taken place along the maximum portion of the shoreline of the island, attributing to the severe level of erosion that has taken place along the shore. Figure 16.10 shows that the negative shifts along the projected shoreline have taken place hugely along the southwestern and western edges of the study area. There have been significant changes between the shorelines of 2017 and the projected shoreline of 2027 in the north-eastern and south-eastern portions. There has also been a small positive shift, detected in the projected shoreline of 2027, regarding the 2017 shoreline, attributable to the deposition along the northern and southern margins of Ghoramara Island. It can be assumed by estimating from the statistics mentioned above and figures that if the rate of change of shoreline and SL rise persists in the same manner, the extent and position of the shoreline in the immediate future may conform to the projected shoreline (2027) shown in Figure 16.11.

16.4 Discussion

As it can be seen in Figure 16.10, that year-wise, Ghoramara Island is steadily shifting to the southeast due to the erosion of the shoreline, which is altering the form of the island. Extensive erosion has occurred in the northwestern part of the island, and slight erosion in the southeast part of the island has caused this apparent lateral change. Table 16.3 lists the changes in the area and perimeter of Ghoramara Island from 1990 to 2017. According to observations, the shoreline shifting of Ghoramara Island began in 1979 (Adarsa et al. 2012).

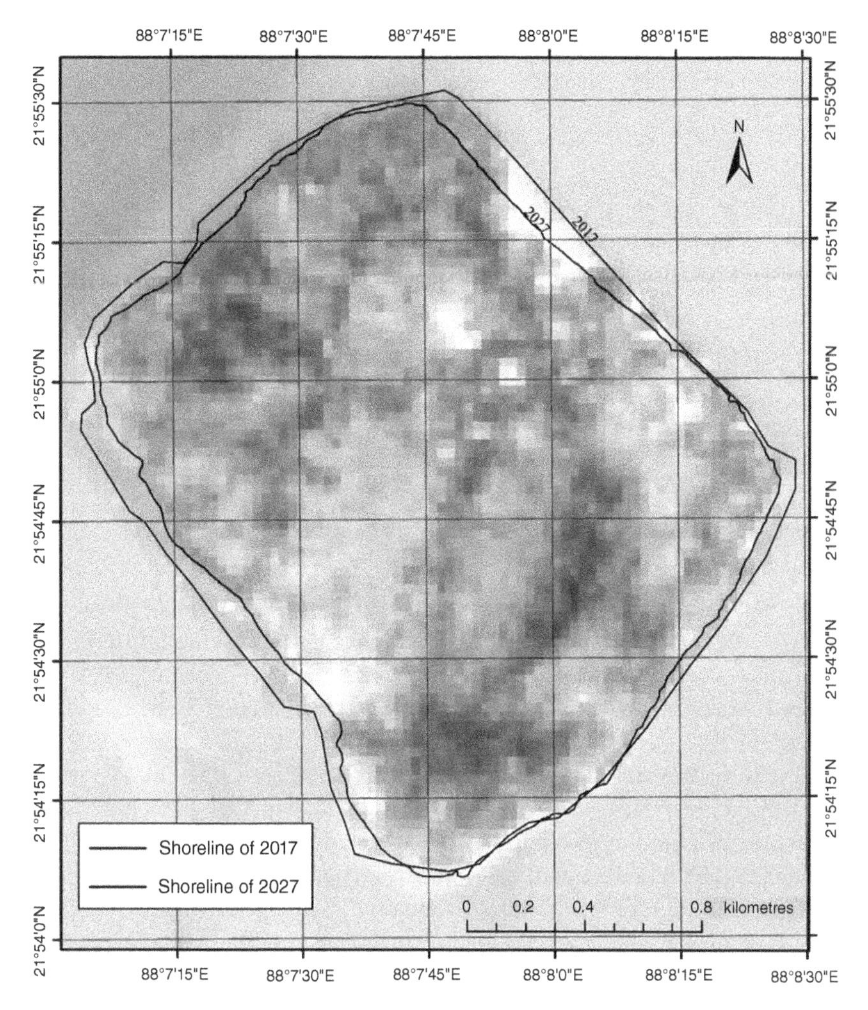

Figure 16.11 The predicted shoreline of 2027 by using the EPR model.

The Hooghly River's Estuarine system is currently encountering high tidal currents and limited freshwater flow. The changes in geomorphology have been seen in Ghoramara, which is mostly a result of changes in estuarine hydrodynamics, all of which are determined by anthropological activities and natural processes (Ghosh et al. 2003). The island system in the estuary of Hooghly, like Ghoramara. Island was safe because there was high availability of freshwater. When the eastward change of River Hooghly's course was caused by a tectonic uplift (Blasco 1977), freshwater flow dropped significantly, and the estuary experienced a water allocation deficit. The barrage of Farakka was built by Kolkata Port Trust (KoPT) in 1975 to resolve the problem but has only marginally improved the situation, but it does not fully solve the problem. Due to the lower inflow of freshwater, siltation in the navigation channels is a big concern. In past years, Ghoramara Island, which was previously accreting farther north, has reportedly eroded substantially. Comprehensive knowledge of coastal processes is needed to improve any current remediation efforts and

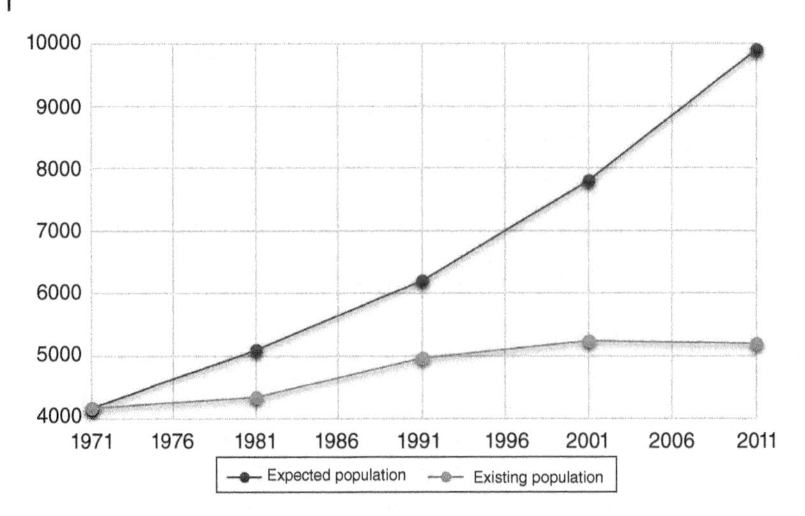

Figure 16.12 Expected and existing population of the Ghoramara Island.

introduce new remedial measures in the region. Figure 16.9 of the current study reveals that the island of Ghoramara has sunk over time; the decline in the area has impacted the island's land use and cover. The cultivated paddy-field area is greatly reduced as well as the uncultivated area is drastically reducing. It used to have a population of 40 thousand on Ghoramara Island. In 2001, the Indian census showed Ghoramara's population is 5236. It is believed that the population has dropped as families were displaced from the island. By 2016, because of the submergence of the island of Lohachara, had 3000 inhabitants and people from Ghoramara Island migrated to other neighbouring islands, including Sagar. Due to the damage to the natural habitat, the residents are forced to leave the island. For the residents of this island, the term 'environmental refugees' was included (AquaBUD 2016).

As per Figure 16.12, the line graphs represent the expected and existing temporal population composition of Ghoramara Island. The graph (Figure 16.12) shows that there really is a substantial discrepancy between the lines indicating the predicted and current populations. The line showing the expected population is growing upwards, successively representing the periodic rise in the island population. Whereas, relative to the expected demographics, the line representing actual population from one period to another refers to a precisely opposite graph of the expected population. As per the line graph (Figure 16.12) depicting the existing demographic, population growth occurred between 1971 and 2001, but at a much slower rate, much lesser than the expected population. Since 2001, as shown in Figure 16.10, the existing population line curved downwards because as population number declined from 5236 in 2000 to 5139 in 2011. This decline in the population of Ghoramara between 2001 and 2011 has a direct correlation with continued erosion of the shoreline and the consequent decrease in the areal extent of Ghoramara. This is forcing the inhabitants to move elsewhere in fear of the submergence of Ghoramara very soon, much like the former submergence of the island of Lochachar.

The rise in the island's vegetative cover is the best solution for controlling erosion. Thus, the plantation of land, which includes the plantation of mangroves, Caesarians, and bamboos, should be executed. The thick growth of bushes and trees on the river's bank not only reduces the flow velocity but also influences the velocity distribution within the river profile. The presence of woody vegetation on both riverbanks of natural and unhindered river channels is not an obstacle to the discharge of floodwaters, as there is adequate space in the river corridor to adapt to the excessive flow. Similarly, at Ghoramara Island, wherein the muddy and sandy bank lands were completely eroded, the vegetated region could endure coastal erosion

16.5 Conclusion

Over the years, multi-resolution satellite imagery and the shoreline using computational approaches have been retrieved surrounding Ghoramara Island. The findings show that shoreline behaviour is quite complex. Ghoramara's shoreline has been severely damaged between 1990 and 2017 when the island experienced significant northeast and northwest erosion. Transect-wise erosion analysis has been deciphered in almost every direction across the island. The pace of erosion around Ghoramara Island is significantly faster than the rate of reduction in the island's area over time. The endpoint predictions, on the other hand, show a significant decline in the entire area of the island. The study's findings will be useful for improved administration and sustainable monitoring of estuarine and coastal islands, and the technique can be replicated on other islands to comprehend the impending consequences of rising SLs and degradation under the present global climate change scenario.

References

Abd El-Kawy, O.R., Rød, J.K., Ismail, H.A., and Suliman, A.S. (2011). Land use and land cover change detection in the western Nile delta of Egypt using remote sensing data. *Applied Geography* 31 (2): 483–494. https://doi.org/10.1016/j.apgeog.2010.10.012.

Adarsa, J., Shamina, S., and Arkoprovo, B. (2012). Morphological change study of Ghoramara Island, Eastern India using multi temporal satellite data. *Research Journal of Recent Sciences 1* (10): 72–81.

AquaBUD (2016). AquaBUD. http://aquabud.blogspot.com/2016/11/the-vanishing-biotope-of-ghoramara.html (accessed 12 February 2020).

Asif, K. (2010). Coastal sediment transport. http://dl.lib.mrt.ac.lk/bitstream/handle/123/2049/ (accessed 12 February 2020).

Bandyopadhyay, S. (2000). Sagardwip: some environmental problems and men. In: *Bish Sataker Sagardwip (20th Century Sagar Island)* (ed. B. Maity) Quarterly, April–June, 17–24. South Pargana, India: Newsman Computer.

BBC (2020). Coastal processes. BBC. https://www.bbc.co.uk/bitesize/guides/zt6r82p/revision/4 (accessed12 February 2021).

Bertacchini, E. and Capra, A. (2010). Map updating and coastline control with very high resolution satellite images: application to Molise and Puglia coasts (Italy). *Italian Journal of Remote Sensing/Rivista Italiana Di Telerilevamento* 42 (2): 103–115. https://doi.org/10.5721/itjrs20104228.

Bheeroo, R.A., Chandrasekar, N., Kaliraj, S., and Magesh, N.S. (2016). Shoreline change rate and erosion risk assessment along the Trou Aux Biches–Mont Choisy beach on the northwest coast of Mauritius using GIS-DSAS technique. *Environmental Earth Sciences* 75 (5): 444. https://doi.org/10.1007/s12665-016-5311-4.

Blasco, F. (1977). *Outline Ecology, Botany and Forestry on the Mangals of the Indian Subcontinent, Wet Coastal Ecosystem*, 241–259. Oxford, UK: Elsevier Scientific.

Caballero, I. and Stumpf, R.P. (2020). Towards routine mapping of shallow bathymetry in environments with variable turbidity: contribution of sentinel-2A/B satellites mission. *Remote Sensing* 12 (3): 451. https://doi.org/10.3390/rs12030451.

Cai, F., Su, X., Liu, J. et al. (2009). Coastal erosion in China under the condition of global climate change and measures for its prevention. *Progress in Natural Science* 19 (4): 415–426. https://doi.org/10.1016/j.pnsc.2008.05.034.

Central Water Commission (2016). Status report on coastal protection & development in India. p. 120. http://cwc.gov.in/CPDAC-Website/Paper_Research_Work/Status_Report_on_Coastal_Protection_and_Development_in_India_2016.pdf (accessed12 February 2020).

Central Water Commission River Management Wing (2003). Guidlines for prepration of coastal protection projects. Central Water Commission. http://old.cwc.gov.in/CPDAC-Website/Guideline/CWC Guidelines for Preparation of Coastal Protection Projects-2003.pdf (accessed 26 November 2020).

Coastal Engineering Research Center (1984). Shore Protection Manual. *US Army Corps of Engineers* 2: 597–603.

Columbia University in the city of New York (2020). Coastal processes. Columbia University in the City of New York. http://www.columbia.edu/~vjd1/coastal_basic.htm (accessed 26 November 2020).

Committee on Climate Change (2018). Managing the environment in a changing climate. Committe on Climate Change.

Department of Environment, Government of West Bengal (2010). Integrated coastal zone mangement project. Department of Environment, Government of West Bengal. http://www.iczmpwb.org/main/coastal_erosion.php (accessed 1 December 2020).

Dolan, R., Fenster, M.S., and Holme, S.H. (1991). Temporal analysis of shoreline recession and accretion. *Journal of Coastal Research* 7 (3): 723–744.

Dolan, R. and Morton, R. (2001). Coastal storms and shoreline change: signal or noise? *Journal of Coastal Research* 17 (3): 714–720.

Farquharson, L.M., Mann, D.H., Swanson, D.K. et al. (2018). Temporal and spatial variability in coastline response to declining sea-ice in northwest Alaska. *Marine Geology* 404 (July): 71–83. https://doi.org/10.1016/j.margeo.2018.07.007.

Fenster, M.S., Dolan, R., and Elder, J.F. (1993). A new method for predicting shoreline positions from historical data. *Journal of Coastal Research* 9 (1): 147–171.

FitzGerald, D.M., Fenster, M.S., Argow, B.A., and Buynevich, I.V. (2008). Coastal impacts due to sea-level rise. *Annual Review of Earth and Planetary Sciences* 36: 601–647. https://doi.org/10.1146/annurev.earth.35.031306.140139.

Geological Survey Ireland (2020). Coastal erosion. Geological Survey Ireland. https://www.gsi.ie/en-ie/geoscience-topics/natural-hazards/Pages/Coastal-Erosion.aspx#:~:text=Abrasion is when rocks and,carried away by the sea.&text=Attrition is when material such,each other wearing them down (accessed 26 November 2020).

Ghosh, T., Bhandari, G., and Hazra, S. (2001). Assessment of landuse/landcover dynamics and shoreline changes of Sagar Island through remote sensing. *22nd Asian Conference on Remote Sensing* (January). p. 43.

Ghosh, T., Bhandari, G., and Hazra, S. (2003). Application of a "bio-engineering" technique to protect Ghoramara Island (Bay of Bengal) from severe erosion. *Journal of Coastal Conservation* 9 (2): 171–178. https://doi.org/10.1652/1400-0350(2003)009[0171:AOABTT]2.0.CO;2.

Himmelstoss, E.A., Henderson, R.E., Kratzmann, M.G., and Farris, A.S. (2018). *Digital Shoreline Analysis System (DSAS) Version 5.0 User Guide*. USGS Numbered Series. Reston,VA: U.S. Geological Survey https://doi.org/10.3133/ofr20181179.

Huang, C., Wylie, B., Yang, L., and Homer, C. (2002). Derivation of a tasselled cap transformation based on Landsat 7 at-satellite reflectance. *International Journal of Remote Sensing* 23 (8): 1741–1748. https://doi.org/10.1080/01431160110106113.

Jayappa, K.S., Mitra, D., and Mishra, A.K. (2006). Coastal geomorphological and land-use and land-cover study of Sagar Island, Bay of Bengal (India) using remotely sensed data. *International Journal of Remote Sensing* 27 (17): 3671–3682. https://doi.org/10.1080/01431160500500375.

Kaliraj, S., Chandrasekar, N., and Magesh, N.S. (2015). Evaluation of coastal erosion and accretion processes along the southwest coast of Kanyakumari, Tamil Nadu using geospatial techniques. *Arabian Journal of Geosciences* 8 (1): 239–253. https://doi.org/10.1007/s12517-013-1216-7.

Kumar Mandal, U., Bikas Nayak, D., Samui, A. et al. (2018). Trend of sea-level-rise in West Bengal Coast. *Indian Journal of Coastal Agricultural Research* 36 (2): 64–73. www.incois.gov.in.

Kundu, S., Mondal, A., Khare, D. et al. (2014). Shifting shoreline of Sagar Island Delta, India. *Journal of Maps* 10 (4): 612–619. https://doi.org/10.1080/17445647.2014.922131.

Lumen (2020). Waves and coastal features. Lumen. https://courses.lumenlearning.com/earthscience/chapter/wave-coastal-features/ (accessed 26 November 2020).

Maiti, S. and Bhattacharya, A.K. (2009). Shoreline change analysis and its application to prediction: a remote sensing and statistics based approach. *Marine Geology* 257 (1–4): 11–23. https://doi.org/10.1016/j.margeo.2008.10.006.

Mangor, K., Drønen, N.K., Kaergaard, K.H., and Kristensen, S.E. (2017). Shoreline management guidelines. https://www.dhigroup.com/upload/campaigns/shoreline/assets/ShorelineManagementGuidelines_Feb2017-TOC.pdf (accessed 25 November 2020).

Marchand, M. (2010). Concepts and science for coastal erosion management. Concise report for policy makers.

Mcsweeny, R. (2020). Explainer: nine' tipping points' that could be triggered by climate change. Carbon Brief. https://www.carbonbrief.org/explainer-nine-tipping-points-that-could-be-triggered-by-climate-change (accessed12 February 2021).

Mohr, M.C. (2001). Site characterization (Issue part V).

Mondal, B. (2015). Rehabilitation; a practical issue in the Sundarban deltaic area with special references to Ghoramara Island, South 24 Parganas, West Bengal, India. *International Journal of Humanities and Social Science Research* 1 (2): 2455–2070. www.ijhssi.org.

Mukhopadhyay, A., Mukherjee, S., Mukherjee, S. et al. (2012). Automatic shoreline detection and future prediction: a case study on Puri coast, Bay of Bengal, India. *European Journal of Remote Sensing* 45 (1): 201–213. https://doi.org/10.5721/EuJRS20124519.

Muthukumarasamy, R., Mukesh, R., Tamilselvi, M. et al. (2013). Shoreline changes using remotesensing andgisenvironment: a case study of Valinokkam to Thoothukudi Area, Tamilnadu, India. *International Journal of Innovative Technology and Exploring Engineering* 2 (6): 72–75.

Nandi, S., Ghosh, M., Kundu, A. et al. (2016). Shoreline shifting and its prediction using remote sensing and GIS techniques: a case study of Sagar Island, West Bengal (India). *Journal of Coastal Conservation* 20 (1): 61–80. https://doi.org/10.1007/s11852-015-0418-4.

Nandy, S. and Bandyopadhyay, S. (2011). Trend of sea level change in the Hugli estuary, India. *Indian Journal of Marine Sciences* 40 (6): 802–812.

Nassar, K., Mahmod, W.E., Fath, H. et al. (2019). Shoreline change detection using DSAS technique: case of North Sinai coast, Egypt. *Marine Georesources and Geotechnology* 37 (1): 81–95. https://doi.org/10.1080/1064119X.2018.1448912.

National Geographic Society (2020). Erosion. National Geographic. https://www.nationalgeographic.org/encyclopedia/erosion/print/#:~:text=Coastal erosion—the wearing away,moving the coastline farther inland (accessed 26 November 2020).

National Ocean Service (2020). Longshore currents. NOAA. https://oceanservice.noaa.gov/education/tutorial_currents/03coastal2.html#:~:text=Rather%2C they arrive at a,called a "longshore current." (accessed12 February 2021).

National Oceanography Centre (2018). Permanent Service for Mean Sea Level, tide gauge data. https://www.psmsl.org/ (accessed 30 December 2020).

Nehra, V. (2016). A study of coastal erosion & its causes, effects and control strategies. *International Journal of Research and Scientific Innovation* 3 (6): 133–135.

Oliver-Smith, A. (2009). Sea level rise and the vulnerability of coastal peoples. InterSecTions (Issue 7). https://www.unisdr.org/files/14028_4097.pdf (accessed 29 November 2020).

Oppenheimer, M. and Glavovic, B. (2019). Sea level rise and implications for low lying islands, coasts and communities IPCC SR ocean and cryosphere. IPCC special report on the ocean and cryosphere in a changing climate [H.-O. Pörtner, D.C. Roberts, V. Masson-Delmotte, P. Zhai, M. Tignor, E. Poloczanska, K. Mintenbeck, M. Nicolai, A. Okem, J. Petzold, B. Rama, N. Weyer (eds.)]. pp. 1–14.

Oyedotun, T.D.T. (2014). Shoreline geometry: DSAS as a tool for historical trend analysis. *Geomorphological Techniques (Online Edition)* 2: 1–12.

Pajak, M.J. and Leatherman, S. (2002). The high water line as shoreline indicator. *Journal of Coastal Research* 18 (2): 329–337.

Paul, A.K. and Bandopadhyay, M.K. (1987). Morphology of Sagar island, a part of Ganga delta. *Journal of the Geological Society of India* 29 (4): 412–423.

Prasad, D.H. and Kumar, N.D. (2014). Coastal erosion studies — a review. *International Journal of Geosciences 5* (March): 341–345.

Prasetya, G. (2001). Chapter 4: Protection from coastal erosion. In: *Coastal Protection in the Aftermath of the Indian Ocean Tsunami: What Role for Forests and Trees? Proceedings of the Regional Technical Workshop, Khao Lak, Thailand, 28–31 August 2006*, 103–132. Food and Agriculture Organization of the United Nations, Regional Office for Asia and the Pacific http://www.fao.org/docrep/010/ag127e/AG127E00.htm.

Reddy, D.V. (2010). *Engineering Geology*. Vikas Publishing House PVT.LTD.

Ryu, J.H., Won, J.S., and Min, K.D. (2002). Waterline extraction from Landsat TM data in a tidal flat a case study in Gomso Bay, Korea. *Remote Sensing of Environment* 83 (3): 442–456. https://doi.org/10.1016/S0034-4257(02)00059-7.

Scott, B. (2005). Coastal changes, rapid. In: *Encyclopedia of Coastal Science* (ed. I. Schwartz), 233–255. Netherlands: Springer.

Senevirathna, E.M.T.K., Edirisooriya, K.V.D., Uluwaduge, S.P., and Wijerathna, K.B.C.A. (2018). Analysis of causes and effects of coastal erosion and environmental degradation in southern coastal belt of Sri Lanka special reference to unawatuna coastal area. *Procedia Engineering* 212: 1010–1017. https://doi.org/10.1016/j.proeng.2018.01.130.

Smithsonian (2020). Currents, waves, and tides. Smithsonian. https://ocean.si.edu/planet-ocean/tides-currents/currents-waves-and-tides (accessed 26 November 2020).

Vargas-T, V.H., Uribe-P, E., Castellanos-A, O.M., and Ríos-R, C.A. (2016). Coastal landforms caused by deposition and erosion along the shoreline between Punta Brava and Punta Betín, Santa Marta, Colombian Caribbean. *Revista de La Academia Colombiana de Ciencias Exactas, Fisicas y Naturales* 40 (157): 664. https://doi.org/10.18257/raccefyn.387.

Zhenye, Z. (2017). Chapter 7 - Coastal erosion. In: *Marine Geo-Hazards in China* (eds. Y. Yincan et al.), 269–296. https://www.sciencedirect.com/science/article/pii/B9780128127261000073. China Ocean Press, Elsevier https://doi.org/10.1016/b978-0-12-812726-1.00007-3.

Index

Urban Ecology and Global Climate Change, First Edition. Edited by Rahul Bhadouria, Shweta Upadhyay, Sachchidanand Tripathi, and Pardeep Singh.